T0255046

Universitext

Universitext is a series of textbooks that presents material from a wide variety of mathematical disciplines at master's level and beyond. The books, often well class-tested by their author, may have an informal, personal even experimental approach to their subject matter. Some of the most successful and established books in the series have evolved through several editions, always following the evolution of teaching curricula, to very polished texts.

Thus as research topics trickle down into graduate-level teaching, first textbooks written for new, cutting-edge courses may make their way into *Universitext*.

More information about this series at
http://www.springer.com/series/223

Universitext

Universitext is a series of textbooks that presents material from a wide variety of mathematical disciplines at master's level and beyond. The books, often well class-tested by their author, may have an informal, personal even experimental approach to their subject matter. Some of the most successful and established books in the series have evolved through several editions, always following the evolution of teaching curricula, to very polished texts.

Thus as research topics trickle down into graduate-level teaching, first textbooks written for new, cutting-edge courses may make their way into *Universitext*.

More information about this series at
http://www.springer.com/series/223

Krzysztof Dębicki • Michel Mandjes

Queues and Lévy Fluctuation Theory

Krzysztof Dębicki
Mathematical Institute
University of Wrocław
Wrocław, Poland

Michel Mandjes
Korteweg-de Vries Institute for
Mathematics
University of Amsterdam
Amsterdam, The Netherlands

ISSN 0172-5939 ISSN 2191-6675 (electronic)
Universitext
ISBN 978-3-319-20692-9 ISBN 978-3-319-20693-6 (eBook)
DOI 10.1007/978-3-319-20693-6

Library of Congress Control Number: 2015945940

Mathematics Subject Classification: Primary 60K25, 60G51; Secondary 90B05

Springer Cham Heidelberg New York Dordrecht London

Printed on acid-free paper

Springer International Publishing AG Switzerland is part of Springer Science+Business Media (www.springer.com)

Preface

After having worked in the domain of Gaussian queues for about a decade, we got the idea to look at similar problems, but now in the context of *Lévy-driven* queues. That step felt as going from hell to heaven: it was not that we did not like Gaussian queues, but in that domain almost everything is incredibly hard, whereas in the Lévy framework so many rather detailed results can be obtained and usually with transparent and clean arguments.

Fluctuation theory for Lévy processes is an intensively studied topic, perhaps owing to its direct applications in finance and risk. Over the past, say, 30 years, a lot of progress has been made, archived in great textbooks, such as Bertoin [43], Kyprianou [146], Sato [193], and the more general book on applied probability and queues by Asmussen [19]. The distinguishing feature of this textbook is that we explicitly draw the connection with queueing theory. To some extent, Lévy-based fluctuation theory and queueing theory have developed autonomously. Our book proves that bringing these branches together opens interesting possibilities for both.

This textbook is a reflection of the courses we have been teaching in Wrocław, Poland, and Amsterdam, the Netherlands, respectively. While Lévy processes had already been part of the curriculum for a while, we felt there was a need for a course that more explicitly paid attention to its fluctuation-theoretic elements and the connection to queues. This course should not only cover the central results (such as the Wiener–Hopf-based results for the running maximum and minimum and in particular the resulting explicit formulae for spectrally one-sided cases) but also, e.g. a detailed analysis of various queueing-related quantities (busy period, workload correlation function, etc.), asymptotic results (explicitly distinguishing between light-tailed and heavy-tailed scenarios), queueing networks, and applications in communication networks and finance (with a specific focus on option pricing). This has resulted in this book, with a twofold target audience. In the first place, the book has been written to teach either master's students or (starting) PhD students. The required background knowledge consists of Markov chains, some (elementary) queueing theory, martingales, and a bit of stochastic integration theory. In addition, the students should be trained in making their way through some lengthy and technical but usually nice (and in the end rewarding) computations. The second target audience consists of researchers with a background in (applied) probability, but not specifically in the material covered in this book, to quickly learn from—when we entered this area, we would have loved it if there had been such a book, and that was precisely the reason why we decided to write it.

We have written this book more or less remotely, each of us locally testing whether the students liked the way we wrote it. It led to many small and several very substantial changes in the setup. We believe that the current form is the most logical and coherent structure that we could come up with. Having said that, there are quite a number of topics that we could have included, but in the end decided to leave out. Book projects are never finished....

This book would not have been written without the great help of many people. At Springer, Joerg Sixt has always been very supportive of our plans and never put any pressure on us. We also thank Søren Asmussen, Peter Glynn, and Tomasz Rolski, senior researchers in our field, for their encouragement in the early stages of the project.

Krzysztof Dębicki would like to thank the coauthors of his 'Lévy papers': Ton Dieker, Abdelghafour Es-Saghouani, Enkelejd Hashorva, Lanpeng Ji, Kamil Kosiński, Tomasz Rolski, and (last but not least) Michel for the joy of the joint work. He is also grateful to his former PhD students Iwona Sierpińska-Tułacz and Kamil Tabiś, for valuable comments on 'Lévy-driven queues' courses that have been taught at the University of Wrocław. He wants to express his special thanks to Enkelejd Hashorva (University of Lausanne)—warm thanks, Enkelejd, for your exceptional hospitality and wise words on maths and life.

Michel Mandjes would like to thank his 'Lévy coauthors' Lars Nørvang Andersen, Jose Blanchet, Onno Boxma, Bernardo D'Auria, Ton Dieker, Abdelghafour Es-Saghouani, Peter Glynn, Jevgenijs Ivanovs, Offer Kella, Kamil Kosiński, Pascal Lieshout, Zbigniew Palmowski, and Tomasz Rolski (besides Krzyś, of course) for the great collaboration over the years. He also would like to extend a special word of thanks to his current PhD students Naser Asghari and Gang Huang, as well as his (former) master's students Krzysztof Bisewski, Sylwester Błaszczuk, Lukáš Drápal, Viktor Gregor, Mariska Heemskerk, Simaitos Šarūnas, Birgit Sollie, Arjun Sudan, Jan Vlachy, Mathijs van der Vlies, and Dorthe van Waarden, who made numerous suggestions for improving the text. A special word of thanks goes to Nicos Starreveld who proofread the manuscript multiple times. Writing this book benefited tremendously from three quiet periods spent in New York City (!): one, in August 2011, hosted by Jose Blanchet at Columbia University, and two, in December 2013 and March 2014, hosted by Mor Armony and Joshua Reed at New York University—many thanks, Jose, Mor, and Josh!

We conclude with a few personal words. I (Krzyś) would like to thank my beloved family: thanks, Asia and Dobroszek, for all the difference you have made in my life. And I (Michel) would like to use this opportunity to express my deep gratitude to my 'home front': thanks, Miranda, Ester, and Chloe, for giving me the opportunity to do what I like most.

Wrocław, Poland
Amsterdam, The Netherlands
December 15, 2014

Krzysztof Dębicki
Michel Mandjes

Contents

Chapter 1
Introduction

The class of Lévy processes consists of all stochastic processes with *stationary* and *independent* increments; here 'stationarity' means that increments corresponding to a fixed time interval are identically distributed, whereas 'independence' refers to the property that increments corresponding to non-overlapping time intervals behave statistically independently. As such, Lévy processes cover several well-studied processes (e.g. Brownian motions and Poisson processes), but also, as this book will show, a wide variety of other processes, with their own specific properties in terms of their path structure. The process' increments being stationary and independent, Lévy processes can be seen as the genuine continuous-time counterpart of the random walk $S_n := \sum_{i=1}^{n} Y_i$, with independent and identically distributed Y_i.

Lévy processes owe their popularity to their mathematically attractive properties as well as their wide applicability: they play an increasingly important role in a broad spectrum of application domains, ranging from finance to biology. Lévy processes were named after the French mathematician Paul Lévy (1886–1971), who played a pioneering role in the systematic analysis of processes with stationary and independent increments. A brief account of the history of Lévy processes (initially simply known as 'processes with stationary and independent increments') and its application fields is given in e.g. Applebaum [12].

Application areas—In *mathematical finance*, Lévy processes are being used intensively to analyze various phenomena. They are for instance suitable when studying credit risk, or for option pricing purposes (see e.g. Cont and Tankov [63]), but play a pivotal role in insurance mathematics as well (see e.g. Asmussen and Albrecher [21]). An attractive feature of Lévy processes, particularly with applications in finance in mind, is that this class is rich in terms of possible path structures: it is perhaps the simplest class of processes that allows sample paths to have continuous parts interspersed with jumps at random epochs.

Another important application domain lies in *operations research* (OR). According to the functional central limit theorem, under mild conditions on the distribution of the increments, a scaled version of discrete-time random walks converges weakly

K. Dębicki, M. Mandjes, *Queues and Lévy Fluctuation Theory*, Universitext,
DOI 10.1007/978-3-319-20693-6_1

to a Brownian motion. In line with this convergence, one can argue that under a suitable scaling and regularity conditions, there is weak convergence of 'classical' GI/G/1 queueing systems (with discrete customers) to a 'queue with Brownian input', usually referred to as *reflected Brownian motion* [217].

A more specific example, in which the limiting process is *not* necessarily Brownian motion, relates to the performance analysis of resources in communication networks. In the mid-1990s it was observed that the distribution of the sizes of documents transferred over the internet is *heavy tailed*: the complementary distribution of the document sizes decays roughly hyperbolically with a tail index such that the mean document size exists, but the corresponding variance is infinite. This entails that under particular conditions the aggregate of traffic generated by many users weakly converges to fractional Brownian motion, but under other conditions there is weak convergence to (a specific class of) Lévy processes (i.e. α-stable Lévy motions); see Mikosch et al. [163], Taqqu et al. [210], or Whitt [217, Chapter 4]. In the latter regime, the performance of the network element can be evaluated by analyzing a queue fed by Lévy input.

Relevance of Lévy-driven queues; their construction; fluctuation theory—The above OR-related considerations underscore the importance of analyzing queues with Lévy input (or *Lévy-driven queues*). It should be noticed, though, that it is not a priori clear what should be understood by such a queue: for instance, in the case that the Lévy process under consideration is a Brownian motion, the input process is not increasing, nor is even a difference of increasing functions (i.e. it is not of finite variation), and therefore the corresponding queue cannot be seen as a storage system in the classical sense. Relying on a description of the queue as the solution of a so-called *Skorokhod problem* [217], however, a formal definition of a Lévy-driven queue can be given; in fact, any stochastic process satisfying some minor regularity assumptions can serve as the input of a queueing system, as argued in e.g. [124]. It is stressed that queues of the 'classical' M/G/1 type (i.e. Poisson arrivals, generally distributed jobs, one server) fit in the framework of Lévy-driven queues. A Lévy-driven queue is also referred to as a *Lévy process reflected at* 0, or a *regulated Lévy process.*

Interestingly, although queues are seemingly absent in the finance applications that we mentioned above, Lévy-driven queues are, in disguise, used intensively in that context as well. The reason for this is that many queueing-related metrics can be expressed in terms of extreme values attained by the driving (i.e. non-reflected) Lévy process. Precisely this knowledge about extremes, a body of results usually referred to as *fluctuation theory*, plays a pivotal role in finance; think for instance, in an insurance context, of the analysis of ruin probabilities, but also of techniques to price various exotic options and to quantify the corresponding sensitivities.

Goal of the book—Having defined Lévy-driven queues, all questions that have been studied for classical queues now have their Lévy counterpart—the high-level goal of this book is to address these issues. For instance, a first question relates to the distribution of the steady-state workload of the queue: imposing the obvious stability criterion, can we explicitly characterize the stationary workload distribution? A second branch of questions relate to the impact of the initial workload; there the focus lies on determining the queue's transient workload, but also various alternative transience-related metrics (such as the busy period and the correlation of the workload process) can be considered. In addition, just as in the world of 'classical' queues, one can think of a variety of variants of the standard Lévy-driven queue: queues with a finite buffer, queues whose input characteristics are affected by the current workload level ('feedback'), queues with vacations and service interruptions, and Lévy-driven polling models. Finally, under specific conditions on the Lévy processes involved, one can let the output of a queue serve as the input for the next queue, and in this way we arrive at the notion of Lévy-driven queueing networks.

The objective of this textbook is to give a systematic account of the literature on Lévy-driven queues. In addition, we also intend to make the reader familiar with the wide set of techniques that has been developed over the past decades. In this survey, techniques that are highlighted include transform-based methods, martingales, rate-conservation arguments, change of measure, importance sampling, large deviations, and numerical inversion.

Complementary reading—A few words on additional recommended literature. In the first place there are the textbooks by Bertoin [43], Kyprianou [146], and Sato [193], which provide a fairly general account of the theory of Lévy processes. All of these have a specific focus, though: they concentrate on fluctuation theory, that is, the theory that describes the extreme values that are attained by the Lévy process under consideration, and which is, as argued above, a topic that is intimately related to Lévy-driven queues. We also mention the book by Applebaum [11], which concentrates more on techniques deriving from stochastic calculus. Asmussen [19, Chapter IX] and Prabhu [179, Chapter 4] provide brief introductions to Lévy-driven queues.

Organization—Our book is organized as follows. Chapters 2, 3, 4, 5, 6, 7, 8, 9, 10, 11, 12, and 13 build up the theory of Lévy-driven queues, whereas Chapters 14 and 15 focus on applications in operations research and finance, respectively; the book concludes in Chapter 16 with a description of numerical techniques. In more detail, the topics addressed in this monograph are the following.

Chapter 2 formalizes the notion of *Lévy-driven queues*; it is argued how in general queues can be defined without assuming that the input process is necessarily non-decreasing. We also define the special class of *spectrally one-sided* Lévy inputs, that is, Lévy processes with either only positive jumps or only negative jumps; we will extensively rely on this notion throughout the survey. In addition, this chapter introduces the class of α-stable Lévy motions.

In Chapter 3 we analyze the *steady-state workload Q*. For spectrally positive input this is done through its Laplace transform, which is a result that dates back to Zolotarev [222] and which is commonly referred to as the 'generalized Pollaczek–Khintchine formula'. The spectrally negative case can be dealt with explicitly, resulting in an exponentially distributed stationary workload. To deal with the case that jumps in both directions are allowed (the *spectrally two-sided* case), we provide a brief introduction to *Wiener–Hopf theory*. We conclude this chapter by presenting explicit results for two specific classes of spectrally two-sided processes: the former is the class in which the jumps have a phase-type distribution, and the second is the class of meromorphic processes.

Then, in Chapter 4 we characterize (in terms of transforms) the distribution of the *transient workload*, that is, the workload Q_t at some time $t \geq 0$, conditional on $Q_0 = x \geq 0$. Again we distinguish between the spectrally two-sided cases (leading to rather explicit expressions) and the general case (as before relying on Wiener–Hopf-type arguments).

Chapter 5 addresses the limiting regime in which the drift of the driving Lévy process is just 'slightly negative', commonly referred to as *heavy traffic*. Resorting to the steady-state and transient results that were derived in the previous chapters, it appears that we observe an interesting dichotomy, in that one should distinguish between two scenarios that show intrinsically different behavior. In the case that the underlying Lévy process has a finite variance, the appropriately scaled workload process tends to a Brownian motion reflected at 0 (i.e. a Lévy-driven queue with Brownian input). If the variance is infinite, on the contrary, we establish convergence to a Lévy-driven queue fed by an α-stable Lévy motion.

Next to the distribution of the (stationary and transient) workload, in queueing theory much attention is paid to the analysis of the *busy period* distribution. The question addressed in Chapter 6 is, given the workload is in stationarity at time 0, how long does it take for the queue to idle? Explicit results in terms of Laplace transforms are presented. The last part of this chapter addresses the distribution of the *minimal value* attained by the workload process in an interval of given length.

Chapter 7 considers another metric that relates to the transient workload, that is, the *workload correlation function*. A variety of techniques are used to analyze the correlation between Q_0 and Q_t, again assuming the queue is in stationarity at time 0. Specifically, the structural result is established that the workload correlation function is positive, decreasing, and convex (as a function of t), relying on the theory of completely monotone functions.

Where the full distribution of Q was uniquely characterized in Chapter 3, Chapter 8 considers the *tail asymptotics of the stationary workload*. Distinguishing between Lévy processes with light and heavy upper tails (as well as an intermediate regime), functions $f(\cdot)$ are identified such that $\mathbb{P}(Q > u)/f(u) \to 1$ as $u \to \infty$ (so-called exact asymptotics). A variety of techniques are used, such as change-of-measure arguments, large deviations, and Tauberian inversion. These techniques also shed light on *how* high buffer levels are achieved.

In Chapter 9 we present *asymptotics related to the transient metrics* that we defined earlier. Again the distinction between Lévy processes with light and heavy tails should be made. We also pay attention to the asymptotics of the joint distribution of the workloads at two different time epochs.

Chapter 10 focuses on *simulating Lévy-driven queues*. Algorithms are presented to (efficiently and accurately) simulate various important classes of Lévy processes and their associated workload processes. In addition, we point out how importance-sampling-based simulation is of great help when estimating rare-event probabilities (and small covariances, associated to the workload at times 0 and t, for t large).

Where the previous chapters considered the standard Lévy-driven queue, Chapter 11 presents results on several variants. In the first place, it is explained how Lévy-driven queues with a *finite buffer* can be constructed and analyzed. After that, we also present results on *feedback queues*, that is, queues in which the current buffer level affects the characteristics of the Lévy input, and *vacation* and *polling* types of models. We also include a short account of queues with *Markov-additive* input; specializing to the spectrally positive case, we present the transform of the stationary workload as well as the corresponding tail asymptotics.

Then, Chapter 12 presents results on Lévy-driven *tandem queues*: the output of the 'upstream queue' serves as input for the 'downstream queue'. For this model the joint steady-state workload is determined, and various special cases are dealt with in more explicit terms (such as the Brownian tandem queue). Also attention is paid to the joint workload asymptotics, that is, the (bivariate) asymptotics corresponding to the event that both workloads grow large.

In Chapter 13 the theory of Chapter 12 is extended to a particular class of Lévy-driven *networks*. Imposing specific conditions on the network structure and the input processes involved, the joint distribution of all workloads can be determined. The techniques featuring here resemble those used to analyze the tandem queue.

In the next two chapters the focus is on applications. First, in Chapter 14 the use of Lévy-driven queues in *OR-type applications* (related to *communication networks*) is pointed out. In particular, it is argued under what conditions and scaling limits will Lévy processes form a natural candidate to model network traffic. These limits involve both aggregation over time (so-called horizontal aggregation) and over the number of network users (vertical aggregation). As a result, the performance of the network nodes can be evaluated by studying the corresponding Lévy-driven queues.

Financial applications are covered by Chapter 15. First a brief survey is given on the specific Lévy processes that are frequently used to model financial processes (such as the evolution of an asset price); special attention is paid to the normal inverse Gaussian process, the variance gamma process, and the generalized tempered stable process (which also covers the CGMY process). Then we explain how Lévy processes can be estimated from data. A substantial part of this chapter focuses on the computation of prices of exotic options, such as the barrier option and the lookback option, whose payoff functions can be expressed in terms of the extreme values (over a given time horizon) that are attained by the price of the underlying asset. The chapter is concluded by an account of the use of Lévy fluctuation theory in non-life insurance.

In Chapter 16 it is shown how fluctuation-theoretic quantities can be numerically evaluated. Many results presented in this book are in terms of transforms, and fast and accurate algorithms are available to numerically invert these. We describe two intrinsically different approaches.

Chapter 17 concludes our textbook. A brief discussion of the current state of the art is given, and we mention a number of topics that need further analysis.

Exercises

Exercise 1.1 Let $(Y_n)_{n\in\{1,2,\ldots\}}$ be a sequence of i.i.d. random variables. Y_1 is defined by

$$Y_1 := \begin{cases} 1 & \text{with probability } p, \\ -k & \text{with probability } q := 1-p, \end{cases}$$

where $p \in (0,1)$.

Consider a sequence of random variables $(X_n)_{n\in\mathbb{N}}$ defined by $X_0 := 0$ and $X_{n+1} := \max\{X_n + Y_{n+1}, 0\}$.

(a) Show that $(X_n)_{n\in\mathbb{N}}$ is an irreducible Markov chain (in discrete time). Give the state space and the transition matrix P.

(b) Which conditions should be fulfilled by the equilibrium distribution $(\pi_n)_{n\in\{0,1,\ldots\}}$, should it exist?

(c) Let X be a random variable on $\{0,1,2,\ldots\}$ distributed according to the equilibrium distribution of $(X_n)_{n\in\mathbb{N}}$; in other words, $\mathbb{P}(X=n)=\pi_n$, as defined above. The probability generating function of X is then given by

$$\phi(z) := \mathbb{E}(z^X) = \sum_{n=0}^{\infty} \pi_n z^n$$

for $z \in [0,1)$. Show that $\phi(z)$ can be written as

$$\phi(z) = \frac{1-\vartheta}{1-\vartheta z}, \text{ for some } \vartheta \in (0,1].$$

How can ϑ be characterized? Deduce an expression for the equilibrium distribution $(\pi_n)_{n\in\{0,1,\ldots\}}$; under which conditions, to be imposed on p and k, does this distribution exist?

(d) Define $\bar{S} := \sup_{n\in\{0,1,\ldots\}} S_n$, where $S_0 := 0$ and $S_n := \sum_{i=1}^n Y_i$, with the Y_i independent and all distributed as Y_1 as introduced above. Also define

$$\varrho := \mathbb{P}(\exists n \in \mathbb{N} : S_n \geq 1).$$

Show that ϱ satisfies $\varrho = p + q\varrho^{k+1}$.

(e) Show that the distributions of $\bar{S}$ and X are equal.

(*Note*: This is a manifestation of 'Reich's principle', which we will treat in detail in Chapter 2; cf. Eqn. (2.5)).

Chapter 2
Lévy Processes and Lévy-Driven Queues

In classical queueing systems, there is the notion of customers (or work) arriving, and subsequently being processed by the server. The class of Lévy processes, being defined as processes with stationary and independent increments, covers processes with highly non-regular trajectories (think for instance of Brownian motion). As a consequence, it is not immediately clear how one should define a *queue with Lévy input*. One of the goals of the present chapter is to introduce a sound notion of Lévy-driven queues.

We do so by first providing an explicit definition of Lévy processes, and then extending the classical definition of a queue to a notion that can be used for general input processes as well (i.e. in principle *any* real-valued stochastic process can serve as input). For more background, we refer the reader e.g. to Applebaum [11], Asmussen [19], Kyprianou [146], and Sato [193].

In Section 2.1, as a first step we introduce notation, to be used throughout this book, together with a number of fundamental properties. As mentioned in Chapter 1, for the special case of one-sided jumps, the results are more explicit. Notation related to such *spectrally one-sided* Lévy processes is given in Section 2.2; this section also includes a number of frequently used Lévy processes. Another important class of Lévy processes, that is, α-stable Lévy motions, is covered by Section 2.3. Finally, in Section 2.4 we present the definition of Lévy-driven queues.

2.1 Infinitely Divisible Distributions, Lévy Processes

We say that a continuous-time real-valued stochastic process $(X_t)_t$ is a Lévy process if it has stationary and independent increments, with $X_0 = 0$ and càdlàg sample paths a.s. (*càdlàg* meaning 'continuous from right, limits from left'). The stationary increments property entails that for given s the distribution of $X_{t+s} - X_t$ is the same irrespective of the value of t, whereas the independent increments property means

K. Dębicki, M. Mandjes, *Queues and Lévy Fluctuation Theory*, Universitext,
DOI 10.1007/978-3-319-20693-6_2

that, for $t \geq 0$, the increment $X_{t+s} - X_s$ is independent of the history of the Lévy process, that is, $(X_u)_{u \leq s}$.

The initial condition $X_0 = 0$ together with the stationary increments property leads, for each $t > 0$, to the equation

$$X_t = \sum_{i=1}^{n} \left(X_{it/n} - X_{(i-1)t/n} \right),$$

in which the increments $X_{it/n} - X_{(i-1)t/n}$ are all distributed as $X_{t/n}$. Moreover, by virtue of the independent increments property, it follows that these increments are also independent. We thus arrive at the following distributional equality, with $X_t^{(i)}$ i.i.d. copies of X_t:

$$X_t \stackrel{\mathrm{d}}{=} \sum_{i=1}^{n} X_{t/n}^{(i)}, \tag{2.1}$$

for any $n \in \mathbb{N}$. In this way we see that, for any t, X_t has an *infinitely divisible* distribution. Indeed, let us recall that a random variable Z is infinitely divisible if for any $n \in \mathbb{N}$ there exist independent and identically distributed (i.i.d.) random variables $Z_{1,n}, \ldots, Z_{n,n}$ such that Z is distributed as $\sum_{m=1}^{n} Z_{m,n}$; see e.g. De Finetti [70]. Conversely, for each infinitely divisible random variable Z there exists a Lévy process $(X_t)_t$ such that $X_1 \stackrel{\mathrm{d}}{=} Z$. This, for example, straightforwardly implies the existence of a Lévy process with Poisson marginals: if Z has a Poisson distribution with mean λ, it is distributed as the sum of n independent Poisson random variables with mean λ/n. Other examples of infinitely divisible distributions are the normal distribution, the negative binomial distribution, and the gamma distribution, as is readily verified.

One can alternatively say that, for any value of t,

$$\xi_t(s) := \log \mathbb{E} e^{isX_t} = t \log \mathbb{E} e^{isX_1} = t\xi(s),$$

for $s \in \mathbb{R}$, where $\xi(s) := \log \mathbb{E} e^{isX_1}$ is referred to as the so-called *Lévy exponent*. This equality is a direct consequence of (2.1), as can be seen as follows. Fixing an $s \in \mathbb{R}$, we find for any two integers m and n that $\xi_m(s) = n\xi_{m/n}(s)$ and $\xi_m(s) = m\xi_1(s)$. Combining these relations, we obtain $\xi_{m/n}(s) = (m/n)\,\xi_1(s) = (m/n)\,\xi(s)$, and hence for all $t \in \mathbb{Q}$ it follows that $\xi_t(s) = t\xi(s)$. By using a limiting argument, it follows immediately that the right continuity of the Lévy process implies that $\xi_t(s) = t\xi(s)$ for any $t \in \mathbb{R}$. As a result, one could informally say that each Lévy process can be associated with an infinitely divisible distribution, and vice versa.

It is immediately seen that the class of Lévy processes contains a number of canonical stochastic processes. In the first place it can be concluded that the *Poisson process* is Lévy. A Poisson process $(X_t)_t$ can be defined as follows: with Y_m i.i.d. exponential random variables with mean $\lambda^{-1} \in (0, \infty)$, we let X_t have the value n

if at the same time $\sum_{m=1}^{n} Y_m \leq t$ and $\sum_{m=1}^{n+1} Y_m > t$. It is well known that X_t has a Poisson distribution with mean λt, and as a consequence,

$$\log \mathbb{E}e^{isX_t} = \log \left(\sum_{n=0}^{\infty} e^{isn} e^{-\lambda t} \frac{(\lambda t)^n}{n!} \right) = \lambda t \left(e^{is} - 1 \right),$$

and hence $(X_t)_t$ is indeed Lévy (with Lévy exponent $\xi(s) = \lambda(e^{is} - 1)$ for $\lambda > 0$). Likewise, we can show that *Brownian motion* without drift is Lévy; here $\xi(s) = -\frac{1}{2}\sigma^2 s^2$ for $\sigma^2 > 0$. In Sections 2.2 and 2.3 we mention various other examples.

It is possible to characterize Lévy processes more specifically: it can be shown that the Lévy exponent $\xi(s)$ is necessarily of the form

$$\xi(s) = isd - \frac{1}{2}s^2\sigma^2 + \int_{-\infty}^{\infty} (e^{isx} - 1 - isx1_{\{|x|<1\}})\Pi(\mathrm{d}x), \tag{2.2}$$

where $d \in \mathbb{R}$, $\sigma \geq 0$, and the spectral measure (or *Lévy measure*) $\Pi(\cdot)$, concentrated on $\mathbb{R} \setminus \{0\}$, satisfies

$$\int_{\mathbb{R}} \min\{x^2, 1\}\Pi(\mathrm{d}x) < \infty.$$

For a proof of this fundamental representation of Lévy processes (or, in fact, a stronger version of it), called in the literature the *Lévy–Khintchine formula*, we refer e.g. to Kyprianou [146, Chapter II].

The triplet (d, σ^2, Π) is commonly referred to as the *characteristic triplet*, as it uniquely defines the underlying Lévy process: every Lévy process has its own specific d, σ^2, and Π. It is noted that in some cases it is possible to extend the domain of $\xi(s)$ to (a subset of) $\mathbb{C}$; we return to this issue in greater detail in Section 2.2 when we speak about Lévy processes with one-sided jumps.

For obvious reasons, we call the first parameter of the characteristic triplet, d, the *deterministic drift*, whereas the term $\frac{1}{2}s^2\sigma^2$ is often referred to as the *Brownian term*. The third term in (2.2) corresponds to the jumps of the Lévy process by the relation that the jumps of size x occur at intensity $\Pi(\mathrm{d}x)$. More precisely, for any bounded interval M such that $0 \notin M$, the sum of the jumps of size within M in the time interval $[0, t)$ is distributed as a compound Poisson random variable with intensity $t \int_M \Pi(\mathrm{d}y)$ and the jump-size distribution

$$\frac{\Pi(\mathrm{d}x)1_{\{x \in M\}}}{\int_M \Pi(\mathrm{d}y)}.$$

Thus, if the jumps are only in the upward (respectively, downward) direction, then the support of Π is concentrated in $(0, \infty)$ (respectively, $(-\infty, 0)$). The process $(X_t)_t$ is of bounded variation if and only if both $\sigma = 0$ and $\int_{-1}^{1} |x|\Pi(\mathrm{d}x) < \infty$; we do not provide details on this, but refer to Kyprianou [146, Section 2.6.1].

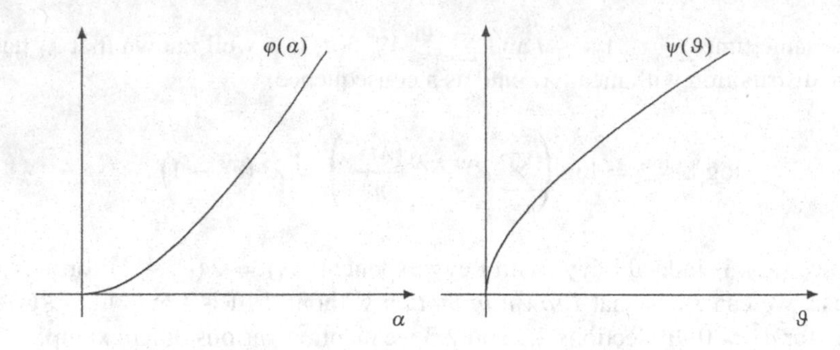

Fig. 2.1 Spectrally positive case: Laplace exponent and its inverse

2.2 Spectrally One-Sided Lévy Processes

Let $(X_t)_{t\geq 0}$ be a Lévy process, as introduced in Section 2.1. Unless stated otherwise, we assume throughout the book that the 'mean drift' $\mathbb{E}X_1$ of the Lévy process is negative, so as to make sure that the corresponding workload process (to be formally introduced in Section 2.4) is stable, thus guaranteeing the existence of a proper stationary workload distribution.

In this monograph, two special cases will often be considered in great detail, that is, spectrally positive and spectrally negative Lévy processes.

The Lévy process has no negative jumps—Here the Lévy process $(X_t)_{t\geq 0}$ has no negative jumps, or is *spectrally positive*; in the sequel this is denoted by $X \in \mathscr{S}_+$. In this case the spectral measure $\Pi(\cdot)$ is concentrated on $(0,\infty)$.

It turns out, in this case, to be convenient to work with the *Laplace exponent*, given by the function $\varphi(\alpha) := \log \mathbb{E}e^{-\alpha X_1}$, rather than the Lévy exponent $\xi(s)$. It is a consequence of the fact that there are only positive jumps that this Laplace exponent is well defined for all $\alpha \geq 0$.

It follows immediately from Hölder's inequality that the Laplace exponent $\varphi(\cdot)$ is convex on $[0,\infty)$; due to the assumption $\mathbb{E}X_1 < 0$, and observing that $\varphi(\cdot)$ has slope $\varphi'(0) = -\mathbb{E}X_1$ at the origin, we conclude that $\varphi(\cdot)$ is increasing on $[0,\infty)$, and hence the inverse $\psi(\cdot)$ of $\varphi(\cdot)$ is well defined on $[0,\infty)$; see Fig. 2.1. In the sequel we also require that X_t is not a *subordinator*, that is, a monotone process; this means that X_1 has probability mass on the negative half-line, which implies that $\lim_{\alpha\to\infty} \varphi(\alpha) = \infty$.

The Lévy process has no positive jumps—In this case the Lévy process $(X_t)_{t\geq 0}$ has no positive jumps, or is *spectrally negative*; throughout this book we denote this by $X \in \mathscr{S}_-$. Now the spectral measure $\Pi(\cdot)$ is concentrated on $(-\infty, 0)$. In this case, we define the *cumulant* $\Phi(\beta) := \log \mathbb{E}e^{\beta X_1}$. This function is well defined and finite for any $\beta \geq 0$ due to the fact that there are no positive jumps. We now rule out that $(X_t)_t$ has decreasing sample paths a.s. Recalling that $\Phi'(0) = \mathbb{E}X_1 < 0$, we see that $\Phi(\beta)$ is *not* a bijection on $[0,\infty)$; we define the *right* inverse through

$$\Psi(q) := \sup\{\beta \geq 0 : \Phi(\beta) = q\}.$$

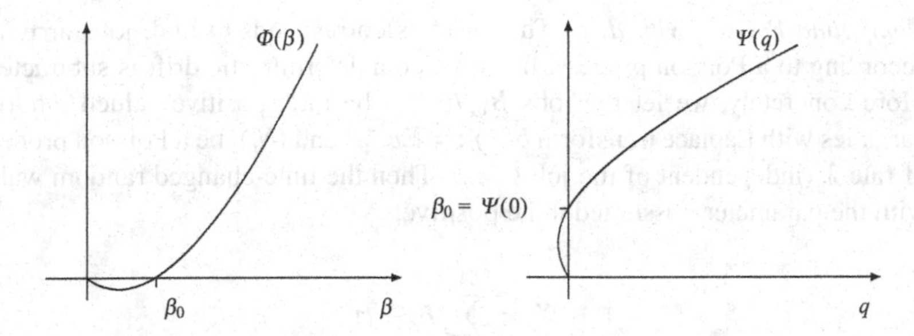

Fig. 2.2 Spectrally negative case: the cumulant and its right inverse

Note that $\beta_0 := \Psi(0) > 0$; this parameter plays a crucial role when analyzing queues with spectrally negative input; see Fig. 2.2.

The Lévy exponent (or the Laplace exponent for $X \in \mathscr{S}_+$, or cumulant for $X \in \mathscr{S}_-$) contains all information about X_1, and hence, due to the infinite divisibility, also about the whole process $(X_t)_t$. For instance, it enables the computation of all moments (provided they exist), as follows. For example, for $X \in \mathscr{S}_+$, we have $\mathbb{E}X_t = -\varphi'(0)\,t$ and $\mathbb{V}\mathrm{ar}\,X_t = \varphi''(0)\,t$ (given that these derivatives are well defined). It is also noted that

$$\varphi'(0) = -d - \int_{[1,\infty)} x\Pi(\mathrm{d}x), \quad \varphi''(0) = \sigma^2 + \int_{(0,\infty)} x^2\Pi(\mathrm{d}x),$$

whereas, for $n = 3, 4, \ldots$,

$$\varphi^{(n)}(0) = (-1)^n \int_{(0,\infty)} x^n \Pi(\mathrm{d}x).$$

We now treat in greater detail a number of examples of spectrally one-sided Lévy processes.

(1) *Brownian motion with drift*. This process has sample paths that are continuous a.s., and is therefore both spectrally positive and spectrally negative. In this case X_t has a normal distribution with mean dt and variance $\sigma^2 t$. It is readily verified that, with U denoting a standard normal random variable, $\mathbb{E}e^{-\alpha X_t} = e^{-\alpha dt}\mathbb{E}e^{-\alpha\sqrt{t}\sigma U}$, and

$$\mathbb{E}e^{-\alpha U} = \int_{-\infty}^{\infty} e^{-\alpha u}\frac{1}{\sqrt{2\pi}}e^{-u^2/2}\mathrm{d}u = e^{\alpha^2/2}\int_{-\infty}^{\infty}\frac{1}{\sqrt{2\pi}}e^{-(u+\alpha)^2/2}\mathrm{d}u = e^{\alpha^2/2}.$$

It is concluded that $\log \mathbb{E}e^{-\alpha X_t} = t(-\alpha d + \frac{1}{2}\alpha^2\sigma^2)$. We write $X \in \mathbb{B}\mathrm{m}(d, \sigma^2)$ when $\varphi(\alpha) = -\alpha d + \frac{1}{2}\alpha^2\sigma^2$. The mean drift of this process is d, which is assumed to be negative (to make sure that $\mathbb{E}X_1 < 0$).

(2) *Compound Poisson with drift*. This process corresponds to i.i.d. jobs arriving according to a Poisson process, from which a deterministic drift is subtracted. More concretely, we let the jobs $B_1, B_2, \ldots$ be i.i.d. positive-valued random variables with Laplace transform $b(\alpha) := \mathbb{E}e^{-\alpha B}$ and $(N_t)_t$ be a Poisson process of rate λ (independent of the job sizes). Then the time-changed random walk, with the parameter r assumed to be positive,

$$X_t = \sum_{i=1}^{N_t} B_i - rt$$

(following the convention that $\sum_{i=1}^{0} B_i := 0$) is a spectrally positive Lévy process which we call a compound Poisson process with drift. We write $X \in \mathbb{CP}(r, \lambda, b(\cdot))$.

It can be verified that

$$\mathbb{E}e^{-\alpha X_t} = e^{r\alpha t} \sum_{n=0}^{\infty} (b(\alpha))^n e^{-\lambda t} \frac{(\lambda t)^n}{n!} = \exp\left(t(r\alpha - \lambda + \lambda b(\alpha))\right).$$

As a consequence, $\varphi(\alpha) = r\alpha - \lambda + \lambda b(\alpha)$. The mean drift of this process is $\mathbb{E}X_1 = \lambda\,\mathbb{E}B - r$, which we assume to be negative to ensure stability.

Clearly, if the depletion rate r were negative, and the jobs were i.i.d. samples from a non-positive distribution (i.e. the jumps were downward), then the resulting process would be spectrally negative.

It is instructive to express the compound Poisson process in terms of a triplet (d, σ^2, Π). Obviously, because of the lack of a Brownian term, $\sigma^2 = 0$. In addition, for the Lévy measure we have $\Pi(\mathrm{d}x) = \lambda\mathbb{P}(B \in \mathrm{d}x)$. It is then readily verified that

$$d = -r + \lambda \int_0^1 x\Pi(\mathrm{d}x).$$

(3) *Gamma process*. This process is characterized by the characteristic triplet (d, σ^2, Π), where $\sigma^2 = 0$ and

$$\Pi(\mathrm{d}x) = \frac{\beta}{x} e^{-\gamma x}\mathrm{d}x \quad \text{for } x > 0, \qquad d = \int_0^1 x\Pi(\mathrm{d}x),$$

for $\gamma, \beta > 0$. From the above formulation it is clear that the jumps of this process are non-negative, that is, the gamma process is spectrally positive. In fact its sample paths are non-decreasing a.s.; we return to this property below.

The Laplace exponent corresponding to the gamma process can be evaluated explicitly, but this requires some non-standard computations. These rely on the well-known *Frullani integral*: for $z \in \mathbb{C}$ with non-positive real part,

$$\beta \log\left(1 - \frac{z}{\gamma}\right) = \int_0^\infty (1 - e^{zx})\frac{\beta}{x}e^{-\gamma x}\mathrm{d}x; \tag{2.3}$$

see e.g. Kyprianou [146, Lemma I.1.7]. The validity of Eqn. (2.3) is a direct consequence of the identity (given that appropriate regularity conditions are imposed on the function $f(\cdot)$)

$$\int_0^\infty \frac{f(ax) - f(bx)}{x}\mathrm{d}x = -\int_0^\infty \int_a^b f'(xy)\mathrm{d}y\,\mathrm{d}x = -\int_a^b \int_0^\infty f'(xy)\mathrm{d}x\,\mathrm{d}y$$
$$= \int_a^b \frac{f(0) - f(\infty)}{y}\mathrm{d}y = (f(0) - f(\infty))\log\frac{b}{a}$$

by picking $f(x) := e^{-x}$, $a = \gamma$, and $b = \gamma - z$.

As a consequence of the above computations, it follows that the corresponding Laplace exponent

$$\varphi(\alpha) = \log \mathbb{E}e^{-\alpha X_1} = -\alpha \int_0^1 x\Pi(\mathrm{d}x) + \int_{-\infty}^\infty (e^{-\alpha x} - 1 + \alpha x 1_{[0,1)}(|x|))\Pi(\mathrm{d}x),$$

can now be rewritten as

$$\int_0^\infty (e^{-\alpha x} - 1)\frac{\beta}{x}e^{-\gamma x}\mathrm{d}x = \beta \log\left(\frac{\gamma}{\gamma + \alpha}\right).$$

From the equation

$$\int_0^\infty \left(\gamma e^{-\gamma x}\frac{(\gamma x)^{\beta t - 1}}{\Gamma(\beta t)}\right) e^{-\alpha x}\mathrm{d}x$$
$$= \left(\frac{\gamma}{\gamma + \alpha}\right)^{\beta t} \int_0^\infty (\gamma + \alpha)e^{-(\gamma+\alpha)x}\frac{((\gamma + \alpha)x)^{\beta t - 1}}{\Gamma(\beta t)}\mathrm{d}x = \left(\frac{\gamma}{\gamma + \alpha}\right)^{\beta t},$$

where $\Gamma(z) := \int_0^\infty e^{-x}x^{z-1}\mathrm{d}x$ denotes the gamma function, it follows that the marginals X_t have a gamma distribution with parameters γ and βt. We write throughout this monograph $X \in \mathbb{G}(\gamma, \beta)$.

The gamma process has interesting qualitative properties. Observe that X_t has the same distribution as the sum of X_s and X_{t-s} (with $s \in (0, t)$), with the latter two random variables being sampled independently, which are both non-negative random variables. From this we conclude that $(X_t)_t$ is a non-decreasing process.

In the second place, it is observed that the gamma process is *not* compound Poisson. This is a consequence of the fact that we cannot write $\Pi(\mathrm{d}x)$ as $\lambda\,\mathbb{P}(B \in \mathrm{d}x)$. To see this, realize that, as a consequence of $\beta/x \cdot e^{-\gamma x}$ being roughly β/x for x close to 0,

$$\int_0^\infty \Pi(\mathrm{d}x) = \int_0^\infty \frac{\beta}{x} e^{-\gamma x} \mathrm{d}x = \infty,$$

and hence it is not possible to properly define a (finite) jump intensity λ. Indeed, the gamma process is a Lévy process with the remarkable property that it has infinitely many jumps (almost surely) in any finite amount of time. We refer to this phenomenon by saying that the gamma process has *small jumps*, or, equivalently, *infinite activity*.
As mentioned above, the gamma process is increasing; to make sure that $\mathbb{E}X_1 < 0$ (so as to guarantee that the corresponding workload process is stable) a negative drift has to be added.

(4) *Inverse Gaussian process.* Like the gamma process, this process is increasing. It is defined as follows. For any $X \in \mathscr{S}_+$, we define the *first passage time*,

$$\tau(x) := \inf\{t \geq 0 : X_t < -x\};$$

this is a notion that will play an important role later in this book. It is straightforward to observe that $e^{-\varphi(\alpha)t}\, e^{-\alpha X_t}$ is a mean-1 martingale [220]: for all $s \leq t$, using the properties of Lévy processes,

$$\begin{aligned}
&\mathbb{E}\left(e^{-\varphi(\alpha)t}\, e^{-\alpha X_t} \mid \{e^{-\varphi(\alpha)u}\, e^{-\alpha X_u} : u \leq s\}\right) \\
&= \mathbb{E}\left(e^{-\varphi(\alpha)t}\, e^{-\alpha X_t} \mid \{X_u : u \leq s\}\right) \\
&= e^{-\varphi(\alpha)s}\, e^{-\alpha X_s}\, \mathbb{E}\left(e^{-\varphi(\alpha)(t-s)}\, e^{-\alpha X_{t-s}}\right) = e^{-\varphi(\alpha)s}\, e^{-\alpha X_s}.
\end{aligned}$$

Considering $X \in \mathbb{B}\mathrm{m}(d, \sigma^2)$, clearly $d < 0$ implies $\tau(x) < \infty$ almost surely. The a.s. continuous sample paths imply that $X_{\tau(x)} = -x$, which, together with 'optional sampling' [220, Chapter A14], leads to

$$\mathbb{E}e^{-\varphi(\alpha)\tau(x)} = e^{-\alpha x}.$$

As a consequence, replacing $\varphi(\alpha)$ by ϑ (and hence α by $\psi(\vartheta)$),

$$\mathbb{E}e^{-\vartheta\tau(x)} = \exp\left(-\left(\frac{d}{\sigma^2} + \sqrt{\left(\frac{d}{\sigma^2}\right)^2 + 2\frac{\vartheta}{\sigma^2}}\right)x\right).$$

Conclude that $\tau(x)$ is an increasing Lévy process (in x); the class of these processes we call *inverse Gaussian*, and we denote it by $\mathbb{IG}(d, \sigma^2)$. Again, to have $\mathbb{E}X_1 < 0$, a negative drift is added. The identification of the spectral measure $\Pi(\cdot)$ is the subject of one of the exercises. The inverse Gaussian process has 'small jumps', too: it experiences an infinite number of jumps (almost surely) over any time interval of finite length.

2.3 α-Stable Lévy Motions

This section focuses on a subclass of Lévy processes that has attracted substantial attention in the literature: α-stable Lévy motions. This class of processes is particularly suitable when modeling various sorts of heavy-tailed phenomena [192].

To introduce α-stable Lévy motions, we first define the class of stable distributions. Here we follow the exposition in Samorodnitsky and Taqqu [192], but various other parameterizations are possible [213]. We say that a random variable Y has a stable distribution if for any $a, b > 0$ there exist $c > 0$ and $d \in \mathbb{R}$ such that

$$aY' + bY'' \stackrel{\mathrm{d}}{=} cY + d,$$

where Y' and Y'' are independent copies of Y. Due e.g. to Bingham et al. [47, Thm. 8.3.2], it turns out that the characteristic function of Y can be written in the form

$$\log \mathbb{E}e^{\mathrm{i}\theta Y} = \begin{cases} -\sigma^\alpha|\theta|^\alpha(1 - \mathrm{i}\beta\mathrm{sign}(\theta)\tan(\pi\alpha/2)) + im\theta, & \alpha \neq 1; \\ -\sigma|\theta|(1 + \mathrm{i}\beta\pi/2\mathrm{sign}(\theta)\log|\theta|) + im\theta, & \alpha = 1; \end{cases}$$

where $\alpha \in (0,2]$, $\beta \in [-1,1]$, $\sigma \in [0,\infty)$, $m \in \mathbb{R}$, and $\mathrm{sign}(x) := 1_{(0,\infty)}(x) - 1_{(-\infty,0)}(x)$. We write that Y is distributed $S_\alpha(\sigma, \beta, m)$.

Let us consider the meaning of the parameters in more detail.

- The parameter α is commonly referred to as the *index of stability*. Later we will observe that α is directly related to the 'heaviness' of the tail distribution. In particular, if $\alpha \in (0,1]$, then $\mathbb{E}|Y| = \infty$ (for $\alpha = 1$ we have the Cauchy distribution). For $\alpha = 2$ we obtain the normal distribution.
- The parameter β is known as the *skewness*. The extreme cases are $\beta = 1$, corresponding to a *totally skewed to the right* distribution, and $\beta = -1$, which corresponds to a *totally skewed to the left* distribution. For $\alpha < 1$, $m = 0$, and $\beta = 1$ (respectively, $\beta = -1$), the support of the distribution is the positive (respectively, negative) half-line, but this is no longer true for $\alpha \geq 1$. The choice of $m = 0$ and $\beta = 0$ leads to a symmetric distribution.
- For obvious reasons, σ is called the *scale parameter*.
- For $\alpha \in (1,2]$, we have $\mathbb{E}Y = m$. This explains why m is called the *shift parameter*.

The following useful property, describing the distribution's tail asymptotics, can be found in e.g. Samorodnitsky and Taqqu [192, p. 16]. As before, $\Gamma(z) := \int_0^\infty e^{-x}x^{z-1}\mathrm{d}x$ denotes the gamma function. Also, $f(x) \sim g(x)$ as $x \to \infty$ means that $f(x)/g(x) \to 1$ as $x \to \infty$.

Proposition 2.1 *Let $Y \stackrel{\mathrm{d}}{=} S_\alpha(\sigma, \beta, m)$ with $\beta \in (-1, 1]$. Then, as $u \to \infty$,*

$$\mathbb{P}(Y > u) \sim C_{\alpha,\sigma}\left(\frac{1+\beta}{2}\right)u^{-\alpha},$$

where

$$C_{\alpha,\sigma} := \begin{cases} \sigma^\alpha(1-\alpha)/\left(\Gamma(2-\alpha)\cos(\pi\alpha/2)\right) & \text{if } \alpha \neq 1; \\ 2\sigma/\pi & \text{if } \alpha = 1. \end{cases}$$

The case $\beta = -1$ has to be treated separately; see e.g. [192, pp. 17–18].

Proposition 2.2 *Let $Y \stackrel{\mathrm{d}}{=} S_\alpha(\sigma, -1, 0)$.*

(i) If $\alpha = 1$, then as $u \to \infty$,

$$\mathbb{P}(Y > u) \sim \frac{1}{\sqrt{2\pi}}\exp\left(-\frac{(\pi/2\sigma)u - 1}{2} - e^{(\pi/2\sigma)u-1}\right).$$

(ii) If $\alpha > 1$, then as $u \to \infty$,

$$\mathbb{P}(Y > u) \sim \frac{1}{\sqrt{2\pi\alpha(\alpha-1)}}\left(\frac{\alpha\hat{\sigma}_\alpha}{u}\right)^{\alpha/(2(\alpha-1))}\exp\left(-(\alpha-1)\left(\frac{u}{\alpha\hat{\sigma}_\alpha}\right)^{\alpha/(\alpha-1)}\right),$$

where

$$\hat{\sigma}_\alpha := \sigma\left(\cos\left(\frac{\pi(2-\alpha)}{2}\right)\right)^{-1/\alpha}.$$

Having defined stable distributions, we can now introduce α-stable Lévy motions, as follows. We say that $(X_t)_t$ is an α-stable Lévy motion if $(X_t)_t$ has stationary and independent increments such that the marginals obey

$$X_t \stackrel{\mathrm{d}}{=} S_\alpha(t^{1/\alpha}, \beta, mt);$$

we write $X \in \mathbb{S}(\alpha, \beta, m)$. From the above we conclude that if $\beta = \pm 1$, then $X \in \mathscr{S}_\pm$.

For given $\alpha \in (0, 2]$ the Lévy measure has the form, for $A, B > 0$,

$$\Pi(\mathrm{d}x) = \left(\frac{A}{(-x)^{\alpha+1}}1_{\{x<0\}} + \frac{B}{x^{\alpha+1}}1_{\{x>0\}}\right)\mathrm{d}x;$$

it is verified that we again have the property of infinitely many jumps in any finite time interval, almost surely.

One could say that α-stable Lévy motions are *self-similar*: picking $m = 0$, and writing $(X_t^{(\alpha)})_t$ to stress the dependence on α, one has that, for $M > 0$,

$$\left(X_{Mt}^{(\alpha)}\right)_t \stackrel{\mathrm{d}}{=} \left(M^{1/\alpha} X_t^{(\alpha)}\right)_t$$

(unless $\alpha = 1$, $\beta \neq 0$). In other words, when zooming in, one sees essentially the same pattern, given that one adjusts the axes in a suitable fashion.

2.4 Lévy-Driven Queues

Having defined Lévy processes, in this section we introduce the notion of queues with Lévy input (or *Lévy-driven queues*). It is noticed, however, that these definitions are by no means restricted to the Lévy framework; based on the formalism defined below, one can define for *any* real-valued stochastic process the corresponding workload process. We provide two types of characterizations.

In the first approach, we define the Lévy-driven queue as the continuous-time counterpart of the classical discrete-time queue. In discrete time, a queue can be described through the well-known Lindley recursion: we have that the workload process (Q_n) satisfies

$$Q_{n+1} = \max\{Q_n + Y_n, 0\},$$

where Y_n is the *net* input to the queue in slot n (i.e. the input minus the amount that can potentially be served). Iterating this recursion, we obtain $Q_{n+1} = \max\{Q_{n-1} + Y_{n-1} + Y_n, Y_n, 0\}$. With $X_n := \sum_{i=0}^n Y_i$, and with $Q_0 = x$ for $x \geq 0$, this eventually leads to

$$Q_n = X_n + \max\left\{x, \max_{0 \leq i \leq n} -X_i\right\}.$$

In this way we have written the workload process $(Q_n)_n$ as a functional of the cumulative net input process $(X_n)_n$, and now the idea is to use the very same functional to define the workload in continuous time.

More concretely, a queue in continuous time can be defined by just taking the continuous-time analogue of the above, so that we obtain

$$Q_t = X_t + \max\{x, L_t\}, \ t \geq 0, \tag{2.4}$$

with

$$L_t := \sup_{0 \leq s \leq t} -X_s = -\inf_{0 \leq s \leq t} X_s;$$

this increasing (and therefore of bounded variation) process L_t is often referred to as *local time* (*at zero*) or a *regulator process*; see e.g. Harrison [108]. Assuming the queue has been running from $-\infty$, one can alternatively write

$$Q_t = \sup_{s \leq t}(X_t - X_s).$$

To ensure the existence of a stationary distribution, it is evident that a stability condition needs to be fulfilled. In the case of input processes $(X_t)_t$ with stationary increments (as is the case in our Lévy context) it needs to be assumed that $\mathbb{E}X_1 < 0$ for the workload process to be stable (which we do throughout this book). If the input process X_t is *reversible*, that is, $(X_{(s-t)-} - X_s)_t \stackrel{\mathrm{d}}{=} (-X_t)_t$ for each given $s > 0$ (which is true in the Lévy case), then we have the following distributional equality for the stationary workload Q, commonly attributed to Reich [182]:

$$Q \stackrel{\mathrm{d}}{=} \sup_{t \geq 0} X_t. \tag{2.5}$$

Above we constructed the Lévy-driven queue in continuous time analogously to its discrete-time counterpart. An alternative way of introducing Lévy-driven queues is by defining them as the solution of a so-called *Skorokhod problem*, as introduced by Skorokhod in [201, 202]; then one commonly says that $(Q_t)_t$ is *the reflection of* $(X_t)_t$ *at* 0. This is done as follows. Let $(L_t^\star)_t$ be a non-decreasing right-continuous process such that the following two requirements are fulfilled.

(A) The workload process $(Q_t)_t$, defined through $Q_0 := x$ and $Q_t := X_t + L_t^\star$, is non-negative for all $t \geq 0$.

(B) $L_t^\star$ can only increase when $Q_t = 0$, that is,

$$\int_0^T Q_t \mathrm{d}L_t^\star = 0, \quad \text{for all } T > 0.$$

Observe that it is natural to impose these conditions on a queueing process. The process $(L_t^\star)_t$ can be informally thought of as the *cumulative idle time process*; then (A) indicates by how much X_t should be increased to obtain Q_t (to account for the effect of the boundary at 0), and (B) entails that it is not possible that at the same time the queue is non-empty and the cumulative idle time grows.

Importantly, it can be proved that the only process satisfying these two conditions is $L_t^\star = \max\{x, L_t\}$, so that $Q_t = X_t + \max\{x, L_t\}$ for $t \geq 0$, where L_t is defined as above; see e.g. Asmussen [19, Prop. IX.2.2] and Robert [185, p. 375]. We conclude that the expression found in this way coincides with the one obtained when taking the continuous counterpart of the discrete-time definition, as in (2.4). For the sake of completeness we include the proof here.

Proposition 2.3 *The process $(L^\star_t)_t$, defined by $L^\star_t := \max\{x, L_t\}$, is the unique solution to the Skorokhod problem (A)–(B).*

Proof There are several ways to prove the statement; we follow the proof of [19, Prop. IX.2.2]. Let $(\bar{L}^\star_t)_t$ be another solution to (A)–(B), and $(\bar{Q}_t)_t$ be the corresponding workload process. Defining $D_t := \bar{L}^\star_t - L^\star_t$, it is our goal to verify that necessarily $D_t \equiv 0$. By applying integration by parts for right-continuous processes of bounded variation, and defining $\Delta D_s := D_s - D_{s-}$,

$$
\begin{aligned}
D_t^2 &= 2\int_0^t D_s \mathrm{d}D_s - \sum_{s\le t}(\Delta D_s)^2 \\
&= 2\int_0^t (\bar{L}^\star_s - L^\star_s)\mathrm{d}\bar{L}^\star_s - 2\int_0^t (\bar{L}^\star_s - L^\star_s)\mathrm{d}L^\star_s - \sum_{s\le t}(\Delta D_s)^2 \\
&= 2\int_0^t (\bar{Q}_s - Q_s)\mathrm{d}\bar{L}^\star_s - 2\int_0^t (\bar{Q}_s - Q_s)\mathrm{d}L^\star_s - \sum_{s\le t}(\Delta D_s)^2,
\end{aligned}
$$

where the last step is due to $X_t = Q_t - L^\star_t = \bar{Q}_t - \bar{L}^\star_t$. Realizing that

$$\int_0^t \bar{Q}_s \mathrm{d}\bar{L}^\star_s = \int_0^t Q^\star \mathrm{d}L^\star_s = 0,$$

it follows that

$$D_t^2 = -2\int_0^t Q_s \mathrm{d}\bar{L}^\star_s - \int_0^t \bar{Q}_s \mathrm{d}L^\star_s - \sum_{s\le t}(\Delta D_s)^2.$$

As Q_s and $\bar{Q}_s$ are non-negative, we conclude that $D_t^2 \le 0$, and therefore $D_t = 0$. □

In the case $X \in \mathbb{CP}(r, \lambda, b(\cdot))$, the queue under study is the well-known M/G/1 queue. We refer to Fig. 2.3 for a pictorial illustration of the evolution of the workload in time, jointly with the $(X_t)_t$ process (where we consider for ease the special case of $Q_0 = 0$ and $r = 1$). It is elementary to verify that in the case that

$$\arg \inf_{0\le s\le t}(X_t - X_s)$$

is smaller than t, this time epoch can be interpreted as the start of the busy period in which t is contained; if it equals t (meaning that X_t is the 'all-time low' of the process so far), then the workload is 0 at time t. It also follows that in this context, the process $L^\star_t$ is the queue's cumulative idle time up to time t.

Importantly, however, we would like to stress that this general notion of a queueing system can be used in settings beyond traditional queues: the process $(X_t)_t$ does not need necessarily to relate to positive quantities of work arriving. In this sense, we now have developed the concept of a queue fed by for instance Brownian

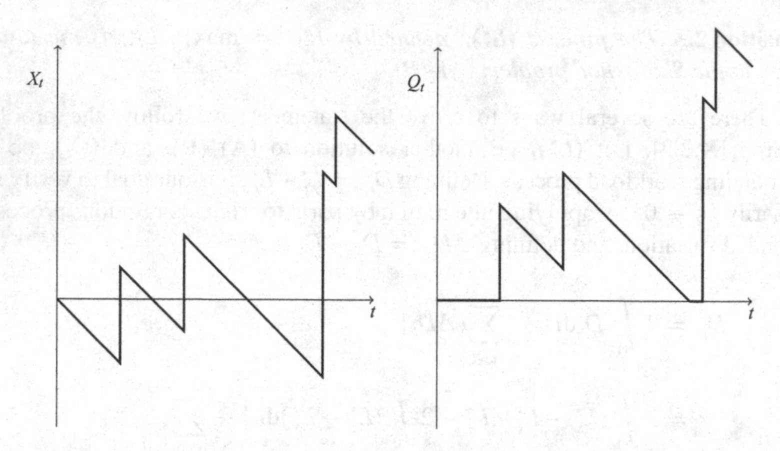

Fig. 2.3 Net input process and workload process for a compound Poisson process

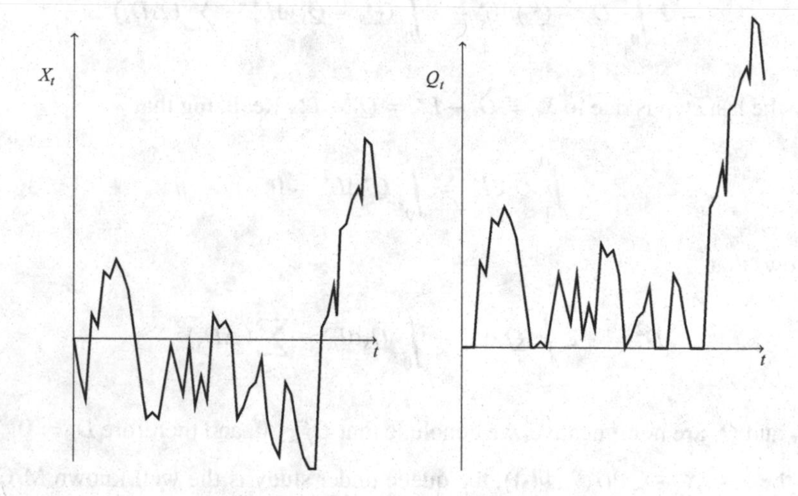

Fig. 2.4 Net input process and workload process for an erratic, 'Brownian-like' process

motion, or any other real-valued continuous-time stochastic process. In the case $X \in \mathbb{B}\mathrm{m}(d, \sigma^2)$, the resulting workload process is often referred to as *reflected* (or *regulated*) *Brownian motion*. We refer to Fig. 2.4 for an illustrative example of such a workload process.

One of the main objectives in this book is the identification of the distribution of the transient workload Q_t and its stationary counterpart $Q := \lim_{t\to\infty} Q_t$. Note that due to (2.5), as $t \uparrow \infty$,

$$\bar{X}_t := \sup_{0\le s\le t} X_s \uparrow \sup_{s\ge 0} X_s \stackrel{\mathrm{d}}{=} Q.$$

Likewise, $(Q_0 \mid Q_{-t} = 0)$ increases to Q as t goes to ∞. In operations research the steady-state workload is the natural performance metric when studying queueing systems that are in operation over long periods of time.

A second frequently used performance measure is the so-called *busy period*, to be denoted by τ, being the time it takes for the queue to drain (starting from time 0):

$$\tau := \inf\{t \geq 0 : Q_t = 0\}.$$

In this book we study the busy period in detail, where we typically assume that the workload is in stationarity at time 0. Several other metrics are analyzed as well, such as the workload correlation function $\mathbb{C}\mathrm{orr}(Q_0, Q_t)$ and the infimum attained by the workload process over a time interval of length t, that is, $\inf_{s\in[0,t]} Q_s$, in both cases assuming the workload is in stationarity at time 0.

Exercises

Exercise 2.1 Prove the Frullani integral equality, Eqn. (2.3), for $z \in \mathbb{C}$ with non-positive real part.

Hint: In the text a rough sketch was provided. Consider first $z \leq 0$. Use that

$$\frac{e^{-\gamma x} - e^{-(\gamma - z)x}}{x} = \int_{\gamma}^{\gamma - z} e^{-yx} \mathrm{d}y,$$

and then change the order of integration. Finally, by analytic extension, show that the formula is valid for $z \in \mathbb{C}$ with non-positive real part.

Exercise 2.2 Consider $X \in \mathbb{IG}(-1, 1)$. Prove that

$$\Pi(\mathrm{d}x) = \frac{1}{\sqrt{2\pi x^3}} e^{-x/2}.$$

Exercise 2.3 Let $X \in \mathbb{S}(\alpha_1, \beta_1, m)$ and $Y \in \mathbb{S}(\alpha_2, \beta_2, m_2)$ be independent.

(a) Check that X_1 is infinitely divisible.

(b) Characterize when $Z_t = X_t + Y_t$ has a stable distribution. Find the parameters of Z_t.

(c) Assume that $m_1 = 0$ and check that X is self-similar, that is, show that

$$(X_{Mt})_t \stackrel{\mathrm{d}}{=} (M^{1/\alpha_1} X_t)_t.$$

(d) Characterize γ for which $\mathbb{E}(X_1)^\gamma < \infty$.

Exercise 2.4 Let $X \stackrel{d}{=} S_\alpha(\sigma, \beta, m)$ with $\alpha \in (1, 2)$. Check that

(a) $aX \stackrel{d}{=} S_\alpha(|a|\sigma, \text{sign}(a)\beta, am)$, for $a \neq 0$;
(b) $-X \stackrel{d}{=} S_\alpha(\sigma, -\beta, -m)$;
(c) X is symmetric if and only if $\beta, m = 0$.

Exercise 2.5 Let $X \stackrel{d}{=} S_\alpha(\sigma, \beta, 0)$ with $\alpha \in (1, 2)$. In addition, we have the processes $X^{(1)} \stackrel{d}{=} S_\alpha(\sigma, 1, 0)$, $X^{(2)} \stackrel{d}{=} S_\alpha(\sigma, -1, 0)$, which we assume to be mutually independent. Check that

$$X \stackrel{d}{=} \left(\frac{1+\beta}{2}\right)^{1/\alpha} X^{(1)} + \left(\frac{1-\beta}{2}\right)^{1/\alpha} X^{(2)}.$$

Exercise 2.6 Prove that the sum of independent compound Poisson processes is a compound Poisson process. Find its parameters.

Exercise 2.7 Let X and Y be two independent Lévy processes; assume Y is increasing.

(a) Show that $(X_{Y_t})_{t\geq 0}$ is a Lévy process as well.
(b) Let X be a (standard) Brownian motion, and $Y \in \mathbb{G}(\beta, \gamma)$. Determine the Lévy exponent of $(X_{Y_t})_{t\geq 0}$.

(*Note*: With a specific choice of the parameters, this process is called a *variance gamma process*; see also Chapter 15.)

Exercise 2.8 Prove Prop. 2.1.

Exercise 2.9 For a given Lévy process X with $\mathbb{E}X_1 < 0$, let Q_0 obey the stationary workload distribution, and let L be the regulator process, with

$$L_t := \sup_{0\leq s\leq t} -X_s = -\inf_{0\leq s\leq t} X_s.$$

Then, according to the definition of the workload process, for $t \geq 0$,

$$Q_t = X_t + \max\{Q_0, L_t\}.$$

Show that $Q_t = \sup_{-\infty<s\leq t}(X_t - X_s)$, and that Q_t is stationary.

Chapter 3
Steady-State Workload

In this chapter we analyze the distribution of the stationary workload Q associated with the workload process $(Q_t)_t$ that was defined in the previous chapter. We first treat (in Section 3.1) the spectrally positive and (in Section 3.2) the spectrally negative case, for which we derive fairly explicit results. We then provide (in Section 3.3) an account of the general case (i.e. the case in which the jumps are not necessarily one sided), relying on Wiener–Hopf theory; in this spectrally two-sided case the results are substantially less clean.

The last two sections of this chapter treat two special cases with two-sided jumps for which the analysis can be done relatively explicitly, owing to specific assumptions imposed on the jumps. In Section 3.4 we consider the queue fed by a compound Poisson input with positive as well as negative jumps (in addition to a drift and a Brownian term), where these jumps have a phase-type distribution. We conclude the chapter in Section 3.5, where we briefly sketch results in the case that the queue's input process is a meromorphic Lévy process.

3.1 Spectrally Positive Case

The objective of this section is to characterize the stationary workload distribution Q of a queue fed by a spectrally positive Lévy process. More specifically, we find an explicit expression for the Laplace transform $\mathbb{E}e^{-\alpha Q}$ in terms of the model primitives $\varphi(\cdot)$ and $\psi(\cdot)$. Our approach is first to derive this expression for queues with compound Poisson input, and then to approximate any spectrally positive process by a compound Poisson. Using the fact that this can be done arbitrarily accurately, we thus find the desired result. We conclude this section by presenting an alternative derivation of the expression for $\mathbb{E}e^{-\alpha Q}$, based on martingale techniques.

As mentioned, we first consider the special case of compound Poisson input; the system under study is then a so-called M/G/1 queue. Jobs arrive according to

K. Dębicki, M. Mandjes, *Queues and Lévy Fluctuation Theory*, Universitext,
DOI 10.1007/978-3-319-20693-6_3

a Poisson process with rate λ, the jobs are i.i.d. and distributed as a non-negative random variable B (independent of the interarrival times), and the system is drained at a constant rate. Calling this depletion rate r, we impose the condition $\lambda\,\mathbb{E}B < r$ so as to guarantee that the queueing system is stable.

First observe that the queue is empty during exponentially distributed periods with mean λ^{-1}: as soon as the workload reaches value 0, it takes this exponentially distributed time before the next job arrives. We let $p_0 := \mathbb{P}(Q = 0)$ be the long-run fraction of time that the system is idle. For any $x > 0$, a *rate conservation argument* (also often referred to as a 'level-crossing argument') yields that the density $f_Q(\cdot)$ (assumed to exist) of the steady-state workload satisfies the equation

$$rf_Q(x) = \lambda \left(\int_{(0,x)} f_Q(y)\mathbb{P}(B > x - y)\mathrm{d}y + p_0\mathbb{P}(B > x) \right).$$

Here the left-hand side represents the 'probability flux' into the set $[0, x)$, whereas the right hand-side is the flux out of $[0, x)$ (where there are two possibilities: crossing level x by a job arriving when the workload is at level $y \in (0, x)$, and crossing level x by a job arriving when the queue is empty). Hence,

$$\begin{aligned} \bar{\kappa}(\alpha) &:= \int_{(0,\infty)} e^{-\alpha x} f_Q(x)\mathrm{d}x \\ &= \frac{1}{r} \int_{(0,\infty)} e^{-\alpha x} \lambda \left(\int_{(0,x)} f_Q(y)\mathbb{P}(B > x - y)\mathrm{d}y + p_0\mathbb{P}(B > x) \right) \mathrm{d}x. \end{aligned} \tag{3.1}$$

Interchanging the integrals, we obtain

$$\begin{aligned} &\int_{(0,\infty)} e^{-\alpha x} \int_{(0,x)} f_Q(y)\mathbb{P}(B > x - y)\mathrm{d}y\,\mathrm{d}x \\ &\quad= \int_{(0,\infty)} e^{-\alpha y} f_Q(y) \int_y^\infty e^{-\alpha(x-y)}\mathbb{P}(B > x - y)\mathrm{d}x\,\mathrm{d}y \\ &\quad= \int_{(0,\infty)} e^{-\alpha y} f_Q(y)\mathrm{d}y \int_0^\infty e^{-\alpha x}\mathbb{P}(B > x)\mathrm{d}x \\ &\quad= \bar{\kappa}(\alpha)\,\frac{1 - b(\alpha)}{\alpha}, \end{aligned}$$

where the last step relies on an elementary integration-by-parts argument. It is now easily verified that (3.1) reduces to

$$r\bar{\kappa}(\alpha) = \lambda\,(\bar{\kappa}(\alpha) + p_0)\,\frac{1 - b(\alpha)}{\alpha}.$$

Realizing that $\kappa(\alpha) := \mathbb{E}e^{-\alpha Q} = p_0 + \bar{\kappa}(\alpha)$ and that $\kappa(\alpha) \to 1$ as $\alpha \downarrow 0$, we conclude that $p_0 = (1 - \lambda\,\mathbb{E}B/r)$, so that we arrive at the following theorem, attributed to Pollaczek [177] and Khintchine [133], usually referred to as the *Pollaczek–Khintchine formula*.

Theorem 3.1 *Let $X \in \mathbb{CP}(r, \lambda, b(\cdot))$. For $\alpha \geq 0$,*

$$\kappa(\alpha) := \mathbb{E}e^{-\alpha Q} = \frac{r\alpha p_0}{r\alpha - \lambda(1 - b(\alpha))} = \frac{\alpha(r - \lambda\,\mathbb{E}B)}{r\alpha - \lambda(1 - b(\alpha))}.$$

Remark 3.1 Let $B_1^{\rm res}, B_2^{\rm res}, \ldots$ be i.i.d. samples from the residual lifetime distribution of B, defined by

$$\mathbb{P}(B^{\rm res} \leq x) = \frac{1}{\mathbb{E}B} \int_0^x \mathbb{P}(B > y) {\rm d}y;$$

from $\mathbb{E}B = \int_0^\infty \mathbb{P}(B > y){\rm d}y$ we know that the right-hand side of the previous display corresponds to a genuine distribution function. Realizing that $b^{\rm res}(\alpha) := \mathbb{E}e^{-\alpha B^{\rm res}} = (1 - b(\alpha))/(\alpha\,\mathbb{E}B)$ (which follows directly, using integration by parts), Thm. 3.1 can alternatively be written as

$$\kappa(\alpha) = \left(1 - \frac{\lambda\,\mathbb{E}B}{r}\right) \sum_{n=0}^{\infty} \left(\frac{\lambda\,\mathbb{E}B}{r}\right)^n (b^{\rm res}(\alpha))^n.$$

As a consequence, with $\varrho := \lambda\,\mathbb{E}B/r$,

$$\mathbb{P}(Q \leq x) = \mathbb{P}\left(\sum_{n=1}^{N} B_n^{\rm res} \leq x\right), \tag{3.2}$$

where $\mathbb{P}(N = n) = (1 - \varrho)\,\varrho^n$. This means that the steady-state workload Q can be interpreted as a geometric number of residuals of the job size B. ◇

Now the idea is to 'bootstrap' our findings for the compound Poisson case to the general spectrally positive case. Our goal is to find an expression for $\kappa(\alpha) = \mathbb{E}e^{-\alpha Q}$ for any $X \in \mathscr{S}_+$, by approximating $\varphi(\alpha)$ by a sequence $\varphi_n(\alpha)$ of terms that correspond to compound Poisson processes, then apply Thm. 3.1 for these compound Poisson processes, and finally take the limit $n \to \infty$.

In the spectrally positive case we have, for a certain d, $\sigma^2 \geq 0$, and measure $\Pi_\varphi(\cdot)$ such that $\int_{(0,\infty)} \min\{1, x^2\}\Pi_\varphi({\rm d}x) < \infty$, that the Laplace exponent reads

$$\varphi(\alpha) = \alpha d + \frac{1}{2}\alpha^2\sigma^2 + \int_{(0,\infty)} (e^{-\alpha x} - 1 + \alpha x\, 1_{\{x \in (0,1)\}})\Pi_\varphi({\rm d}x).$$

Now define, for a sequence ε_n such that $\varepsilon_n \downarrow 0$ as $n \to \infty$, the approximating Laplace exponent

$$\varphi_n(\alpha) := \left(d + \int_{\varepsilon_n}^1 x\Pi_\varphi(\mathrm{d}x) + \frac{\sigma^2}{\varepsilon_n}\right)\alpha + \frac{\sigma^2}{\varepsilon_n^2}(e^{-\alpha\varepsilon_n} - 1) + \int_{\varepsilon_n}^\infty (e^{-\alpha x} - 1)\Pi_\varphi(\mathrm{d}x). \tag{3.3}$$

It is easily verified that $\varphi_n(\alpha) \to \varphi(\alpha)$ as $n \to \infty$, whereas, for all $n \in \mathbb{N}$, $\varphi_n'(0) = \varphi'(0)$.

Importantly, $\varphi_n(\alpha)$, as given in (3.3), is *the Laplace exponent of a compound Poisson process*. This is seen as follows. The drift term of this compound Poisson process is

$$d_n := d + \int_{\varepsilon_n}^1 x\Pi_\varphi(\mathrm{d}x) + \frac{\sigma^2}{\varepsilon_n} > 0.$$

Then, the term $\sigma^2/\varepsilon_n^2 \cdot (e^{-\alpha\varepsilon_n} - 1)$ can be interpreted as the contribution of a Poisson stream (arrival rate $\lambda_{1,n} := \sigma^2/\varepsilon_n^2$) of jobs of deterministic size $\beta_{1,n} := \varepsilon_n$. Finally,

$$\int_{\varepsilon_n}^\infty (e^{-\alpha x} - 1)\Pi_\varphi(\mathrm{d}x) = \Pi_\varphi([\varepsilon_n, \infty)) \int_{\varepsilon_n}^\infty (e^{-\alpha x} - 1)\frac{\Pi_\varphi(\mathrm{d}x)}{\Pi_\varphi([\varepsilon_n, \infty))},$$

which is the contribution of a Poisson stream (arrival rate $\lambda_{2,n} := \Pi_\varphi([\varepsilon_n, \infty))$) of jobs, whose sizes are i.i.d. samples from a 'truncated distribution'. This distribution has density $\Pi_\varphi(\mathrm{d}x)/\Pi_\varphi([\varepsilon_n, \infty))$, for $x \geq \varepsilon_n$, and mean

$$\beta_{2,n} := \int_{\varepsilon_n}^\infty x\frac{\Pi_\varphi(\mathrm{d}x)}{\Pi_\varphi([\varepsilon_n, \infty))}.$$

Let Q_n be the steady-state workload of the queue fed by a compound Poisson process with Laplace exponent $\varphi_n(\alpha)$. Due to $\varphi_n(\alpha) \to \varphi(\alpha)$ it is conceivable that $\mathbb{E}e^{-\alpha Q_n} \to \mathbb{E}e^{-\alpha Q}$. From Thm. 3.1, we find that $\mathbb{E}e^{-\alpha Q_n}$ equals

$$\alpha(d_n - \lambda_{1,n}\beta_{1,n} - \lambda_{2,n}\beta_{2,n}) \Bigg/ \left(d_n\alpha - \frac{\sigma^2}{\varepsilon_n^2}(1 - e^{-\alpha\varepsilon_n}) - \int_{\varepsilon_n}^\infty (1 - e^{-\alpha x})\Pi_\varphi(\mathrm{d}x)\right).$$

It is a matter of straightforward calculus now to show that

$$\mathbb{E}e^{-\alpha Q_n} \to \frac{\alpha\varphi'(0)}{\varphi(\alpha)} \quad \text{as } n \to \infty; \tag{3.4}$$

the convergence follows from straightforward algebra. In other words, under the proviso that we can prove that $\mathbb{E}e^{-\alpha Q_n} \to \mathbb{E}e^{-\alpha Q}$, we have established the following result. Thm. 3.2 is often attributed to Zolotarev [222]; it is sometimes referred to as the *generalized Pollaczek–Khintchine formula.*

Theorem 3.2 *Let $X \in \mathscr{S}_+$. For $\alpha \geq 0$,*

$$\kappa(\alpha) := \mathbb{E}e^{-\alpha Q} = \frac{\alpha\varphi'(0)}{\varphi(\alpha)}.$$

The convergence $\mathbb{E}e^{-\alpha Q_n} \to \mathbb{E}e^{-\alpha Q}$ is a technical issue that lies beyond the scope of this textbook; we refer to [139, 207] for related results.

Thm. 3.2 provides us with the Laplace–Stieltjes transform of the random variable under consideration, but it is noticed that there are powerful techniques to numerically invert these transforms. Besides the classical contribution by Abate and Whitt [2], we wish to draw attention to novel ideas developed by den Iseger, reported on in [79]; we return to this topic in Chapter 16.

Alternative proofs of Thm. 3.2 rely on martingale techniques, most notably the celebrated *Kella–Whitt martingale* [130]; see also [146, Section 4.4] and [19, Section IX.3]. With

$$L_t(x) := \max\{0, L_t - x\} = \max\left\{0, -\inf_{0\leq s\leq t} X_s - x\right\},$$

it can be shown using stochastic integration theory that, for $X \in \mathscr{S}_+$,

$$K_t := \varphi(\alpha)\int_0^t e^{-\alpha Q_s}\mathrm{d}s + e^{-\alpha x} - e^{-\alpha Q_t} - \alpha L_t(x)$$

is a martingale. Below we provide the skeleton of the proof of this claim; after this proof we show how the martingale $(K_t)_t$ can be applied to obtain a compact derivation of the generalized Pollaczek–Khintchine formula.

Proposition 3.1 *$(K_t)_t$ is a martingale.*

Proof (sketch) Consider an adapted continuous process $(Y_t)_t$ that we assume to be of locally bounded variation, and define the process $Z_t := x + X_t + Y_t$. In addition, we introduce the processes $M_t := e^{-\alpha X_t}e^{-t\varphi(\alpha)}$ (which we have proved to be a martingale), and $B_t := e^{-\alpha Y_t}e^{t\varphi(\alpha)}$.

From stochastic integration theory [117], it is known that

$$\bar{K}_t := \int_0^t B_{s-}\mathrm{d}M_s$$

is a local martingale. Then we apply integration by parts. Using that $(Y_t)_t$ is continuous, we have

$$M_tB_t - M_0B_0 = \int_0^t M_{s-}\mathrm{d}B_s + \int_0^t B_{s-}\mathrm{d}M_s = \int_0^t M_{s-}\mathrm{d}B_s + \bar{K}_t,$$

and also

$$\bar{K}_t = M_t B_t - M_0 B_0 - \int_0^t M_s \mathrm{d}B_s.$$

Now plugging in the definitions, we obtain

$$\bar{K}_t = e^{\alpha(x-Z_t)} - 1 - \int_0^t e^{-\alpha X_s} e^{-s\varphi(\alpha)} \mathrm{d}\left(e^{-\alpha Y_s} e^{s\varphi(\alpha)}\right).$$

Realizing that

$$\mathrm{d}\left(e^{-\alpha Y_s} e^{s\varphi(\alpha)}\right) = \left(e^{-\alpha Y_s} e^{s\varphi(\alpha)}\right) \cdot (-\alpha\, \mathrm{d}Y_s + \varphi(\alpha)\, \mathrm{d}s),$$

it is seen that $\bar{K}_t = -e^{\alpha x} \check{K}_t$, with

$$\check{K}_t := \left(\varphi(\alpha) \int_0^t e^{-\alpha Z_s} \mathrm{d}s + e^{-\alpha x} - e^{-\alpha Z_t} - \alpha \int_0^t e^{-\alpha Z_s} \mathrm{d}Y_s\right).$$

In other words, $(\check{K}_t)_t$ is a local martingale.

Now take for $(Y_t)_t$ the process $(L_t(x))_t$, so that $Z_t = Q_t$ (with initial condition $Q_0 = x$). Notice that the above requirements for $(Y_t)_t$ are fulfilled; it is continuous due to the fact that $X \in \mathscr{S}_+$, and in addition it is non-decreasing. As $L_t(x)$ (as a function of t) only increases when $Q_t = 0$, we have

$$\int_0^t e^{-\alpha Q_s} \mathrm{d}L_s(x) = L_t(x).$$

It now follows that $(K_t)_t$ is indeed a local martingale. It is actually even a martingale [19, Lemma IX.3.3]; we do not prove this property. □

As mentioned earlier, we now use the martingale $(K_t)_t$ to prove the generalized Pollaczek–Khintchine formula. Assume that the queue is in stationarity at time 0. Stopping the martingale at time 1, realizing that the martingale has mean 0, and using that the stationarity of $(Q_t)_t$ implies that $\mathbb{E} \int_0^1 e^{-\alpha Q_s} \mathrm{d}s = \mathbb{E}e^{-\alpha Q}$, we obtain the identity

$$0 = \mathbb{E}K_1 = \varphi(\alpha)\mathbb{E}e^{-\alpha Q} + \mathbb{E}e^{-\alpha Q} - \mathbb{E}e^{-\alpha Q} - \alpha \mathbb{E}L_1(Q),$$

so that

$$\mathbb{E}e^{-\alpha Q} = \frac{\alpha\, \mathbb{E}L_1(Q)}{\varphi(\alpha)}.$$

Now realizing that $\mathbb{E}e^{-\alpha Q} \to 1$ as $\alpha \downarrow 0$, we retrieve Thm. 3.2. In passing, we have shown that, in stationarity, the 'mean amount of local time per time unit' equals $\varphi'(0)$.

Example 3.1 Consider the case of reflected Brownian motion, that is, we suppose $X \in \mathbb{B}\mathrm{m}(d, \sigma^2)$ for some $d < 0$. Then, with $\nu := -2d/\sigma^2 > 0$,

$$\mathbb{E}e^{-\alpha Q} = \frac{\alpha\varphi'(0)}{\varphi(\alpha)} = \frac{\nu}{\nu + \alpha}.$$

We conclude that the steady-state workload in a Brownian queue has an exponential distribution with mean $1/\nu$. Observe that the steady-state workload of this queue has no atom at 0. ◊

Example 3.2 Consider the case of $X \in \mathbb{S}(\alpha, 1, -r)$ with $\alpha \in (1, 2)$ and $r > 0$; recall that $X \in \mathscr{S}_+$ as the underlying stable distribution is totally skewed to the right. Then, using that according e.g. to Furrer [96, Prop. 2.25],

$$\varphi(s) = rs + \frac{1}{\cos(\pi(\alpha/2 - 1))} s^\alpha,$$

one can invert the transform of Thm. 3.2 to obtain [96, Prop. 3.3]

$$\mathbb{P}(Q > u) = \sum_{n=0}^{\infty} \frac{(-r\cos(\pi(\alpha/2 - 1)))^n}{\Gamma(1 + (\alpha - 1)n)} u^{(\alpha-1)n}.$$

It is concluded that Q has a so-called *Mittag–Leffler distribution*. ◊

Thm. 3.2 reveals all moments of the steady-state workload Q, and in particular its mean and variance:

$$\mu := \mathbb{E}Q = -\frac{\mathrm{d}}{\mathrm{d}\alpha} \frac{\alpha\varphi'(0)}{\varphi(\alpha)}\bigg|_{\alpha\downarrow 0} = \frac{\varphi''(0)}{2\varphi'(0)}, \tag{3.5}$$

and similarly,

$$v := \mathbb{V}\mathrm{ar}\, Q = \frac{1}{4}\left(\frac{\varphi''(0)}{\varphi'(0)}\right)^2 - \frac{1}{3}\frac{\varphi'''(0)}{\varphi'(0)}, \tag{3.6}$$

provided that these objects are well defined. As a general rule, Q has a finite nth moment if and only if X_1 has a finite $(n + 1)$st moment; for details on this property, we refer to Kyprianou [146, Exercise 7.1].

3.2 Spectrally Negative Case

For spectrally negative input, the reasoning is substantially simpler. First observe that $e^{\beta_0 X_t}$ is a martingale, with $\beta_0 := \Psi(0) > 0$; this is a direct consequence of the fact that $\Phi(\beta_0) = 0$.

Note that, by virtue of 'Reich's identity' (see Eqn. (2.5)), we have

$$\mathbb{P}(Q \geq u) = \mathbb{P}\left(\sup_{t\geq 0} X_t \geq u\right) = \mathbb{P}(\exists t \geq 0 : X_t \geq u).$$

Now consider the stopping time $\sigma(u) := \inf\{t \geq 0 : X_t \geq u\}$; obviously, the event $\{\sigma(u) < \infty\}$ coincides with $\{\exists t \geq 0 : X_t \geq u\}$. 'Optional sampling' (see e.g. Williams [220, Chapter A14]) thus gives, for any positive u,

$$1 = e^{\beta_0 X_0} = \mathbb{E}\left(e^{\beta_0 X_{\sigma(u)}} 1_{\{\sigma(u)<\infty\}}\right) = e^{\beta_0 u}\mathbb{P}(\sigma(u) < \infty) = e^{\beta_0 u}\mathbb{P}(Q \geq u);$$

here we use that, due to the fact that there are no jumps in the upward direction, given a certain level $u > 0$ is reached, it is attained with equality. We conclude that $\mathbb{P}(Q > u) = \mathbb{P}(Q \geq u) = e^{-\beta_0 u}$.

Theorem 3.3 *Let $X \in \mathscr{S}_-$. Then Q is exponentially distributed with mean $1/\beta_0$.*

It is noted that a similar argument can be used for the waiting time in the classical G/M/1 queue; it entails that the waiting time distribution in that model is exponential (with an atom at 0).

It can be intuitively understood that Q is exponentially distributed. Recall that, according to (2.5), Q is distributed as $\sup_{t\geq 0} X_t$. Observe that, due to the fact that there are no positive jumps, the *running supremum process* $\bar{X}_t := \sup_{0\leq s\leq t} X_s$ attains any value between 0 and $\sup_{t\geq 0} X_t$. It is seen that, using that the increments of X are stationary and independent,

$$\mathbb{P}\left(\sup_{t\geq 0} X_t > x + y \,\middle|\, \sup_{t\geq 0} X_t > x\right) = \mathbb{P}\left(\sup_{t\geq 0} X_t > y\right),$$

which implies that $\sup_{t\geq 0} X_t$ is memoryless.

3.3 Spectrally Two-Sided Case

In this section, we consider the stationary workload in the situation that the input process is not necessarily spectrally one sided. We use, as we did earlier, 'Reich's identity' (see Eqn. (2.5)), in the sense that we analyze the all-time supremum attained by the Lévy process $(X_t)_t$. The body of theory related to these results is known as *Wiener–Hopf theory*. In this monograph we do not provide an in-depth treatment of Wiener–Hopf results, but restrict ourselves to a brief introduction. For a detailed description we refer e.g. to Kyprianou [146, Chapter 6], and for a compact but fairly complete account, to Kyprianou [147]; see also [102, 187], as well as the textbooks [43, 193].

In this section we highlight the main results from Wiener–Hopf theory. We do so by predominantly concentrating on the discrete-time case (i.e. the situation of a discrete-time random walk); we state the associated Wiener–Hopf result and sketch the crucial elements of the proof. The continuous-time case can be dealt with essentially analogously (albeit that a number of technical complications have to be addressed); we restrict ourselves to just stating the main result in this continuous-time case.

Discrete time—Consider the random walk $S_n := \sum_{i=1}^n Y_i$, with the Y_i being i.i.d., distributed as a generic random variable Y. Let $\bar{S}_n$ be the running maximum process:

$$\bar{S}_n := \sup_{i\in\{1,\ldots,n\}} S_i;$$

G_n denotes the (first) epoch at which that running maximum is attained. Let T be an (independent) geometric random variable, that is, $\mathbb{P}(T = k) = p(1-p)^k$, for some $p \in (0, 1)$ and $k \in \{0, 1, \ldots\}$. Our goal is to study the joint distribution of the pairs $(\bar{S}_T, G_T)$ and $(S_T - \bar{S}_T, T - G_T)$.

- First consider the pair $(\bar{S}_T, G_T)$. Observe that the number of record values attained before T is a *geometric* random variable; let us denote this number by N. It follows that both $\bar{S}_T$ and G_T can be written as the sum of N i.i.d. non-negative random variables. It is an elementary exercise that such 'geometric sums' of i.i.d. random variables are infinitely divisible (to this end, realize that a geometric random variable with success probability p can be written as the sum of n negative-binomial random variables with parameters n^{-1} and p). It is concluded that $\bar{S}_T$ and G_T are infinitely divisible.
- Then observe that $(S_T - \bar{S}_T, T - G_T)$ is independent of $(\bar{S}_T, G_T)$. This can be intuitively understood from the fact that, by virtue of the memoryless property of the geometric distribution, neither the position of the running maximum (i.e. $\bar{S}_T$) nor the epoch at which this is attained (i.e. G_T) has any impact on the amount by which the process has gone down between G_T and T (i.e. $\bar{S}_T - S_T$), nor the time elapsed until the 'killing epoch' T (i.e. $T - G_T$).
- It can then be seen that $S_T - \bar{S}_T$ has the same distribution as the running *minimum* process (draw a picture!); in addition, $T - G_T$ corresponds to the time epoch that this minimum is attained. Using the same argumentation as before, these two random variables are infinitely divisible.
- It is a straightforward computation that, with $s \in (0, 1]$ and $\alpha \in \mathbb{R}$,

$$\mathbb{E}\, s^T e^{\alpha i S_T} = \frac{p}{1 - (1-p)s\,\mathbb{E}e^{\alpha i Y}}. \tag{3.7}$$

At the same time,

$$\begin{aligned}
&\exp\left(-\int_{-\infty}^{\infty}\sum_{n=1}^{\infty}\frac{1}{n}(1-s^n e^{\alpha \mathrm{i} x})(1-p)^n \mathbb{P}(S_n \in \mathrm{d}x)\right) \\
&= \exp\left(-\sum_{n=1}^{\infty}\frac{1}{n}(1-s^n \mathbb{E}e^{\alpha \mathrm{i} S_n})(1-p)^n\right) \\
&= \exp\left(-\sum_{n=1}^{\infty}\frac{1}{n}\left((1-p)^n - \left((1-p)s\mathbb{E}e^{\alpha \mathrm{i} Y}\right)^n\right)\right) \\
&= \exp\left(\log p - \log\left(1 - s(1-p)\,\mathbb{E}e^{\alpha \mathrm{i} Y}\right)\right),
\end{aligned}$$

which evidently equals (3.7).

- Hence (S_T, T) can be written as the sum of two independent terms, that is, $(\bar{S}_T, G_T)$ and $(S_T - \bar{S}_T, T - G_T)$, which are both infinitely divisible. Also, observe that $\bar{S}_T$ is non-negative and $S_T - \bar{S}_T$ is non-positive. As a result, applying standard Wiener–Hopf arguments,

$$\mathbb{E}\, s^{G_T} e^{\alpha \mathrm{i} \bar{S}_T} = \exp\left(-\int_0^{\infty}\sum_{n=1}^{\infty}\frac{1}{n}(1-s^n e^{\alpha \mathrm{i} x})(1-p)^n \mathbb{P}(S_n \in \mathrm{d}x)\right),$$

and

$$\mathbb{E}\, s^{T-G_T} e^{\alpha \mathrm{i}(S_T - \bar{S}_T)} = \exp\left(-\int_{-\infty}^{0}\sum_{n=1}^{\infty}\frac{1}{n}(1-s^n e^{\alpha \mathrm{i} x})(1-p)^n \mathbb{P}(S_n \in \mathrm{d}x)\right).$$

We have thus found the distribution of the running maximum $\bar{S}_T$, and, in fact, a set of more refined results as well, such as the distribution of the running maximum (i.e. $\bar{S}_T$) jointly with the value of the process at the killing epoch (i.e. S_T). Intuitively, by letting the success probability $p \downarrow 0$, the random variable T becomes infinitely large. This means that we obtain the distribution of the all-time supremum $\bar{S}$ by inserting $p = 0$ (and obviously $s = 1$) in the expression above; we thus obtain

$$\mathbb{E}e^{\alpha \mathrm{i} \bar{S}} = \exp\left(-\int_0^{\infty}\sum_{n=1}^{\infty}\frac{1}{n}(1-e^{\alpha \mathrm{i} x})\mathbb{P}(S_n \in \mathrm{d}x)\right). \tag{3.8}$$

The stationary workload being distributed as the all-time maximum $\bar{X}$ of the driving Lévy process $(X_t)_t$, our objective in this section is to find the continuous-time counterpart of (3.8).

Continuous time—Let us therefore proceed with analyzing the continuous-time setting; from now on T is exponentially distributed with mean $1/\vartheta$, independently of the Lévy process $(X_t)_t$. Trivially, for $\beta \geq 0$ and $\alpha \in \mathbb{R}$,

$$\mathbb{E}e^{-\beta T+\alpha i X_T} = \frac{\vartheta}{\vartheta+\beta-\log \mathbb{E}e^{\alpha i X_1}}. \tag{3.9}$$

But on the other hand, using the Frullani integral identity (2.3), we also have

$$\begin{aligned}
&\exp\left(-\int_0^\infty \int_{-\infty}^\infty \frac{1}{t}\left(e^{-\vartheta t}-e^{-(\vartheta+\beta)t}e^{\alpha i x}\right)\mathbb{P}(X_t\in \mathrm{d}x)\mathrm{d}t\right)\\
&= \exp\left(-\int_0^\infty \frac{1}{t}\left(e^{-\vartheta t}-e^{-(\vartheta+\beta)t}\mathbb{E}e^{\alpha i X_t}\right)\mathrm{d}t\right)\\
&= \exp\left(-\int_0^\infty \frac{1}{t}\left(e^{-\vartheta t}-e^{-(\vartheta+\beta-\log \mathbb{E}e^{\alpha i X_1})t}\right)\mathrm{d}t\right)\\
&= \frac{\vartheta}{\vartheta+\beta-\log \mathbb{E}e^{\alpha i X_1}}.
\end{aligned}$$

Define the following two functions that play a crucial role in Lévy fluctuation theory:

$$k(\vartheta,\alpha) := \exp\left(-\int_0^\infty \int_{(0,\infty)} \frac{1}{t}\left(e^{-t}-e^{-\vartheta t-\alpha x}\right)\mathbb{P}(X_t\in \mathrm{d}x)\mathrm{d}t\right); \tag{3.10}$$

$$\bar{k}(\vartheta,\beta) := \exp\left(-\int_0^\infty \int_{(-\infty,0)} \frac{1}{t}\left(e^{-t}-e^{-\vartheta t+\beta x}\right)\mathbb{P}(X_t\in \mathrm{d}x)\mathrm{d}t\right). \tag{3.11}$$

Mimicking the line of reasoning for the discrete-time random walk case, we obtain the following result; a graphical illustration is provided in Fig. 3.1.

Theorem 3.4 *The pairs $(\bar{X}_T, G_T)$ and $(X_T-\bar{X}_T, T-G_T)$ are independent. Also, with $k(\cdot\,,\cdot)$ as defined in (3.10), for $\beta \geq 0$ and $\alpha \in \mathbb{R}$,*

$$\begin{aligned}
\mathbb{E}e^{-\beta G_T+\alpha i \bar{X}_T} &= \frac{k(\vartheta+\beta,-\alpha i)}{k(\vartheta,0)}\\
&= \exp\left(-\int_0^\infty \int_0^\infty \frac{1}{t}\left(e^{-\vartheta t}-e^{-(\vartheta+\beta)t}e^{\alpha i x}\right)\mathbb{P}(X_t\in \mathrm{d}x)\mathrm{d}t\right),
\end{aligned}$$

and, with $\bar{k}(\cdot\,,\cdot)$ as defined in (3.11), for $\beta \geq 0$ and $\alpha \in \mathbb{R}$,

$$\begin{aligned}
\mathbb{E}e^{-\beta(T-G_T)+\alpha i(X_T-\bar{X}_T)} &= \frac{\bar{k}(\vartheta+\beta,\alpha i)}{\bar{k}(\vartheta,0)}\\
&= \exp\left(-\int_0^\infty \int_{-\infty}^0 \frac{1}{t}\left(e^{-\vartheta t}-e^{-(\vartheta+\beta)t}e^{\alpha i x}\right)\mathbb{P}(X_t\in \mathrm{d}x)\mathrm{d}t\right).
\end{aligned}$$

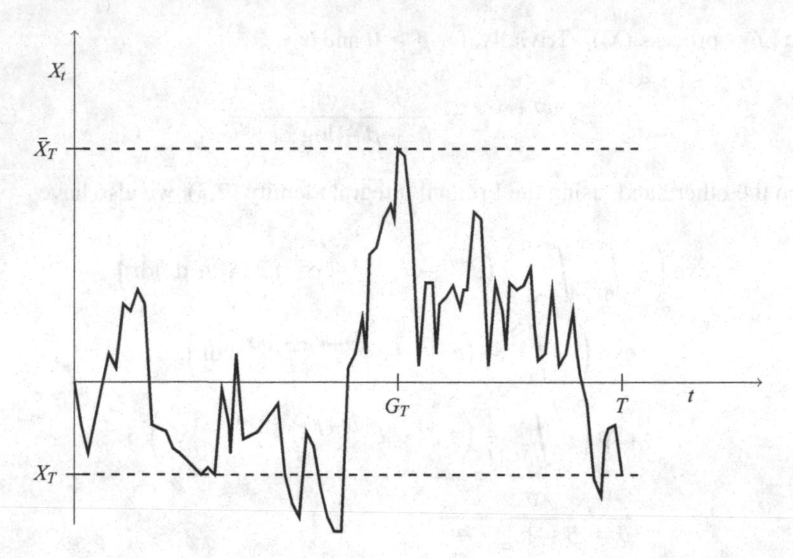

Fig. 3.1 Graphical representation of the Wiener–Hopf result; the maximum value $\bar{X}_T$ is attained at time G_T. The pairs $(\bar{X}_T, G_T)$ and $(X_T - \bar{X}_T, T - G_T)$ are independent, with T being exponentially distributed

This theorem also provides us with the joint transform of the all-time supremum $\bar{X} := \lim_{t\to\infty} \bar{X}_t$ and the corresponding epoch $G := \lim_{t\to\infty} G_t$, simply by taking $\vartheta = 0$ in the formulas.

It is noted that in the spectrally one-sided cases, the function $k(\cdot,\cdot)$ can be computed explicitly. For $X \in \mathscr{S}_-$, we have $k(q,\beta) = 1/(\Psi(q)+\beta)$, whereas for $X \in \mathscr{S}_+$,

$$k(\vartheta,\alpha) = \frac{\psi(\vartheta)-\alpha}{\vartheta-\varphi(\alpha)}.$$

Also, for $X \in \mathscr{S}_-$,

$$\bar{k}(q,\beta) = \frac{\Psi(q)-\beta}{q-\Phi(\beta)},$$

and for $X \in \mathscr{S}_+$, we have $\bar{k}(\vartheta,\alpha) = 1/(\psi(\vartheta)+\alpha)$.

We can now use Thm. 3.4, so as to obtain an expression for $\mathbb{E}e^{-\alpha Q}$, by plugging in $\beta = 0$, and by letting the parameter ϑ go to 0. We thus obtain the following result.

Theorem 3.5 *Let X be a general Lévy process. For $\alpha \geq 0$,*

$$\mathbb{E}e^{-\alpha Q} = \exp\left(-\int_0^\infty \int_{(0,\infty)} \frac{1}{t}\,(1-e^{-\alpha x})\,\mathbb{P}(X_t \in \mathrm{d}x)\mathrm{d}t\right) = \frac{k(0,\alpha)}{k(0,0)}.$$

It is noted that in the case that $(X_t)_t$ is compound Poisson, there is a minor subtlety that needs to be taken into account, essentially relating to the case that at time t *no* jobs have arrived, for which this formula needs to be slightly adapted; we refer for details to Kyprianou [146, pp. 167–168].

Example 3.3 Suppose $X \in \mathbb{B}\text{m}(d, \sigma^2)$ for some $d < 0$; we recover the result of Example 3.1. For ease we consider the case that $d = -1$ and $\sigma^2 = 1$, but the general case works analogously. We have to evaluate

$$\exp\left(-\int_0^\infty \int_0^\infty \frac{1}{t}\,(1-e^{-\alpha x})\,\frac{1}{\sqrt{2\pi t}}\exp\left(-\frac{(x+t)^2}{2t}\right)\mathrm{d}x\,\mathrm{d}t\right).$$

First change the order of integration, and then apply [146, Exercise 1.6(iii)], to obtain

$$\mathbb{E}e^{-\alpha Q} = \exp\left(-\int_0^\infty \frac{1}{x}\left(e^{-2x}-e^{-x(2+\alpha)}\right)\mathrm{d}x\right).$$

Then, with the Frullani integral identity, the last expression can be rewritten as $2/(2+\alpha)$, as desired: Q is exponentially distributed with mean $\frac{1}{2}$. ◇

Example 3.4 Interestingly, the result stated in Thm. 3.5 allows us to deal with specific situations that are *not* in $\mathscr{S}_+$ or $\mathscr{S}_-$. Consider for instance the queue with compound Poisson input; the jobs arrive according to a Poisson process with rate λ, and the job sizes are i.i.d. samples from a normal distribution with mean $d < 0$ and variance σ^2. In this case, we have to evaluate, by conditioning on the number of jobs that have entered the system in $(0, t)$,

$$\exp\left(-\int_0^\infty \int_0^\infty \frac{1}{t}\,(1-e^{-\alpha x})\sum_{k=1}^\infty \frac{e^{-\lambda t}(\lambda t)^k}{k!}\frac{1}{\sqrt{2\pi k\sigma^2}}\exp\left(-\frac{(x-kd)^2}{2k\sigma^2}\right)\mathrm{d}x\,\mathrm{d}t\right);$$

note that the $k = 0$ term can be omitted. First perform the integration over t; it leads to (recognize the gamma function!)

$$\exp\left(-\sum_{k=1}^\infty \int_0^\infty (1-e^{-\alpha x})\frac{1}{\sqrt{2\pi k\sigma^2}}\frac{1}{k}\exp\left(-\frac{(x-kd)^2}{2k\sigma^2}\right)\mathrm{d}x\right)$$
$$= \exp\left(-\sum_{k=1}^\infty \frac{1}{k}\left(\Psi_{\rm N}\left(-\frac{d\sqrt{k}}{\sigma}\right) - \left(e^{-\alpha d+\frac{1}{2}\alpha^2\sigma^2}\right)^k \Psi_{\rm N}\left(-\frac{d\sqrt{k}}{\sigma}+\sigma\alpha\sqrt{k}\right)\right)\right),$$

with $\Psi_{\rm N}(\cdot)$ denoting the complementary distribution function of a standard normal random variable. It is observed that λ cancels (which could be a priori expected; why?); a similar procedure can be executed in the case that a deterministic drift is added, but then λ obviously does not cancel. ◇

Phase-type jumps in one direction—As demonstrated earlier in this section, for $X \in \mathscr{S}_+$ or $X \in \mathscr{S}_-$, the transform $\mathbb{E}e^{-\alpha Q}$ can be evaluated in an explicit form in terms of the model primitives (i.e. in terms of the functions $\varphi(\cdot)$, $\psi(\cdot)$ for $X \in \mathscr{S}_+$, and the functions $\Phi(\cdot)$ and $\Psi(\cdot)$ for $X \in \mathscr{S}_-$). Wiener–Hopf theory shows that this is *not* possible if the jumps are not one sided (except for particular special cases; see e.g. Example 3.4): if the driving Lévy process is not spectrally one sided, one has not succeeded in expressing $\mathbb{E}e^{-\alpha Q}$ explicitly in terms of the Lévy exponent of the input process (and related quantities). Instead, Thm. 3.5 expresses the Laplace transform of the stationary workload in terms of a double integral involving the density $\mathbb{P}(X_t \in dx)$, which is for many Lévy processes not known in closed form.

There are spectrally two-sided Lévy processes in which rather explicit analysis is possible, though: those for which either the upward jumps or the downward jumps have a *phase-type* distribution. We start our exposition by giving the definition of phase-type distributions; see Asmussen [19, Section III.4] for more background.

Definition 3.1 Let $(J_t)_t$ denote a continuous-time Markov jump process on the finite state space $\{1,\dots,n\}\cup\dagger$, where states $1,\dots,n$ are transient and $\dagger$ is absorbing. Let (the row vector) $\boldsymbol{a}$ denote a distribution on $\{1,\dots,n\}$, to be interpreted as the initial distribution. Defining ζ as the first entrance time of the process $(J_t)_t$ to state $\dagger$, a non-negative random variable P is said to be of phase type if $\mathbb{P}(\zeta \le t) = \mathbb{P}(P \le t)$, for all $t \ge 0$. The transition matrix of $(J_t)_t$ is given by

$$\begin{pmatrix} T & \boldsymbol{t} \\ \boldsymbol{0}^{\mathrm{T}} & 0 \end{pmatrix},$$

where $\boldsymbol{t} := -T\boldsymbol{1}$, and $\boldsymbol{0}$ and $\boldsymbol{1}$ denote an n-dimensional all-zeros vector and an n-dimensional all-ones vector, respectively.

Phase-type random variables, to be informally thought of as sums and mixtures of independent exponentially distributed random variables, are particularly useful because they may serve as accurate approximations of general distributions on the positive half-line. They have the attractive property that they are at the same time relatively easy to work with. Let us proceed by providing a number of classical examples of phase-type distributions.

- *Erlang distribution.* In this case the phase-type random variable P corresponds to the sum of $n \in \{1,2,\dots\}$ independent exponentially distributed random variables, say each having mean $1/\beta > 0$. In the language of the above definition, we have $a_1 = 1$, $T_{i,i+1} = -T_{ii} = \beta$ for $i = 1,\dots,n-1$, $T_{nn} = -t_n = \beta$, while the other entries of $\boldsymbol{a}$, T, and $\boldsymbol{t}$ are 0. Figure 3.2 pictorially illustrates the Erlang distribution (characterized by the parameters n and β).
- *Hyperexponential distribution.* Here the phase-type random variable P corresponds to probability a_i to an exponentially distributed random variable with mean $1/\beta_i$, for $i = 1,\dots,n$; we obviously require $\sum_{i=1}^n a_i = 1$. This distribution can, in the notation introduced above, be represented by $t_i = -T_{ii} = \beta_i$ for

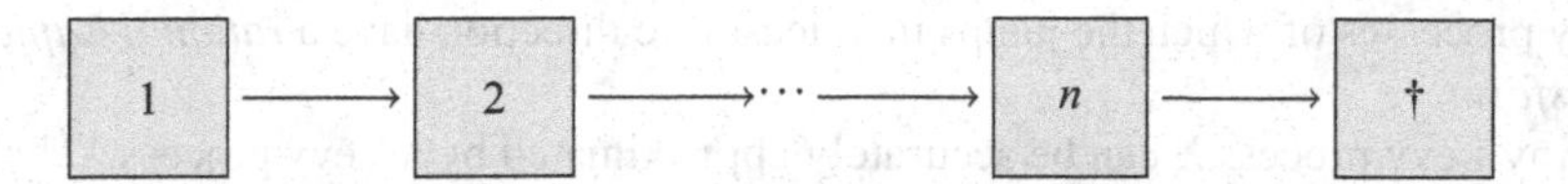

Fig. 3.2 Graphical representation of an Erlang distribution

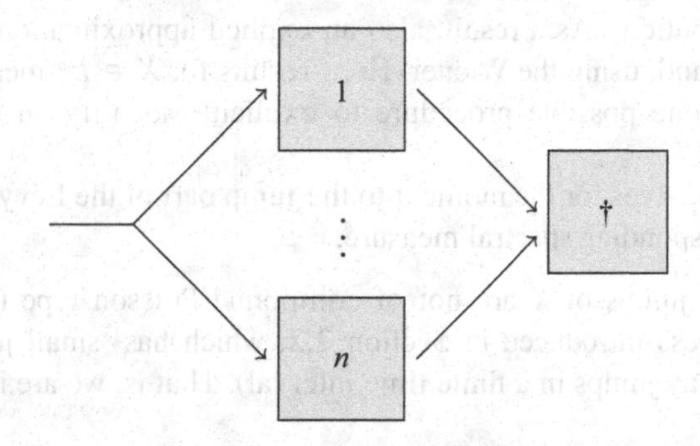

Fig. 3.3 Graphical representation of a hyperexponential distribution

$i = 1, \ldots, n$, while all other entries of T are 0. Figure 3.3 provides an illustration of this distribution (characterized by the n-dimensional vectors $\boldsymbol{a}$ and $\boldsymbol{\beta}$).

As mentioned above, phase-type distributions have several attractive properties. Using standard Markov-chains theory, it is readily verified that the distribution is $\mathbb{P}(P \leq x) = 1 - \boldsymbol{a}e^{Tx}\mathbf{1}$; the nth moment is $\mathbb{E}(P^n) = (-1)^n n!\,\boldsymbol{a}T^{-n}\mathbf{1}$; the Laplace transform is

$$\mathbb{E}e^{-\alpha P} = \boldsymbol{a}\,(\alpha I - T)^{-1}\boldsymbol{t}.$$

We have already noted that phase-type distributions have the potential to provide accurate approximations of any distribution on the positive half-line. More precisely, the class of phase-type distributions is *dense* (in the sense of weak convergence) in the set of all probability distributions on the positive half-line; see Asmussen [19, Thm. III.4.2]. In fact even the smaller class of mixtures of Erlang distributions already has this property; in this case P corresponds to an Erlang distribution with parameters n_k and β_k with probability p_k, for $k = 1, \ldots, m$.

In the sequel we let $\mathscr{P}_+$ be the class of Lévy processes whose jumps in the upward direction are of phase type; likewise, $\mathscr{P}_-$ are the Lévy processes with phase-type downward jumps. We denote by $\mathscr{P}$ the class of Lévy processes whose jumps in one direction are of phase type, that is, $\mathscr{P} := \mathscr{P}_+ \cup \mathscr{P}_-$. As we mentioned above, for $X \in \mathscr{P}$ the Wiener–Hopf factors can still be given relatively explicitly, that is, in terms of the roots of a given equation; the key results of Wiener–Hopf theory for $X \in \mathscr{P}$ are presented in e.g. [22, 149, 150]. We do not include these results in this book, but the case of $X \in \mathscr{P}_+ \cap \mathscr{P}_-$, that is, phase-type jumps in *both* directions, is treated in detail in the Section 3.4. As an aside we remark that, in great generality, the results for $X \in \mathscr{P}$ carry over to a larger class of processes, that is, the class of

Lévy processes of which the jumps in at least one direction have a *rational Laplace transform.*

Any Lévy process X can be accurately approximated by a Lévy process $X^{(\mathrm{app})} \in \mathscr{P}$; in fact, the 'denseness result' mentioned above entails that the 'fit' can be made arbitrarily precise (which is typically achieved by increasing the dimension n of the phase-type distribution). As a result, also an explicit approximation $\mathbb{E}e^{-\alpha Q^{(\mathrm{app})}}$ for $\mathbb{E}e^{-\alpha Q}$ can be found, using the Wiener–Hopf results for $X \in \mathscr{P}$ mentioned above. We now sketch one possible procedure to explicitly identify an approximating process $X^{(\mathrm{app})} \in \mathscr{P}$.

We restrict ourselves for the moment to the jump part of the Lévy process X; let $\Pi(\cdot)$ be the corresponding spectral measure.

(A) Suppose the jumps of X are not of compound Poisson type (recall e.g. the gamma process introduced in Section 2.2, which has 'small jumps', that is, infinitely many jumps in a finite time interval). That is, we are in the situation that

$$\int_{-\infty}^{\infty} \Pi(\mathrm{d}x) = \infty.$$

It is evident that, for any $\varepsilon > 0$, we can write $X = X^{(1)} + X^{(2)}$, where the spectral measure of $X^{(1)}$, say $\Pi^{(1)}$, equals the restriction of $\Pi(\cdot)$ to $(-\varepsilon, \varepsilon)$, and the spectral measure of $X^{(2)}$, say $\Pi^{(2)}$, equals the restriction of $\Pi(\cdot)$ to $(-\infty, -\varepsilon] \cup [\varepsilon, \infty)$. Clearly $X^{(2)}$ is of compound Poisson type, which is covered by step (B), so we are left with dealing with $X^{(1)}$.
To this end, define

$$\mu_\varepsilon := \begin{cases} \int_{-\varepsilon}^{\varepsilon} x\Pi(\mathrm{d}x) \text{ if } \int_{-\infty}^{\infty} |x|\Pi(\mathrm{d}x) < \infty, \\ 0 \qquad\qquad \text{otherwise,} \end{cases} \qquad \sigma_\varepsilon := \int_{-\varepsilon}^{\varepsilon} x^2 \Pi(\mathrm{d}x).$$

Then, due to Asmussen and Rosiński [30], if $\Pi(x)$ has the form $L(x)/|x|^{\alpha+1}$ for $x \to 0$, the function $L(\cdot)$ being slowly varying at 0 and $\alpha \in (0, 2)$, then, as $\varepsilon \downarrow 0$,

$$\left(\frac{X^{(1)}(t) - \mu_\varepsilon t}{\sigma_\varepsilon}\right)_t \xrightarrow{\mathrm{d}} (B_t)_t, \tag{3.12}$$

where '$\xrightarrow{\mathrm{d}}$' stands for weak convergence in the space $D[0,1]$ equipped with the uniform metric, and $(B_t)_t$ denotes standard Brownian motion. As a consequence, we can approximate $X^{(1)}$ by a Brownian motion with drift: $X_t^{(1)} \approx \mu_\varepsilon t + \sigma_\varepsilon B_t$, for some small $\varepsilon > 0$.

(B) Based on step (A), we can approximate our general Lévy process by the sum of a drift, a Brownian motion, and a compound Poisson process; we have eliminated the 'small jumps'. To generate an approximating Lévy process

$X^{(\mathrm{app})}$ in the class $\mathscr{P}$, we have to approximate either the upward jumps or the downward jumps of the Poisson process by phase-type jumps; as mentioned above, in principle this can be done arbitrarily accurately. There are various algorithms to generate phase-type fits; see e.g. [28, 91, 112].

Observe that after performing the steps (A) and (B) we have found an (accurately) approximating Lévy process $X^{(\mathrm{app})}$ in $\mathscr{P}$; hence the transform $\mathbb{E}e^{-\alpha Q}$ can be approximated by $\mathbb{E}e^{-\alpha Q^{(\mathrm{app})}}$, which can be found in (semi-)explicit form. It is stressed that there is an obvious trade-off between accuracy and computational effort needed. The smaller ε, the better the approximation of the small jump process $X^{(2)}$ by a Brownian motion with drift, but the larger the arrival rate λ_ε of the compound Poisson process $X^{(1)}$. The approach described above resembles the procedure proposed by Jeannin and Pistorius in [116]; there the jumps in both directions are approximated by means of a *generalized hyperexponential* model.

3.4 Spectrally Two-Sided Case: Phase-Type Jumps

In the previous section we briefly touched on results related to the case $X \in \mathscr{P}_+ \cup \mathscr{P}_-$ (i.e. phase-type jumps in *at least* one direction). This section provides a more detailed account of the case $X \in \mathscr{P}_+ \cap \mathscr{P}_-$ (i.e. phase-type jumps in *both* directions). We follow the exposition of Asmussen [20], to which we refer for a complete treatment.

The Lévy exponent for $X \in \mathscr{P}_+ \cap \mathscr{P}_-$ reads

$$\xi(s) = \mathrm{i}sd - \frac{1}{2}s^2\sigma^2 + \int_{-\infty}^{0} \left(e^{\mathrm{i}sx} - 1\right) \Pi_-(\mathrm{d}x) + \int_{0}^{\infty} \left(e^{\mathrm{i}sx} - 1\right) \Pi_+(\mathrm{d}x),$$

where

$$\Pi_-(\mathrm{d}x) = \lambda_- f_{B_-}(-x)\,\mathrm{d}x, \quad \Pi_+(\mathrm{d}x) = \lambda_+ f_{B_+}(x)\,\mathrm{d}x.$$

Here $\lambda_- > 0$ (respectively, $\lambda_+ > 0$) is the Poisson arrival rate of negative (respectively, positive) jumps. The downward jumps, the absolute values of which have density $f_{B_-}(\cdot)$, are assumed to be of phase type; using the terminology of Definition 3.1, the underlying random variable B_- is characterized by the dimension n^-, the initial distribution $\boldsymbol{a}^-$, and the transition rate matrix T^-. Likewise, the upward jumps are distributed as a random variable B_+, corresponding to a phase-type distribution with density $f_{B_+}(\cdot)$; this distribution is characterized by dimension n^+, initial distribution $\boldsymbol{a}^+$, and transition rate matrix T^+.

In the sequel it turns out to be convenient to work with the cumulant $\check{\xi}(s) := \xi(-is)$, given by

$$\check{\xi}(s) = ds + \frac{1}{2}s^2\sigma^2 +$$
$$\lambda_-\left(a_-\left(sI^- - T^-\right)^{-1}t^- - 1\right) + \lambda_+\left(a_-\left(-sI^+ - T^+\right)^{-1}t^+ - 1\right), \quad (3.13)$$

where I^- and I^+ are identity matrices of dimension n^- and n^+, respectively. Relying on expression (3.13), it follows that $\check{\xi}(s)$ can be written as $\check{\xi}_n(s)/\check{\xi}_d(s)$, where the functions in both the numerator and the denominator are polynomials in s. The degree of the denominator is clearly $n^- + n^+$ (use e.g. Cramer's rule). Focusing on the case that $\sigma^2 > 0$, which we assume throughout the rest of this section, the degree of the numerator is $n^- + n^+ + 2$; the case that $\sigma^2 = 0$ can be dealt with analogously, but is left out for brevity.

The primary objective of this section is to evaluate $\ell(u \mid \vartheta) := \mathbb{P}(\bar{X}_T \geq u)$, for T being exponentially distributed with mean $1/\vartheta$ (independently of the driving Lévy process X). Having identified an expression for $\ell(u \mid \vartheta)$, we have found $k(\vartheta, \alpha)/k(\vartheta, 0) = \mathbb{E}e^{-\alpha\bar{X}_T}$ as well; to see this realize that, applying integration by parts,

$$\int_0^\infty e^{-\alpha u}\ell(u \mid \vartheta)\mathrm{d}u = \int_0^\infty \ell(u \mid \vartheta)\mathrm{d}\left(-\frac{e^{-\alpha u}}{\alpha}\right) = \frac{1 - \mathbb{E}e^{-\alpha\bar{X}_T}}{\alpha}$$
$$= \frac{1}{\alpha}\left(1 - \frac{k(\vartheta, \alpha)}{k(\vartheta, 0)}\right).$$

Two-sided exit—To finally be able to evaluate $\ell(u \mid \vartheta)$, we first set up a procedure to determine the two-sided exit probability

$$\ell(u, v) := \mathbb{P}\left(X_{\tau[u,v)} \geq v\right), \quad (3.14)$$

where $\tau[u, v) := \inf\{t \geq 0 : X_t \notin [u, v)\}$, for $u \leq 0 \leq v$. This procedure can then be used to develop an algorithm to determine $\ell(u) := \mathbb{P}(\bar{X} \geq 0)$, and finally $\ell(u \mid \vartheta) = \mathbb{P}(\bar{X}_T \geq u)$. We have decided to first concentrate on the two-sided exit probability (3.14) for two reasons. First, the ideas behind the procedure to determine (3.14) are a natural step when developing the analogous procedure for the one-sided counterpart

$$\ell(u) = \mathbb{P}(\exists t \geq 0 : X_t \geq u).$$

Second, the quantity $\ell(u, v)$ is of independent interest, as it plays an important role in the theory of *finite-buffer queues*; see Section 11.1.

Define the following events. For $i = 1, \ldots, n^+$ we let $E_i^+(u, v)$ correspond to the event that the interval $[u, v)$ is first left because there is a time epoch t such that

$X_t > v$ while the phase-type distribution is in state i. Likewise, for $j = 1,\dots,n^-$ in the event $E_i^+(u,v)$ the interval $[u,v)$ is first left because there is a time epoch t such that $X_t < u$ while the phase-type distribution is in state j. The event $E_0^+(u,v)$ corresponds to the event that the interval $[u,v)$ is first left due to a t such that $X_t = v$ (i.e. u is exceeded not because of a jump, but due to the Brownian part of the Lévy process); $E_0^-(u,v)$ is defined analogously. We define the probabilities $p_i^+(u,v) := E_i^+(u,v)$ and $p_i^-(u,v) := E_i^-(u,v)$, for $i = 0,\dots,n^+$ and $j = 0,\dots,n^-$. Notice that these are $n^- + n^+ + 2$ unknowns that we wish to identify.

We introduce the following (mean-0) martingale:

$$\check{\xi}(s)\int_0^t e^{sX_r}\mathrm{d}r + 1 - e^{sX_t}$$

(where verification of this process being a martingale is standard). Note that this is a variant of the Kella–Whitt martingale that we introduced for the spectrally positive case, but focusing on the Lévy process $(X_t)_t$ rather than the workload process $(Q_t)_t$ (and therefore not taking reflection at 0 into account). The idea is that we apply 'optional sampling', with the stopping time $\tau[u,v)$; for considerations regarding the justification of this procedure, we refer to the in-depth treatment in [20]. We thus arrive at

$$0 = \check{\xi}(s)\left(\int_0^{\tau[u,v)} e^{sX_r}\mathrm{d}r\right) + 1$$
$$- \sum_{i=0}^{n^+}\mathbb{E}\left(e^{sX_{\tau[u,v)}}1_{\{E_i^+(u,v)\}}\right) - \sum_{j=0}^{n^-}\mathbb{E}\left(e^{sX_{\tau[u,v)}}1_{\{E_j^-(u,v)\}}\right).$$

Now realize that for $i = 0$ the random variable $X_{\tau[u,v)}$ on the event $E_i^+(u,v)$ equals the deterministic quantity v; for $i = 1,\dots,n^+$ this v should be increased by the phase-type random variable B_+ but now with initial distribution $\boldsymbol{e}_i^+$ rather than $\boldsymbol{a}^+$; the initial distribution $\boldsymbol{e}_i^+$ corresponds to starting in state i with probability 1. A similar reasoning applies to $E_j^-(u,v)$, with $j = 0,\dots,n^-$. As a consequence, with

$$b_i^+(s) := \boldsymbol{e}_i^+(-sI^+ + T^+)^{-1}\boldsymbol{t}^+, \quad b_j^-(s) := \boldsymbol{e}_j^-(sI^- + T^-)^{-1}\boldsymbol{t}^-,$$

we arrive at

$$0 = \check{\xi}(s)\left(\int_0^{\tau[u,v)} e^{sX_r}\mathrm{d}r\right) + 1$$
$$- e^{sv}\left(p_0^+(u,v) + \sum_{i=1}^{n^+} b_i^+(s)p_i^+(u,v)\right) - e^{su}\left(p_0^-(u,v) + \sum_{j=1}^{n^-} b_j^-(s)p_j^-(u,v)\right).$$

From Ivanovs et al. [114] it now follows that there are (in the complex plane) $n^- + n^+ + 2$ solutions to the equation $\check{\xi}(s) = 0$, say $s^{(1)}, \dots, s^{(n^-+n^++2)}$ (one of which equals 0). Hence we obtain the following $n^- + n^+ + 2$ equations, solving equally many unknowns: for $k = 1, \dots, n^- + n^+ + 2$,

$$1 = e^{s^{(k)}v}\left(p_0^+(u,v) + \sum_{i=1}^{n^+} b_i^+(s^{(k)})p_i^+(u,v)\right) + e^{s^{(k)}u}\left(p_0^-(u,v) + \sum_{j=1}^{n^-} b_j^-(s^{(k)})p_j^-(u,v)\right).$$

Finally, $\ell(u,v)$ follows from

$$\ell(u,v) = p_0^+(u,v) + \sum_{i=1}^{n^+} p_i^+(u,v).$$

One-sided exit—We now consider the one-sided exit probability, exploiting the techniques deployed above. Let $\sigma(u)$ be the first time the level $u > 0$ is exceeded (realize this is a defective random variable!), and define the stopping time $\sigma_t(u) := \min\{t, \sigma(u)\}$. Also introduce $p_{i,t}(u)$, for $i = 1, \dots, n^+$, as the probabilities that u is exceeded before time t while the jump (which has phase-type distribution, and is distributed as the random variable B_+) is in state i. In addition, $p_{0,t}(u)$ is the probability that level u is exceeded before time t without an 'overshoot' (i.e. u is exceeded due to the Brownian part: $\sigma(u) < t$ and $X_{\sigma(u)} = u$). Application of 'optional sampling' yields

$$0 = \check{\xi}(s)\left(\int_0^{\sigma_t(u)} e^{sX_r}\mathrm{d}r\right) + 1 - e^{su}\left(p_{0,t}(u) + \sum_{i=1}^{n^+} b_i^+(s)p_{i,t}(u)\right) - \mathbb{E}\left(e^{sX_t}1_{\{t<\sigma(u)\}}\right). \quad (3.15)$$

Appealing again to [114], it follows that there are $n^+ + 1$ roots with a positive real part, which we call $\bar{s}^{(1)}, \dots, \bar{s}^{(n^++1)}$; it can be argued [20, Section 4] that (3.15) applies to all these roots. Again the number of unknowns equals the number of equations: we have for $k = 1, \dots, n^+ + 1$, when sending $t \to \infty$,

$$1 = e^{\bar{s}^{(k)}u}\left(p_0(u) + \sum_{i=1}^{n^+} b_i^+(\bar{s}^{(k)})p_i(u)\right),$$

where $p_i(u) := p_{i,\infty}(u)$, for $i = 0, \ldots, n^+$. Finally, it is observed that we have now identified the steady-state workload distribution:

$$\ell(u) := \mathbb{P}(\bar{X} \geq u) = \mathbb{P}(Q \geq u) = \sum_{i=0}^{n^+} p_i(u).$$

Now that we have identified the distribution of the all-time supremum, we proceed by focusing on $\ell(u \,|\, \vartheta) := \mathbb{P}(\bar{X}_T \geq u)$, where we recall that T is exponentially distributed with mean $1/\vartheta$. To this end, the idea is to work with the (mean-0) martingale

$$\left(\breve{\xi}(s) - \vartheta\right) \int_0^t e^{sX_r - \vartheta r} \mathrm{d}r + 1 - e^{sX_t - \vartheta t}, \tag{3.16}$$

and again we apply 'optional sampling' with stopping time $\sigma_t(u)$. From [114] it follows that there are $n^+ + 1$ roots to the equation $\breve{\xi}(s) = \vartheta$ that have a positive real part, for any $\vartheta > 0$, say $\bar{s}^{(1)}(\vartheta), \ldots, \bar{s}^{(n^+ + 1)}(\vartheta)$.

Analogously to the events we introduced before, let $E_i(u)$ be the event in which level u is crossed due to an upward jump, while the underlying phase-type random variable is in state i (for $i = 1, \ldots, n^+$), and $E_0(u)$ be its counterpart in which crossing level u is due to the Brownian part of the Lévy process (i.e. $X_{\sigma(u)} = u$). Observe that

$$\ell(u \,|\, \vartheta) = \mathbb{P}(\sigma(u) \leq T) = \mathbb{E}\left(e^{-\vartheta \sigma(u)} 1_{\{\sigma(u) < \infty\}}\right) = \sum_{i=0}^{n^+} \ell_i(u \,|\, \vartheta),$$

where

$$\ell_i(u \,|\, \vartheta) := \mathbb{E}\left(e^{-\vartheta \sigma(u)} 1_{\{\sigma(u) < \infty, E_i(u)\}}\right).$$

Now the $\ell_i(u \,|\, \vartheta)$ can be determined as before. 'Optional sampling' with stopping time $\sigma_t(u)$ applied to Eqn. (3.16) (after subsequently taking $t \to \infty$ and plugging in the roots $\bar{s}^{(1)}(\vartheta), \ldots, \bar{s}^{(n^+ + 1)}(\vartheta)$) leads to the following $n^+ + 1$ equations:

$$1 = e^{\bar{s}^{(k)}(\vartheta)\, u} \left(\ell_0(u \,|\, \vartheta) + \sum_{i=1}^{n^+} b_i^+(\bar{s}^{(k)}(\vartheta)) \ell_i(u \,|\, \vartheta) \right).$$

From this system of equations, it is easily seen that the $\ell_i(u \mid \vartheta)$ can be written as linear combinations of the $w_k(u) := \exp(-\bar{s}^{(k)}(\vartheta)\, u)$: for a (known) square matrix $M_{ik}(\vartheta)$ (where i runs from 0 to n^+ and k from 1 to $n^+ + 1$) we have for $i = 0, \ldots, n^+$,

$$\ell_i(u \mid \vartheta) = \sum_{k=1}^{n^+ + 1} M_{ik}(\vartheta) w_k(u),$$

leading to

$$\mathbb{E}e^{-\alpha \bar{X}_T} = \frac{k(\vartheta, \alpha)}{k(\vartheta, 0)} = 1 - \sum_{i=0}^{n^+} \sum_{k=1}^{n^+ + 1} M_{ik}(\vartheta) \frac{\alpha}{\alpha + \bar{s}^{(k)}(\vartheta)}.$$

An expression for $\bar{k}(q, \beta)/\bar{k}(q, 0)$ follows similarly.

We remark that in the above analysis we continuously assumed that all roots involved are *simple*; for an in-depth analysis of the consequences of non-unique roots, we refer to D'Auria et al. [67].

3.5 Spectrally Two-Sided Case: Meromorphic Processes

In this section we study the case that the process X corresponds to a so-called *meromorphic* Lévy process; Wiener–Hopf theory for this class of processes has attracted substantial attention during the past decade. We do not include all proofs and details, but instead we restrict ourselves to presenting the most important concepts and results. For a full treatment we refer to Kuznetsov et al. [145] and its predecessor [144].

In Section 3.4 we saw that the class of Lévy processes with phase-type jumps, that is, $\mathscr{P}_+ \cap \mathscr{P}_-$, is relatively easy to work with, in that the Wiener–Hopf factors (and related quantities) could be given in a fairly explicit way. It is noted, however, that class fails to incorporate 'small jumps': the models considered are of the compound Poisson type, possibly with an additional Brownian term. The class of meromorphic Lévy processes remedies this deficiency: it has the potential to include small jumps (for an example of this see the β-class in Section 15.1), while the Wiener–Hopf factors still allow a relatively explicit form. It is noted that, at a structural level, there is a strong level of similarity between the framework described in the present section on the one hand, and that of the phase-type jumps highlighted in Section 3.4 on the other hand.

As in the previous section, it is convenient to consider the cumulant $\check{\xi}(s) := \xi(-is)$. For the class of meromorphic Lévy processes, denoted by $\mathscr{M}$, we have

$$\check{\xi}(s) = ds + \frac{1}{2}s^2\sigma^2 + \int_{-\infty}^{0} (e^{sx} - 1 - sx)\, \Pi_-(\mathrm{d}x) + \int_{0}^{\infty} (e^{sx} - 1 - sx)\, \Pi_+(\mathrm{d}x),$$

where

$$\Pi_-(\mathrm{d}x) = \left(\sum_{n=1}^{\infty} a_n^- b_n^- e^{b_n^- x}\right)\mathrm{d}x, \quad \Pi_+(\mathrm{d}x) = \left(\sum_{n=1}^{\infty} a_n^+ b_n^+ e^{-b_n^+ x}\right)\mathrm{d}x,$$

with $a_n^-, a_n^+, b_n^-, b_n^+$ all positive, the sequences b_n^- and b_n^+ both increasing in n, in such a way that $b_n^- \to \infty$ as well as $b_n^+ \to \infty$ as $n \to \infty$. It is noted that, as explained in detail in [145, Section 2], there is an interesting link between the class of meromorphic Lévy processes and the concept of completely monotone functions (where we mention, as an aside, that the latter plays a key role in Chapter 7 of this textbook). It is readily checked that $\check{\xi}(s)$ can be alternatively written as

$$\check{\xi}(s) = ds + \frac{1}{2}s^2\sigma^2 + s^2\sum_{n=1}^{\infty}\frac{a_n^-}{b_n^-(b_n^- + s)} + s^2\sum_{n=1}^{\infty}\frac{a_n^+}{b_n^+(b_n^+ - s)}.$$

We thus conclude that there are poles at the locations $-b_n^-, b_n^+$, for $n = 1, 2, \ldots$.

Now consider (again in line with the setup of the previous section) the roots of the equation $\check{\xi}(s) = \vartheta$, for a given $\vartheta > 0$. [145, Thm. 1(v)] entails that these roots are real and interlace with the poles b_n^-, b_n^+. More precisely, calling the positive roots $s_n^+ \equiv s_n^+(\vartheta)$ and the negative roots $s_n^- \equiv s_n^-(\vartheta)$, for $n = 1, 2, \ldots$, we have the ordering

$$\cdots - b_2^- < -s_2^- < -b_1^- < -s_1^- < 0 < s_1^+ < b_1^+ < s_2^+ < b_2^+ < \cdots.$$

[145, Thm. 1(v)] also states that, for all $\vartheta > 0$,

$$\vartheta - \check{\xi}(s) = \vartheta\left(\prod_{n=1}^{\infty}\frac{1 - (s/s_n^+(\vartheta))}{1 - s/b_n^+}\right)\left(\prod_{n=1}^{\infty}\frac{1 + (s/s_n^-(\vartheta))}{1 + s/b_n^-}\right).$$

Observing that, with T being exponentially distributed with mean $1/\vartheta$, we have $\mathbb{E}e^{sX_T} = \vartheta/(\vartheta - \check{\xi}(s))$, it is found that, for $\alpha \geq 0$,

$$\mathbb{E}e^{-\alpha\bar{X}_T} = \frac{k(\vartheta,\alpha)}{k(\vartheta,0)} = \left(\prod_{n=1}^{\infty}\frac{1 + \alpha/b_n^+}{1 + (\alpha/s_n^+(\vartheta))}\right),$$

and likewise, for $\beta \geq 0$,

$$\mathbb{E}e^{-\beta(\bar{X}_T - X_T)} = \frac{\bar{k}(\vartheta,\beta)}{\bar{k}(\vartheta,0)} = \left(\prod_{n=1}^{\infty}\frac{1 + \beta/b_n^-}{1 + (\beta/s_n^-(\vartheta))}\right).$$

The transform $\mathbb{E}e^{-\alpha Q}$ of the stationary workload now follows from the above expression for $\mathbb{E}e^{-\alpha\bar{X}_T}$ by passing ϑ to 0.

In Section 15.1 we describe the so-called β-class of Lévy processes; this class, but also several other examples of Lévy processes in the class $\mathscr{M}$, can be found in [145, Section 3].

Exercises

Exercise 3.1 Prove for $\alpha \geq 0$ that $\varphi_n(\alpha) \to \varphi(\alpha)$ as $n \to \infty$, with $\varphi_n(\cdot)$ given in Eqn. (3.3). Also, prove the convergence in (3.4).

Exercise 3.2 Let X be the sum of $\mathbb{B}\mathrm{m}(d, \sigma^2)$ and $\mathbb{CP}(r, \lambda, b(\cdot))$ (which we assume to be independent), where the jumps of the compound Poisson process stem from an Erlang distribution with parameters n and μ. Under what condition is the queue stable? Suppose this stability condition is fulfilled; what is the Laplace–Stieltjes transform of Q? Does this distribution have an atom in 0?

Exercise 3.3 Let $X \in \mathbb{CP}(r, \lambda, b(\cdot))$, with $r < 0$ and the random variable B taking values in $(-\infty, 0]$ only: a compound Poisson with *positive* drift, and *negative* jumps. (Recall that in the definition of $\mathbb{CP}(r, \lambda, b(\cdot))$ the drift was *subtracted*, so that a positive drift corresponds to $r < 0$.)

(a) Under what condition is the corresponding Lévy-driven queue stable?
(b) Let $f_Q(\cdot)$ be the density of the stationary workload. Show that this density satisfies, for $x > 0$,

$$-rf_Q(x) = \lambda \int_{(x,\infty)} f_Q(y)\mathbb{P}(B \leq x - y)\mathrm{d}y.$$

(c) Does this distribution have an atom in 0?
(d) Define $b_-(\beta) := \mathbb{E}\, e^{\beta B}$. Show that there is a unique positive β_0 such that $-r\beta_0 - \lambda + \lambda b_-(\beta_0) = 0$.
(e) Combine your answers to questions (b) and (d) to prove that

$$\frac{1 - b_-(\beta_0)}{\beta_0} f_Q(x) = \int_{-\infty}^{0} f_Q(x - y)\mathbb{P}(B \leq y)\mathrm{d}y.$$

(f) Use the representation $b_-(\beta) = \int_{-\infty}^{0} e^{\beta x}\mathbb{P}(B \in \mathrm{d}x)$ and integration by parts to obtain, for all $x > 0$,

$$f_Q(x) \int_{-\infty}^{0} e^{\beta_0 y}\mathbb{P}(B \leq y)\mathrm{d}y = \int_{-\infty}^{0} f_Q(x - y)\mathbb{P}(B \leq y)\mathrm{d}y,$$

and show that, for all $x > 0$ and $y \leq 0$,

$$f_Q(x)e^{\beta_0 y} = f_Q(x - y).$$

(g) Argue that $f_Q(x) = \beta_0 e^{-\beta_0 x}$, for $x \geq 0$.
(h) Show that this answer is in line with Thm. 3.3.

Exercise 3.4 Suppose that $X \in \mathbb{CP}(r, \lambda, b(\cdot))$, with the jumps being distributed exponentially with mean $1/\vartheta$. Let Q be the corresponding stationary workload.

(a) Derive explicit expressions for $\mathbb{E}\, Q$ and $\mathbb{V}\mathrm{ar}\, Q$.
(b) Find the distribution of N and B_1^{res} in the representation

$$\mathbb{P}(Q \leq x) = \mathbb{P}\left(\sum_{n=1}^{N} B_n^{\mathrm{res}} \leq x\right).$$

Exercise 3.5 Let $X \in \mathbb{B}\mathrm{m}(d, \sigma^2)$. Prove that the stationary workload distribution exists only if $d < 0$.

Exercise 3.6 Let X be spectrally negative. Prove that $e^{\beta_0 X_t}$, with $\beta_0 := \Psi(0)$, is a martingale.

Exercise 3.7 Let X be a Lévy process with $\mathbb{E}X_1 > 0$. Show that the queue fed by X is *unstable*, that is,

$$\lim_{t\to\infty} Q_t = \infty$$

a.s., regardless of the value of Q_0.
Hint: Realize that Q_t increases in Q_0, and that $Q_t \geq X_t$.

Exercise 3.8 Let $X \in \mathbb{S}(\alpha, -1, -r)$, with $\alpha \in (1, 2)$ and $r > 0$. Find the distribution of the steady-state workload Q of a queue fed by X.

Exercise 3.9 Let X correspond to a Poisson(λ^+) stream of upward jumps with an exponentially distributed size with mean $1/\mu^+$, superimposed by a Poisson(λ^-) stream of downward jumps with an exponentially distributed size with mean $1/\mu^-$. Determine the distribution of Q.

Chapter 4
Transient Workload

This chapter focuses on characterizing the transient workload. The structure is the same as that of the previous chapter: in terms of Laplace–Stieltjes transforms, we subsequently address the transient workload for the spectrally positive case, the spectrally negative case, and the general (i.e. spectrally two-sided) case. Notice that in this chapter the requirement $\mathbb{E}X_1 < 0$ is not needed: the notion of transient workload is well defined without assuming the underlying queueing system is stable.

4.1 Spectrally Positive Case

In this section we characterize the workload at time t, for the spectrally positive case, in terms of a so-called *double transform*, that is, we find an expression for

$$\mathbb{E}_x e^{-\alpha Q_T} := \mathbb{E}\left(e^{-\alpha Q_T} \mid Q_0 = x\right) = \int_0^\infty \vartheta e^{-\vartheta t}\mathbb{E}\left(e^{-\alpha Q_t} \mid Q_0 = x\right) \mathrm{d}t;$$

in other words, we consider the transform of the transient workload after an exponential amount of time (with mean ϑ^{-1}). We present three approaches that find an expression for this double transform, which uniquely defines the distribution of Q_t, conditional on $Q_0 = x$.

In a first approach, we use an argument reminiscent of the level-crossing procedure that we introduced for the steady-state workload, and which is due e.g. to Beneš [39] and Takács [208]. As in the previous chapter, we first focus on $X \in \mathbb{CP}(r, \lambda, b(\cdot))$; for ease we normalize time such that $r = 1$ (which can be done without losing any generality). Define $F_t(y)$ as the probability that Q_t does not exceed y:

$$F_t(y) := \mathbb{P}(Q_t \leq y \mid Q_0 = x).$$

K. Dębicki, M. Mandjes, *Queues and Lévy Fluctuation Theory*, Universitext,
DOI 10.1007/978-3-319-20693-6_4

Now consider the event that at time $t+\Delta t$ the workload is at most y, for some $y > 0$. This means that (i) either the workload was below $y + \Delta t$ at time t, and there was no job arriving between t and $t + \Delta t$, or (ii) the workload was $z \in [0, y)$ at time t, but between t and $t + \Delta t$ a job arrived of size at most $y - z$; events corresponding to more than one arrival have a probability that is $o(\Delta t)$. This reasoning yields the following equation:

$$\begin{aligned} F_{t+\Delta t}(y) &= F_t(y + \Delta t)(1 - \lambda\, \Delta t) \\ &\quad + \lambda\, \Delta t \left(\int_0^y f_t(z)\mathbb{P}(B \leq y - z)\mathrm{d}z + F_t(0)\mathbb{P}(B \leq y) \right) + o(\Delta t). \end{aligned}$$

Subtracting $F_t(y)$ from both sides, dividing the whole equation by Δt, and then letting $\Delta t \downarrow 0$, leads to the partial differential equation

$$\frac{\partial}{\partial t} F_t(y) = f_t(y) - \lambda F_t(y) + \lambda \left(\int_0^y f_t(z)\mathbb{P}(B \leq y - z)\mathrm{d}z + F_t(0)\mathbb{P}(B \leq y) \right), \tag{4.1}$$

for $y > 0$, with $f_t(\cdot) := F_t'(\cdot)$ denoting the density of Q_t.

The next step is to convert this partial differential equation into an explicit expression for the double transform. To make the notation more compact, we introduce $\kappa_T(\alpha) := \mathbb{E}_x e^{-\alpha Q_T}$, and

$$\bar{\kappa}_T(\alpha) := \int_0^\infty \vartheta e^{-\vartheta t} \int_{(0,\infty)} e^{-\alpha y} f_t(y)\mathrm{d}y\, \mathrm{d}t = \kappa_T(\alpha) - \mathbb{P}(Q_T = 0).$$

The basic idea is now to take the double transform of the whole partial differential equation (4.1). We do this term by term. To this end, first notice that standard calculus yields that

$$\int_0^\infty \vartheta e^{-\vartheta t} \int_{(0,\infty)} e^{-\alpha y} \left(\frac{\partial}{\partial t} F_t(y) \right) \mathrm{d}y\, \mathrm{d}t = \frac{\vartheta}{\alpha} \left(\kappa_T(\alpha) - e^{-\alpha x} \right); \tag{4.2}$$

this is found by first interchanging the order of the integrals, and then using the identity

$$\int_0^\infty \vartheta e^{-\vartheta t} \left(\frac{\partial}{\partial t} F_t(y) \right) \mathrm{d}t = -\vartheta 1_{\{y>x\}} + \vartheta^2 \int_0^\infty e^{-\vartheta t} F_t(y)\mathrm{d}t,$$

which can be obtained by an elementary integration-by-parts argument. Similarly, we find

$$\int_0^\infty \vartheta e^{-\vartheta t} \int_{(0,\infty)} e^{-\alpha y} F_t(y)\mathrm{d}y\, \mathrm{d}t = \frac{1}{\alpha}\, \kappa_T(\alpha), \tag{4.3}$$

and also

$$\int_0^\infty \vartheta e^{-\vartheta t} \int_{(0,\infty)} e^{-\alpha y} \left(\int_0^y f_t(z)\mathbb{P}(B \le y - z)\mathrm{d}z + F_t(0)\mathbb{P}(B \le y) \right) \mathrm{d}y\, \mathrm{d}t$$
$$= \frac{b(\alpha)}{\alpha}\kappa_T(\alpha). \tag{4.4}$$

Upon combining Eqns. (4.2), (4.3), and (4.4), we thus obtain the following equation in $\kappa_T(\alpha)$ and $\bar{\kappa}_T(\alpha)$:

$$\frac{\vartheta}{\alpha}\left(\kappa_T(\alpha) - e^{-\alpha x}\right) = \bar{\kappa}_T(\alpha) - \frac{\lambda}{\alpha}\kappa_T(\alpha) + \frac{\lambda b(\alpha)}{\alpha}\kappa_T(\alpha).$$

Now using $\bar{\kappa}_T(\alpha) = \kappa_T(\alpha) - \mathbb{P}(Q_T = 0)$, and recalling the expression for the Laplace exponent in the compound Poisson case (i.e. $\varphi(\alpha) = \alpha - \lambda + \lambda b(\alpha)$), we can isolate $\kappa_T(\alpha)$:

$$\kappa_T(\alpha) = \frac{\vartheta}{\vartheta - \varphi(\alpha)}\left(e^{-\alpha x} - \frac{\alpha}{\vartheta}\mathbb{P}(Q_T = 0)\right).$$

Now the double transform has been expressed in terms of the model primitives (i.e. the Laplace exponent $\varphi(\cdot)$), apart from the term $\mathbb{P}(Q_T = 0)$. Observe that $\mathbb{P}(Q_T = 0)$ is a function of ϑ and x only (i.e. independent of α); call this function $G(\vartheta, x)$. As $\kappa_T(\alpha)$ is a transform, we should have that for all (α, ϑ) for which the denominator vanishes (i.e. $\alpha = \psi(\vartheta)$), the numerator vanishes too (as otherwise the transform equals ∞). As a consequence, we have that for all $x \ge 0$,

$$e^{-\psi(\vartheta)x}\frac{\vartheta}{\psi(\vartheta)} = G(\vartheta, x).$$

We have now identified an explicit expression for the double transform $\kappa_T(\alpha)$ for the case of compound Poisson input.

As before, the next step is to approximate any $X \in \mathscr{S}_+$ by a compound Poisson process. This procedure eventually yields the following result.

Theorem 4.1 *Let $X \in \mathscr{S}_+$, and let T be exponentially distributed with mean $1/\vartheta$, independently of X. For $\alpha \ge 0$, $x \ge 0$,*

$$\mathbb{E}_x e^{-\alpha Q_T} = \vartheta \int_0^\infty e^{-\vartheta t}\mathbb{E}_x e^{-\alpha Q_t}\mathrm{d}t = \frac{\vartheta}{\vartheta - \varphi(\alpha)}\left(e^{-\alpha x} - \frac{\alpha}{\psi(\vartheta)}e^{-\psi(\vartheta)x}\right).$$

Remark 4.1 It can be seen that this transform which characterizes the workload's time-dependent behavior is in line with the generalized Pollaczek–Khintchine formula (see Thm. 3.2), in at least two ways. First we can check what happens when we let $\vartheta \downarrow 0$: then the exponentially distributed time T lies infinitely far in the future. Indeed,

$$\lim_{\vartheta\downarrow 0}\frac{\vartheta}{\vartheta-\varphi(\alpha)}\left(e^{-\alpha x}-\frac{\alpha}{\psi(\vartheta)}e^{-\psi(\vartheta)x}\right)=\frac{\alpha}{\varphi(\alpha)}\cdot\lim_{\vartheta\downarrow 0}\frac{\vartheta}{\psi(\vartheta)}=\frac{\alpha\varphi'(0)}{\varphi(\alpha)},$$

using that $\vartheta/\psi(\vartheta)\to 1/\psi'(0)=\varphi'(0)$ as $\vartheta\downarrow 0$.

Second we can check what the transform is of the workload after an exponentially distributed amount of time, *starting in the workload's stationary distribution*; obviously this transform should correspond to the stationary distribution, too. This turns out to be indeed the case:

$$\begin{aligned}\int_0^\infty \mathbb{E}_x e^{-\alpha Q_T}\mathbb{P}(Q_0\in \mathrm{d}x) &= \int_0^\infty\left(\frac{\vartheta}{\vartheta-\varphi(\alpha)}\left(e^{-\alpha x}-\frac{\alpha}{\psi(\vartheta)}e^{-\psi(\vartheta)x}\right)\right)\mathbb{P}(Q_0\in \mathrm{d}x)\\ &= \frac{\vartheta}{\vartheta-\varphi(\alpha)}\left(\mathbb{E}e^{-\alpha Q_0}-\frac{\alpha}{\psi(\vartheta)}\mathbb{E}e^{-\psi(\vartheta)Q_0}\right)\\ &= \frac{\vartheta}{\vartheta-\varphi(\alpha)}\left(\frac{\alpha\varphi'(0)}{\varphi(\alpha)}-\frac{\alpha}{\psi(\vartheta)}\frac{\psi(\vartheta)\varphi'(0)}{\vartheta}\right)=\frac{\alpha\varphi'(0)}{\varphi(\alpha)},\end{aligned}$$

as expected. ◇

Remark 4.2 Thm. 4.1 can also be used to determine the joint distribution of the workload at time 0 and at an exponential time T, assuming that the workload is in stationarity at time 0. Relying on the arguments used in Remark 4.1, in self-evident notation,

$$\mathbb{E}e^{-\bar\alpha Q_0-\alpha Q_T}=\frac{\vartheta\varphi'(0)}{\vartheta-\varphi(\alpha)}\left(\frac{\alpha+\bar\alpha}{\varphi(\alpha+\bar\alpha)}-\frac{\alpha}{\psi(\vartheta)}\frac{\psi(\vartheta)+\bar\alpha}{\varphi(\psi(\vartheta)+\bar\alpha)}\right).$$

As is easily verified, inserting $\alpha=0$ or $\bar\alpha=0$ yields the transform of the steady-state workload, as identified in Thm. 3.2. ◇

A second approach to identifying the double transform associated to the transient workload works as follows. Similarly to the first approach, it addresses the compound Poisson case first, and then the usual procedure is used to extend the result to the general spectrally positive case. Define $\kappa^x(\alpha):=\mathbb{E}_x e^{-\alpha Q_T}$. Starting at 0, one should distinguish between (i) the clock T expiring before the first jump of the compound Poisson process, and vice versa, and (ii) whether or not the buffer has become empty. One thus obtains

$$\begin{aligned}\kappa^x(\alpha) &= \int_0^\infty\int_0^x \lambda e^{-(\lambda+\vartheta)y}\kappa^{x-y+z}(\alpha)\mathrm{d}y\,\mathrm{d}\mathbb{P}(B\le z)\\ &\quad+\frac{\vartheta}{\vartheta+\lambda-\alpha}(e^{-\alpha x}-e^{-(\lambda+\vartheta)x})\\ &\quad+\frac{\lambda}{\vartheta+\lambda}e^{-(\lambda+\vartheta)x}\int_0^\infty\kappa^z(\alpha)\mathrm{d}\mathbb{P}(B\le z)+\frac{\vartheta}{\vartheta+\lambda}e^{-(\lambda+\vartheta)x};\end{aligned}\tag{4.5}$$

for instance, the last term corresponds to the scenario that no job arrives before the exponential clock T expires. It is a lengthy though elementary verification that $\kappa^x(\alpha) = Ke^{-kx} + Le^{-\ell x}$ satisfies this equation, when picking $k = \alpha$, $\ell = \psi(\vartheta)$, $K = \vartheta/(\vartheta - \varphi(\alpha))$, and $L = -K\alpha/\psi(\vartheta)$, which corresponds to the solution stated in Thm. 4.1. As we mentioned above, approximating the spectrally positive process by a compound Poisson process leads to the result.

A third approach, which we detail now, relies on application of the Kella–Whitt martingale $(K_t)_t$—see e.g. Asmussen [19, Thm. IX.3.10] and Kella et al. [125], and the brief account in Section 3.1. Let T again be exponentially distributed with mean $1/\vartheta$, which is obviously a stopping time. 'Optional sampling' thus provides us with

$$0 = \mathbb{E}K_0 = \mathbb{E}K_T = \varphi(\alpha)\int_0^\infty\int_0^t \vartheta e^{-\vartheta t}e^{-\alpha Q_s}\mathrm{d}s\,\mathrm{d}t - e^{-\alpha x} - \mathbb{E}_x e^{-\alpha Q_T} - \alpha\mathbb{E}L_T(x).$$

The first term of the right-hand side can alternatively be written as

$$\varphi(\alpha)\int_0^\infty\int_s^\infty \vartheta e^{-\vartheta t}e^{-\alpha Q_s}\mathrm{d}t\mathrm{d}s = \frac{\varphi(\alpha)}{\vartheta}\int_0^\infty \vartheta e^{-\vartheta s}e^{-\alpha Q_s}\mathrm{d}s = \frac{\varphi(\alpha)}{\vartheta}\,\mathbb{E}_x e^{-\alpha Q_T}.$$

Now $\mathbb{E}_x e^{-\alpha Q_T}$ can be solved, and we obtain an expression in which the unknown term $\mathbb{E}L_T(x)$ appears in the numerator, and in which the denominator equals $\vartheta - \varphi(\alpha)$. Then use the fact that the root of the denominator (i.e. $\alpha = \psi(\vartheta)$) should be a root of the numerator as well, as otherwise the transform equals ∞ at $\alpha = \psi(\vartheta)$. This enables us to solve $\mathbb{E}L_T(x)$, and finally we obtain the result of Thm. 4.1.

The special case of $X \in \mathbb{B}\mathrm{m}(d, \sigma^2)$ can be solved explicitly. We state without proof that

$$\mathbb{P}(Q_t \le y \mid Q_0 = x) = 1 - \Phi_{\mathrm{N}}\left(\frac{-y + x + dt}{\sigma\sqrt{t}}\right) - e^{2dy/\sigma^2}\Phi_{\mathrm{N}}\left(\frac{-y - x - dt}{\sigma\sqrt{t}}\right), \tag{4.6}$$

with $\Phi_{\mathrm{N}}(\cdot) := 1 - \Psi_{\mathrm{N}}(\cdot)$ denoting the distribution function of a standard normal random variable; see e.g. Harrison [108, p. 49].

Above we pointed out how to characterize the workload distribution after an exponentially distributed amount of time. This procedure can be extended to cover phase-type [19, Section III.4] amounts of time; we demonstrate how such a procedure works by considering, as an example, the case that T is distributed as $T_1 + T_2$, with T_1 and T_2 independent, and T_i exponentially distributed with mean $1/\vartheta_i$, for $i = 1, 2$.

Suppose we wish to evaluate the joint distribution of the workloads at T_1 and $T_1 + T_2$, by considering the transform

$$\mathbb{E}_x\, e^{-\alpha_1 Q_{T_1} - \alpha_2 Q_{T_1+T_2}}.$$

Conditioning on the value of Q_{T_1}, this quantity can be rewritten as

$$\int_0^\infty e^{-\alpha_1 y}\, \mathbb{E}_y e^{-\alpha_2 Q_{T_2}}\, \mathbb{P}_x\left(Q_{T_1} \in \mathrm{d}y\right),$$

which, due to Thm. 4.1, reads

$$\int_0^\infty e^{-\alpha_1 y}\left(\frac{\vartheta_2}{\vartheta_2 - \varphi(\alpha_2)}\left(e^{-\alpha_2 y} - \frac{\alpha_2}{\psi(\vartheta_2)} e^{-\psi(\vartheta_2) y}\right)\right) \mathbb{P}_x\left(Q_{T_1} \in \mathrm{d}y\right).$$

Observe that the above display can be interpreted as

$$\frac{\vartheta_2}{\vartheta_2 - \varphi(\alpha_2)}\left(\mathbb{E}_x\, e^{-(\alpha_1+\alpha_2) Q_{T_1}} - \frac{\alpha_2}{\psi(\vartheta_2)} \mathbb{E}_x\, e^{-(\alpha_1+\psi(\vartheta_2)) Q_{T_1}}\right).$$

Again applying Thm. 4.1, we obtain

$$\frac{\vartheta_2}{\vartheta_2 - \varphi(\alpha_2)}\left(\frac{\vartheta_1}{\vartheta_1 - \varphi(\alpha_1+\alpha_2)}\left(e^{-(\alpha_1+\alpha_2)x} - \frac{\alpha_1+\alpha_2}{\psi(\vartheta_1)} e^{-\psi(\vartheta_1)x}\right)\right.$$
$$\left. - \frac{\alpha_2}{\psi(\vartheta_2)} \frac{\vartheta_1}{\vartheta_1 - \varphi(\alpha_1+\psi(\vartheta_2))}\left(e^{-(\alpha_1+\psi(\vartheta_2))x} - \frac{\alpha_1+\psi(\vartheta_2)}{\psi(\vartheta_1)} e^{-\psi(\vartheta_1)x}\right)\right).$$

From this formula we can compute the transform of the workload after an Erlang(2)-distributed time. To this end, we set $\vartheta_1 = \vartheta_2 = \vartheta$, so that the univariate transform $\mathbb{E}_x\, e^{-\alpha Q_{T_1+T_2}}$ equals

$$\left(\frac{\vartheta}{\vartheta - \varphi(\alpha)}\right)^2 \left(e^{-\alpha x} - \frac{\alpha}{\psi(\vartheta)} e^{-\psi(\vartheta)x}\right)$$
$$+ \frac{\vartheta}{\vartheta - \varphi(\alpha)} \alpha\vartheta \cdot \lim_{\eta \to \vartheta}\left(\frac{e^{-\psi(\eta)x}}{\psi(\eta)} - \frac{e^{-\psi(\vartheta)x}}{\psi(\vartheta)}\right) \Big/ (\eta - \vartheta).$$

A straightforward application of 'L'Hôpital' yields, with T having an Erlang(2) distribution, that the transform $\mathbb{E}_x\, e^{-\alpha Q_T}$ can be rewritten as

$$\left(\frac{\vartheta}{\vartheta - \varphi(\alpha)}\right)^2 \left(e^{-\alpha x} - \frac{\alpha}{\psi(\vartheta)} e^{-\psi(\vartheta)x}\right) - \frac{\alpha\vartheta^2}{\vartheta - \varphi(\alpha)} \frac{\psi'(\vartheta) e^{-\psi(\vartheta)x}(1 + x\psi(\vartheta))}{(\psi(\vartheta))^2}.$$

Letting $\vartheta \downarrow 0$, Thm. 3.2 is recovered, as is readily verified.

Evidently, the case of T having an Erlang(n) distribution, for $n \in \{3, 4, \ldots\}$, can be dealt with analogously (but the expressions become cumbersome). For 'large' n, and replacing ϑ by ϑn, such a procedure yields an approximation for the transform of the transient workload after a *deterministic* amount of time $1/\vartheta$.

4.2 Spectrally Negative Case

We now turn to the spectrally negative case. In this case computation of the double transform $\mathbb{E}_x e^{-\alpha Q_T}$, for a given value of $x \geq 0$, turns out to be infeasible. To resolve this, the idea now is to consider a transform not only with respect to time t, but in addition with respect to the initial position x. The main result of this section is that we uniquely characterize the transient workload distribution by finding an explicit expression for the resulting *triple transform*, with T representing an exponentially distributed random variable with mean q^{-1}:

$$\int_0^\infty e^{-\beta x}\mathbb{E}_x e^{-\alpha Q_T}\mathrm{d}x = \int_0^\infty \int_0^\infty q e^{-qt} e^{-\beta x}\mathbb{E}_x e^{-\alpha Q_t}\mathrm{d}x\,\mathrm{d}t$$

$$= \int_0^\infty \int_0^\infty \int_0^\infty q e^{-qt} e^{-\beta x} e^{-\alpha y}\mathbb{P}(Q_t \in \mathrm{d}y)\,\mathrm{d}x\,\mathrm{d}t,$$

in terms of the model primitives $\Phi(\cdot)$ and $\Psi(\cdot)$.

Following the setup of Kyprianou [146, Chapter VIII], we first introduce, for spectrally negative Lévy processes, families of functions $W^{(q)}(\cdot)$ and $Z^{(q)}(\cdot)$ as follows. Let $W^{(q)}(x)$, with $q \geq 0$, be a strictly increasing and continuous function whose Laplace transform satisfies, for $x \geq 0$,

$$\int_0^\infty e^{-\beta x} W^{(q)}(x)\mathrm{d}x = \frac{1}{\Phi(\beta) - q}, \qquad \beta > \Psi(q), \tag{4.7}$$

and $W^{(q)}(x) = 0$ for negative x; such a function exists, as follows from [146, Thm. 8.1(i)]. In addition,

$$Z^{(q)}(x) := 1 + q\int_0^x W^{(q)}(y)\mathrm{d}y. \tag{4.8}$$

The functions $W^{(q)}(\cdot)$ and $Z^{(q)}(\cdot)$ are usually referred to as the *q-scale functions*; for their numerical evaluation, see e.g. [188, 206].

As mentioned, our objective is to analyze the transient workload distribution. A first characterization is the following. With T being exponentially distributed with mean q^{-1}, we have the density of Q_T, given that $Q_0 = x$:

$$\mathbb{P}_x(Q_T \in \mathrm{d}y) = \left(e^{-\Psi(q)y}\Psi(q)Z^{(q)}(x) - qW^{(q)}(x-y)\right)\mathrm{d}y; \tag{4.9}$$

see [175, Eqn. (19)], and also [84].

Considering the case that $q \downarrow 0$, we find that, due to $Z^{(0)}(x) \equiv 1$ for all $x \geq 0$, Q_T has an exponential distribution with mean $1/\Psi(0)$—this was to be expected, because this limiting regime corresponds to the stationary situation. Likewise, we obtain that Q_T equals x with probability 1 as $q \to \infty$.

The validity of Eqn. (4.9) results from the following line of reasoning.

- First it is observed that, using a reversibility argument,

$$(Q_T \mid Q_0 = 0) = X_T - \inf_{0 \le s \le T} X_s \stackrel{\mathrm{d}}{=} \sup_{0 \le s \le T} X_s =: \bar{X}_T.$$

Now $\bar{X}_T$ has an exponential distribution with mean $1/\Psi(q)$, which can be seen as follows. Observe that $e^{\Psi(q)X_t - qt}$ is a martingale, and recall the stopping time $\sigma(x) = \inf\{t \ge 0 : X_t \ge x\}$ (see Section 3.2) as the (potentially defective) random variable corresponding to the first epoch that X_t exceeds x. 'Optional sampling' yields

$$\mathbb{E}\left(e^{-q\sigma(x)} 1_{\{\sigma(x)<\infty\}}\right) = e^{-\Psi(q)x}.$$

Using the obvious duality of the events $\{\sigma(x) \le t\}$ and $\{\bar{X}_t \ge x\}$, we find that

$$\begin{aligned}
\mathbb{E}\left(e^{-q\sigma(x)} 1_{\{\sigma(x)<\infty\}}\right) &= \int_0^\infty e^{-qt} \mathbb{P}(\sigma(x) \in \mathrm{d}t) \\
&= \int_0^\infty q e^{-qt} \mathbb{P}(\sigma(x) \le t) \mathrm{d}t \\
&= \int_0^\infty q e^{-qt} \mathbb{P}(\bar{X}_t \ge x) \mathrm{d}t = \mathbb{P}(\bar{X}_T \ge x),
\end{aligned}$$

and hence, as desired, in obvious notation,

$$\mathbb{P}_0(Q_T \in \mathrm{d}x) = \mathbb{P}(\bar{X}_T \in \mathrm{d}x) = \Psi(q) e^{-\Psi(q)x} \mathrm{d}x. \tag{4.10}$$

- Define $X'_t := -X_t$. The goal of this step is to verify that

$$\mathbb{P}(\bar{X}'_T \in \mathrm{d}x) = \frac{q}{\Psi(q)} W^{(q)}(\mathrm{d}x) - q W^{(q)}(x) \mathrm{d}x. \tag{4.11}$$

Relation (4.11) follows by first combining the facts that (i) $X' \in \mathscr{S}_+$, and (ii) the distributional quality

$$\bar{X}'_T \stackrel{\mathrm{d}}{=} X'_T - \inf_{0 \le s \le T} X'_s,$$

which can be interpreted as the workload of a queue fed by $-X$, started empty, at time T. Then we can use Thm. 4.1 to compute the Laplace transform $\mathbb{E}_0 e^{-\alpha \bar{X}'_T}$ of $\bar{X}'_T$. Using the definition of the q-scale function, we can also compute the Laplace transform

$$\int_0^\infty e^{-\alpha x}\left(\frac{q}{\Psi(q)}W^{(q)}(\mathrm{d}x) - qW^{(q)}(x)\mathrm{d}x\right).$$

After some calculus, it turns out that both transforms coincide. This means that we have shown (4.11).

- We now find an explicit expression for $\mathbb{P}_x(Q_T \in \mathrm{d}y, T < \tau)$, with τ defined as the stopping time $\inf\{t \geq 0 : Q_t = 0\}$; see e.g. [42, 205] and [146, Thm. 8.7]. With X' as defined above, it holds that

$$\mathbb{P}_x(Q_T \in \mathrm{d}y, T < \tau) = \mathbb{P}((\bar{X}'_T - X'_T) - \bar{X}'_T \in \mathrm{d}y - x, \bar{X}'_T \leq x). \tag{4.12}$$

From Wiener–Hopf theory (see Section 3.3), we know that $\bar{X}'_T - X'_T$ and $\bar{X}'_T$ are independent. Elementary manipulations, and applying (4.10) and (4.11), yields that (4.12) equals

$$\begin{aligned}\int_{z=x-y}^{x} &\Psi(q)e^{-\Psi(q)(y+z-x)}\left(\frac{q}{\Psi(q)}W^{(q)}(\mathrm{d}z) - qW^{(q)}(z)\mathrm{d}z\right)\mathrm{d}y\\ &= q\left(e^{-\Psi(q)y}W^{(q)}(x) - W^{(q)}(x-y)\right)\mathrm{d}y,\end{aligned}$$

using integration by parts.

- Using the strong Markov property,

$$\mathbb{P}_x(Q_T \in \mathrm{d}y) = \mathbb{P}_x(Q_T \in \mathrm{d}y, T < \tau) + \mathbb{P}(\tau(x) < T)\mathbb{P}_0(Q_T \in \mathrm{d}y); \tag{4.13}$$

where $\tau(x)$ is, as throughout this monograph, defined as the first epoch that the process $(X_t)_t$ (started at 0) drops below $-x$. Recalling that T is exponentially distributed with mean q^{-1}, it is readily verified that $\mathbb{P}(\tau(x) < T) = \mathbb{E}e^{-q\tau(x)}$. Integrating (4.13) over positive y now yields

$$\begin{aligned}1 &= \int_0^\infty q\left(e^{-\Psi(q)y}W^{(q)}(x) - W^{(q)}(x-y)\right)\mathrm{d}y + \mathbb{E}e^{-q\tau(x)}\\ &= \frac{q}{\psi(q)}W^{(q)}(x) - \int_0^x qW^{(q)}(y)\mathrm{d}y + \mathbb{E}e^{-q\tau(x)},\end{aligned}$$

and hence, using the relation between $Z^{(q)}(\cdot)$ and $W^{(q)}(\cdot)$,

$$\mathbb{E}e^{-q\tau(x)} = Z^{(q)}(x) - \frac{q}{\Psi(q)}W^{(q)}(x). \tag{4.14}$$

Plugging all expressions into (4.13) then yields density (4.9), as desired.

We now derive a second characterization of the transient workload. It is a matter of straightforward calculus to show that Eqn. (4.9) leads to an explicit expression for the triple transform of the transient workload Q_t:

$$\int_0^\infty e^{-\beta x}\mathbb{E}_x e^{-\alpha Q_T}\mathrm{d}x = I_1(\alpha,\beta,q) - I_2(\alpha,\beta,q), \tag{4.15}$$

where the integrals $I_1(\alpha,\beta,q)$ and $I_2(\alpha,\beta,q)$ are given by

$$I_1(\alpha,\beta,q) := \int_0^\infty\int_0^\infty qe^{-\beta x}e^{-\alpha y}e^{-\Psi(q)y}\frac{\Psi(q)}{q}Z^{(q)}(x)\mathrm{d}x\,\mathrm{d}y,$$

$$I_2(\alpha,\beta,q) := \int_0^\infty\int_0^\infty qe^{-\beta x}e^{-\alpha y}W^{(q)}(x-y)\mathrm{d}x\,\mathrm{d}y.$$

It turns out that it is possible to compute $I_1(\alpha,\beta,q)$ and $I_2(\alpha,\beta,q)$ explicitly in terms of $\Phi(\cdot)$ and $\Psi(\cdot)$. Using (4.7) and (4.8), we obtain

$$\begin{aligned}
I_1(\alpha,\beta,q) &= \frac{\Psi(q)}{\Psi(q)+\alpha}\int_0^\infty e^{-\beta x}Z^{(q)}(x)\mathrm{d}x \\
&= \frac{\Psi(q)}{\Psi(q)+\alpha}\left(\frac{1}{\beta}+\int_0^\infty\int_y^\infty qW^{(q)}(y)e^{-\beta x}\mathrm{d}x\,\mathrm{d}y\right) \\
&= \frac{\Psi(q)}{\Psi(q)+\alpha}\frac{1}{\beta}\left(1+\frac{q}{\Phi(\beta)-q}\right) = \frac{\Psi(q)}{\Psi(q)+\alpha}\frac{1}{\beta}\frac{\Phi(\beta)}{\Phi(\beta)-q}.
\end{aligned}$$

Likewise,

$$I_2(\alpha,\beta,q) = \int_0^\infty qe^{-(\alpha+\beta)y}\frac{1}{\Phi(\beta)-q}\mathrm{d}y = \frac{q}{\alpha+\beta}\frac{1}{\Phi(\beta)-q}.$$

This leads to the following result that uniquely characterizes the distribution of Q_t, conditional on $Q_0 = x$, in terms of the model primitives.

Theorem 4.2 *Let $X \in \mathscr{S}_-$, and let T be exponentially distributed with mean $1/q$, independently of X. For $\alpha \geq 0$ and $\beta > 0$,*

$$\int_0^\infty e^{-\beta x}\mathbb{E}_x e^{-\alpha Q_T}\mathrm{d}x = \frac{1}{\beta}\left(\frac{\Psi(q)}{\Psi(q)+\alpha}+\frac{q}{\Phi(\beta)-q}\frac{\Psi(q)-\beta}{\Psi(q)+\alpha}\frac{\alpha}{\alpha+\beta}\right).$$

Remark 4.3 It is a trivial exercise to verify that the joint distribution of the workload at times 0 and T, given the workload is in stationarity at time 0, equals

$$\mathbb{E}e^{-\bar{\alpha}Q_0-\alpha Q_T} = \int_0^\infty \beta_0 e^{-(\bar{\alpha}+\beta_0)x}\mathbb{E}_x e^{-\alpha Q_T}\mathrm{d}x,$$

which further simplifies to

$$\frac{\beta_0}{\bar{\alpha}+\beta_0}\left(\frac{\Psi(q)}{\Psi(q)+\alpha}+\frac{q}{\Phi(\bar{\alpha}+\beta_0)-q}\frac{\Psi(q)-\bar{\alpha}-\beta_0}{\Psi(q)+\alpha}\frac{\alpha}{\alpha+\bar{\alpha}+\beta_0}\right).$$

It is readily checked that by inserting $\alpha = 0$ or $\bar{\alpha} = 0$ we obtain the exponential stationary distribution, as expected. ◇

As in the spectrally positive case, we can characterize the workload in the spectrally negative case at a phase-type distributed time. With T_i having an exponential distribution with mean $1/q_i$ $(i = 1, 2)$, independent of each other and the Lévy process X, we now point out how to determine

$$\int_0^\infty e^{-\beta x}\mathbb{E}_x e^{-\alpha Q_{T_1+T_2}}\,\mathrm{d}x = \int_0^\infty e^{-\beta x}\,\mathbb{E}_y e^{-\alpha Q_{T_2}}\,\mathbb{P}_x(Q_{T_1}\in \mathrm{d}y)\,\mathrm{d}x. \tag{4.16}$$

To this end, first observe that, relying on (4.9),

$$\begin{aligned}\mathbb{E}_y e^{-\alpha Q_{T_2}} &= \int_0^\infty e^{-\alpha z}\left(e^{-\Psi(q_2)z}\Psi(q_2)Z^{(q_2)}(y) - q_2 W^{(q_2)}(y-z)\right)\mathrm{d}z\\ &= \frac{\Psi(q_2)Z^{(q_2)}(y)}{\alpha+\Psi(q_2)} - q_2\int_0^\infty e^{-\alpha z}W^{(q_2)}(y-z)\mathrm{d}z.\end{aligned}$$

Using (4.9) again, we conclude that (4.16) equals $J_1 - J_2 - J_3 + J_4$, with the $J_i \equiv J_i(\alpha, \beta, q_1, q_2)$, for $i = 1, \ldots, 4$, given by

$$J_1 := \int_0^\infty\int_0^\infty e^{-\beta x}\left(e^{-\Psi(q_1)y}\Psi(q_1)Z^{(q_1)}(x)\right)\left(\frac{\Psi(q_2)Z^{(q_2)}(y)}{\Psi(q_2)+\alpha}\right)\mathrm{d}y\,\mathrm{d}x,$$

$$J_2 := \int_0^\infty\int_0^\infty e^{-\beta x}\left(q_1 W^{(q_1)}(x-y)\right)\left(\frac{\Psi(q_2)Z^{(q_2)}(y)}{\Psi(q_2)+\alpha}\right)\mathrm{d}y\,\mathrm{d}x,$$

$$J_3 := \int_0^\infty\int_0^\infty e^{-\beta x}\left(e^{-\Psi(q_1)y}\Psi(q_1)Z^{(q_1)}(x)\right)\left(q_2\int_0^\infty e^{-\alpha z}W^{(q_2)}(y-z)\right)\mathrm{d}z\,\mathrm{d}y\,\mathrm{d}x,$$

and

$$J_4 := \int_0^\infty\int_0^\infty e^{-\beta x}\left(q_1 W^{(q_1)}(x-y)\right)\left(q_2\int_0^\infty e^{-\alpha z}W^{(q_2)}(y-z)\right)\mathrm{d}z\,\mathrm{d}y\,\mathrm{d}x.$$

When evaluating these integrals, the following identity is extensively used (and is derived similarly to the calculation of $I_1(\alpha, \beta, q)$):

$$\int_0^\infty e^{-\beta x}Z^{(q)}(x)\mathrm{d}x = \frac{1}{\beta}\frac{\Phi(\beta)}{\Phi(\beta)-q} = \frac{\Phi(\beta)}{\beta}\int_0^\infty e^{-\beta x}W^{(q)}(x)\mathrm{d}x.$$

After substantial calculus,

$$J_1 = \frac{1}{\beta}\frac{\Psi(q_2)}{\Psi(q_2)+\alpha}\frac{\Phi(\beta)}{\Phi(\beta)-q_1}\frac{q_1}{q_1-q_2},$$

$$J_2 = \frac{1}{\beta}\frac{\Psi(q_2)}{\Psi(q_2)+\alpha}\frac{\Phi(\beta)}{\Phi(\beta)-q_2}\frac{q_1}{\Phi(\beta)-q_1},$$

$$J_3 = \frac{1}{\beta}\frac{\Psi(q_1)}{\Psi(q_1)+\alpha}\frac{\Phi(\beta)}{\Phi(\beta)-q_1}\frac{q_2}{q_1-q_2},$$

and

$$J_4 = \frac{1}{\alpha+\beta}\frac{q_1}{\Phi(\beta)-q_1}\frac{q_2}{\Phi(\beta)-q_2}.$$

Now consider the special case that $q_1 = q_2 = q$, so that $T_1 + T_2$ has an Erlang(2) distribution. It turns out that in that case $J_1 - J_2 - J_3 + J_4$ equals, by applying 'L'Hôpital' again,

$$\frac{1}{\beta}\frac{\Phi(\beta)}{\Phi(\beta)-q}\frac{\Psi(q)-q\Psi'(q)}{\Psi(q)+\alpha} - \frac{1}{\beta}\frac{\Psi(q)}{\Psi(q)+\alpha}\frac{\Phi(\beta)}{\Phi(\beta)-q}\frac{q}{\Phi(\beta)-q}$$
$$+\frac{1}{\alpha+\beta}\left(\frac{q}{\Phi(\beta)-q}\right)^2.$$

Observe that when inserting $q = 0$ one obtains the transform corresponding to the exponential steady-state distribution (with mean $1/\beta_0$), that is,

$$\int_0^\infty e^{-\beta x}\mathbb{E}_x e^{-\alpha Q_{T_1+T_2}}\mathrm{d}x = \frac{1}{\beta}\frac{\Psi(0)}{\Psi(0)+\alpha} = \frac{1}{\beta}\frac{\beta_0}{\beta_0+\alpha},$$

as desired.

4.3 Spectrally Two-Sided Case

The Wiener–Hopf results, as presented in Section 3.3, facilitate an analysis of the transform of the transient workload for the case that the underlying process X is not necessarily in $\mathscr{S}_+$ or $\mathscr{S}_-$; we follow the line of reasoning of Kella and Mandjes [126].

As before, let T be an exponentially distributed random variable with mean $1/\vartheta$. We focus on computing the triple transform

$$\int_0^\infty e^{-\beta x}\mathbb{E}_x e^{\alpha i Q_T}\mathrm{d}x = \int_0^\infty\int_0^\infty \vartheta e^{-\vartheta t}e^{-\beta x}\mathbb{E}_x e^{\alpha i Q_t}\mathrm{d}x\,\mathrm{d}t. \tag{4.17}$$

With, as before, $X'_t := -X_t$, we have that $Q_T = -X'_T + \max\{x, \bar{X}'_T\}$. It means that we can rewrite expression (4.17) as

$$\int_0^\infty e^{-\beta x} \int_{y=0}^x \int_{z=-\infty}^\infty e^{\alpha i(-z+x)} \mathbb{P}(\bar{X}'_T \in dy, X'_T \in dz) dx$$
$$+ \int_0^\infty e^{-\beta x} \int_{y=x}^\infty \int_{z=-\infty}^\infty e^{\alpha i(-z+y)} d\mathbb{P}(\bar{X}'_T \in dy, X'_T \in dz) dx.$$

Interchanging the order of integration, this can be rewritten as

$$\frac{1}{\beta} \frac{\alpha i}{\beta - \alpha i} \mathbb{E} e^{(\alpha i - \beta)\bar{X}'_T - \alpha i X'_T} + \frac{1}{\beta} \mathbb{E} e^{\alpha i \bar{X}'_T - \alpha i X'_T}$$
$$= \frac{1}{\beta} \left(\frac{\alpha i}{\beta - \alpha i} \mathbb{E} e^{-\beta \bar{X}'_T} + 1 \right) \mathbb{E} e^{\alpha i (\bar{X}'_T - X'_T)}.$$

Using the results presented in Section 3.3, this can be written in terms of the Wiener–Hopf factors, as follows.

Theorem 4.3 *Let X be a general Lévy process, and let T be exponentially distributed with mean $1/\vartheta$, independently of X. For $\alpha, \beta \geq 0$,*

$$\int_0^\infty e^{-\beta x} \mathbb{E}_x e^{-\alpha Q_T} dx = \frac{1}{\beta} \left(1 - \frac{\alpha}{\alpha + \beta} \frac{\bar{k}(\vartheta, \beta)}{\bar{k}(\vartheta, 0)} \right) \frac{k(\vartheta, \alpha)}{k(\vartheta, 0)}.$$

Notice that when sending ϑ to 0, we indeed obtain the transform of the stationary workload, that is,

$$\int_0^\infty \beta e^{-\beta x} \mathbb{E}_x e^{-\alpha Q} dx = \frac{k(0, \alpha)}{k(0, 0)},$$

and by sending ϑ to ∞, as expected,

$$\int_0^\infty \beta e^{-\beta x} \mathbb{E}_x e^{-\alpha Q_0} dx = \frac{\beta}{\beta + \alpha}.$$

As a second sanity check, one could verify whether the result is in line with Thms. 4.1 and 4.2, which address the spectrally one-sided situations. For $X \in \mathscr{S}_+$, Thm. 4.1 yields that

$$\int_0^\infty e^{-\beta x} \mathbb{E}_x e^{-\alpha Q_T} dx = \frac{1}{\alpha + \beta} \frac{\vartheta}{\vartheta - \varphi(\alpha)} - \frac{\alpha}{\psi(\vartheta)} \frac{1}{\psi(\vartheta) + \beta} \frac{\vartheta}{\vartheta - \varphi(\alpha)},$$

whereas Thm. 4.3 leads to

$$\int_0^\infty e^{-\beta x}\mathbb{E}_x e^{-\alpha Q_T}\mathrm{d}x = \frac{1}{\beta}\left(1-\frac{\alpha}{\alpha+\beta}\frac{\psi(\vartheta)}{\psi(\vartheta)+\beta}\right)\frac{\vartheta}{\psi(\vartheta)}\frac{\psi(\vartheta)-\alpha}{\vartheta-\varphi(\alpha)}.$$

It takes a bit of algebra to conclude that these expressions coincide. For $X \in \mathscr{S}_-$, it is straightforward to verify that Thm. 4.2 coincides with Thm. 4.3, and therefore we leave out the underlying computations.

Remark 4.4 Let $\check{Q}$ be the distribution of Q_T, with Q_0 sampled from an exponential distribution (independent of anything else) with mean $1/\beta$. Then Thm. 4.3 entails that

$$\begin{aligned}\mathbb{E}e^{-\alpha\check{Q}} &= \left(1-\frac{\alpha}{\alpha+\beta}\frac{\bar{k}(\vartheta,\beta)}{\bar{k}(\vartheta,0)}\right)\frac{k(\vartheta,\alpha)}{k(\vartheta,0)}\\ &= \left(1-\frac{\bar{k}(\vartheta,\beta)}{\bar{k}(\vartheta,0)}+\frac{\beta}{\alpha+\beta}\frac{\bar{k}(\vartheta,\beta)}{\bar{k}(\vartheta,0)}\right)\frac{k(\vartheta,\alpha)}{k(\vartheta,0)}. \end{aligned} \quad (4.18)$$

This decomposition says that $\check{Q}$ can be interpreted as the sum of two independent random variables, say, $\check{Q} = \check{Q}_1 + \check{Q}_2$. One of them, say $\check{Q}_1$, corresponds to the factor $k(\vartheta,\alpha)/k(\vartheta,0)$, and the underlying random variable $\bar{X}_T$ (i.e. the maximum attained up to time T), or, equivalently, Q_T with $Q_0 = 0$. The other term, $\check{Q}_2$, is the contribution due to the fact that the queue does not start empty at time 0, but rather starts according to the exponential distribution with mean $1/\beta$. The random variable corresponding to this term has value 0 with probability

$$\check{p} := 1-\frac{\bar{k}(\vartheta,\beta)}{\bar{k}(\vartheta,0)},$$

and is sampled from an exponential distribution with mean $1/\beta$ with probability $1-\check{p}$. ◇

As mentioned, Thm. 4.3 enables the evaluation of the transform of the workload after an exponential amount of time, if the initial workload has been independently sampled from an exponential distribution. The next question is, what can be done if Q_0 is *not* exponentially distributed? Below we describe what procedure can be followed in the case that Q_0 is sampled from a (finite) *mixture of Erlang distributions.* Realizing that [19, Thm. III.4.2] the class of mixtures of Erlang distributions is dense (in the sense of weak convergence) in the set of all probability distributions on the positive half-line (even if the 'shape parameters' β_k are required to be identical), we can approximate any distribution of Q_0 arbitrarily closely by such a mixture of Erlangs.

Let p_k (with $k \in \{0, \ldots, m\}$) be non-negative real numbers summing to 1. For $k \in \{1, \ldots, m\}$, introduce parameters $n_k \in \{0, 1, \ldots\}$ and $\beta_k > 0$. Let the density of Q_0 be given, for $x > 0$, by

$$g(x) = \sum_{k=1}^{m} p_k \beta_k^{n_k+1} \frac{x^{n_k}}{n_k!} e^{-\beta_k x},$$

with an additional probability $p_0 \in [0, 1]$ in 0. We obviously have

$$\mathbb{E}e^{-\alpha Q_T} = p_0 \mathbb{E}_0 e^{-\alpha Q_T} + \int_{(0,\infty)} g(x) \mathbb{E}_x e^{-\alpha Q_T} \mathrm{d}x.$$

The first term on the right-hand side obviously equals $p_0 k(\vartheta, \alpha)/k(\vartheta, 0)$, so let us focus on the second term. Based on the ideas that we used for the exponential case, for this second term we can write

$$\begin{aligned}
&\int_{(0,\infty)} g(x) \mathbb{E}_x e^{\mathrm{i}\alpha Q_T} \mathrm{d}x \\
&= \sum_{k=1}^{m} p_k \int_0^\infty \beta_k^{n_k+1} \frac{x^{n_k}}{n_k!} e^{-\beta_k x} \int_{y=0}^{x} \int_{z=-\infty}^{\infty} e^{\alpha \mathrm{i}(-z+x)} \mathbb{P}(\bar{X}_T' \in \mathrm{d}y, X_T' \in \mathrm{d}z) \mathrm{d}x \\
&+ \sum_{k=1}^{m} p_k \int_0^\infty \beta_k^{n_k+1} \frac{x^{n_k}}{n_k!} e^{-\beta_k x} \int_{y=x}^{\infty} \int_{z=-\infty}^{\infty} e^{\alpha \mathrm{i}(-z+y)} \mathbb{P}(\bar{X}_T' \in \mathrm{d}y, X_T' \in \mathrm{d}z) \mathrm{d}x. \quad (4.19)
\end{aligned}$$

Recall the identities

$$\int_y^\infty x^n e^{-\gamma x} \mathrm{d}x = \frac{n!}{\gamma^{n+1}} \sum_{\ell=0}^{n} \frac{(\gamma y)^\ell}{\ell!} e^{-\gamma y}, \quad \int_0^y x^n e^{-\gamma x} \mathrm{d}x = \frac{n!}{\gamma^{n+1}} \left(1 - \sum_{\ell=0}^{n} \frac{(\gamma y)^\ell}{\ell!} e^{-\gamma y}\right).$$

As a result, the first term in (4.19), with $\alpha_k := \beta_k - \alpha \mathrm{i}$ (and upon interchanging the order of the integrals), reduces to

$$\sum_{k=1}^{m} p_k \left(\frac{\beta_k}{\alpha_k}\right)^{n_k+1} \left(\sum_{\ell=0}^{n_k} \int_{z=-\infty}^{\infty} \int_{y=0}^{\infty} \frac{(\alpha_k y)^\ell}{\ell!} e^{-\alpha \mathrm{i} z} e^{-\alpha_k y} \mathbb{P}(\bar{X}_T' \in \mathrm{d}y, X_T' \in \mathrm{d}z) \right),$$

which can be interpreted as

$$\sum_{k=1}^{m} p_k \left(\frac{\beta_k}{\alpha_k}\right)^{n_k+1} \left(\sum_{\ell=0}^{n_k} \frac{\alpha_k^\ell}{\ell!} \mathbb{E}\left((\bar{X}_T')^\ell e^{-\alpha_k \bar{X}_T'} e^{-\alpha i X_T'}\right)\right)$$

$$= \sum_{k=1}^{m} p_k \left(\frac{\beta_k}{\alpha_k}\right)^{n_k+1} \left(\sum_{\ell=0}^{n_k} \frac{\alpha_k^\ell}{\ell!} \mathbb{E} e^{\alpha i(\bar{X}_T' - X_T')}\, \mathbb{E}\left((\bar{X}_T')^\ell e^{-\beta_k \bar{X}_T'}\right)\right)$$

$$= \frac{k(\vartheta, -\alpha i)}{k(\vartheta, 0)} \sum_{k=1}^{m} p_k \left(\frac{\beta_k}{\alpha_k}\right)^{n_k+1} \left(\sum_{\ell=0}^{n_k} \frac{\alpha_k^\ell}{\ell!} (-1)^\ell \frac{\bar{k}^{(\ell)}(\vartheta, \beta_k)}{\bar{k}(\vartheta, 0)}\right),$$

where $\bar{k}^{(\ell)}(\vartheta, \beta)$ is the ℓth derivative of $\bar{k}(\vartheta, \beta)$ with respect to β; the first equality uses the fact that $\bar{X}_T' - X_T'$ and $\bar{X}_T$ are independent. Likewise, the second term in (4.19) equals

$$\sum_{k=1}^{m} p_k \left(\int_{z=-\infty}^{\infty} \int_{y=0}^{\infty} \left(e^{-\alpha i z} e^{\alpha i y} - \sum_{\ell=0}^{n_k} \frac{\beta_k^\ell}{\ell!} y^\ell e^{-\alpha i z} e^{-\alpha_k y}\right) \mathrm{d}\mathbb{P}(\bar{X}_T' \in \mathrm{d}y, X_T' \in \mathrm{d}z)\right)$$

$$= \sum_{k=1}^{m} p_k \left(\mathbb{E} e^{\alpha i(\bar{X}_T' - X_T')} - \sum_{\ell=0}^{n_k} \frac{\beta_k^\ell}{\ell!} \mathbb{E}\left((\bar{X}_T')^\ell e^{-\alpha_k \bar{X}_T'} e^{-\alpha i X_T'}\right)\right)$$

$$= \frac{k(\vartheta, -\alpha i)}{k(\vartheta, 0)} \sum_{k=1}^{m} p_k \left(1 - \sum_{\ell=0}^{n_k} \frac{\beta_k^\ell}{\ell!} (-1)^\ell \frac{\bar{k}^{(\ell)}(\vartheta, \beta_k)}{\bar{k}(\vartheta, 0)}\right).$$

Combining the above we obtain, with $\gamma_{k,\ell} := \beta_k^\ell - \beta_k^{n_k+1}(\alpha + \beta_k)^{\ell - n_k - 1}$, the generalization of (4.18):

$$\mathbb{E} e^{-\alpha Q_T} = \left(p_0 + \sum_{k=1}^{m} p_k \left(1 - \sum_{\ell=0}^{n_k} \frac{(-1)^\ell}{\ell!} \gamma_{k,\ell} \frac{\bar{k}^{(\ell)}(\vartheta, \beta_k)}{\bar{k}(\vartheta, 0)}\right)\right) \frac{k(\vartheta, \alpha)}{k(\vartheta, 0)};$$

with $m = 1$, $p_1 = 1$ (and hence $p_0 = 0$), $\beta_1 = \beta$, and $n_k = 0$, Eqn. (4.18) is recovered.

We conclude this chapter by pointing out how the results presented above can be used to evaluate another metric, that is, the mean amount of time the workload process has spent above a given level. To this end, we first introduce the random variable $V(t, u) \in [0, t]$ by setting

$$V(t, u) := \int_0^t 1_{\{Q_s \geq u\}} \mathrm{d}s.$$

Our objective is to analyze the double transform of $\mathbb{E}_x V(t,u)$, that is,

$$\int_0^\infty e^{-\alpha u} \int_0^\infty \vartheta e^{-\vartheta t} \mathbb{E}_x V(t,u) \mathrm{d}t \, \mathrm{d}u = \int_0^\infty e^{-\alpha u} \int_0^\infty \vartheta e^{-\vartheta t} \int_0^t \mathbb{P}_x(Q_s \geq u) \mathrm{d}s \, \mathrm{d}t \, \mathrm{d}u,$$

which, after swapping the two inner integrals, reduces to

$$\int_0^\infty e^{-\alpha u} \int_0^\infty e^{-\vartheta s} \mathbb{P}_x(Q_s \geq u) \mathrm{d}s \, \mathrm{d}u.$$

Again interchanging the order of the integrals, and applying integration by parts, we obtain

$$\int_0^\infty e^{-\alpha u} \int_0^\infty \vartheta e^{-\vartheta t} \mathbb{E}_x V(t,u) \mathrm{d}t \, \mathrm{d}u = \frac{1 - \mathbb{E}_x e^{-\alpha Q_T}}{\alpha \vartheta},$$

with T being exponentially distributed with mean $1/\vartheta$. Then the transform (with respect to the initial position x) of this expression can be phrased in terms of the functions $k(\cdot,\cdot)$ and $\bar{k}(\cdot,\cdot)$ relying on Thm. 4.3.

Exercises

Exercise 4.1 Verify Eqns. (4.2), (4.3), and (4.4), and use them to compute $\kappa_T(\alpha)$.

Exercise 4.2 Verify Eqn. (4.5).

Exercise 4.3 Let $X \in \mathscr{S}_+$ and T be exponentially distributed with mean $1/\vartheta$. Prove that

$$\mathbb{P}(Q_T = 0) = \left(\lim_{\alpha \to \infty} \frac{\alpha}{\varphi(\alpha)} \right) \frac{\vartheta}{\psi(\vartheta)} e^{-\psi(\vartheta) x}.$$

Use this to find the steady-state probability $\mathbb{P}(Q = 0)$, and show that for the case of $X \in \mathbb{CP}(r, \lambda, b(\cdot))$ this gives the well-known expression

$$\mathbb{P}(Q = 0) = 1 - \frac{\lambda \, \mathbb{E}B}{r}.$$

Exercise 4.4 Verify Eqn. (4.6) using Thm. 4.1.

Exercise 4.5 Verify Eqn. (4.15).

Exercise 4.6 Let $X \in \mathscr{S}_-$, and let T be exponentially distributed with mean $1/q$, independently of X. Determine

$$\int_0^\infty \beta_0 e^{-\beta_0 x} \mathbb{E}_x e^{-\alpha Q_T} \mathrm{d}x.$$

Interpret the result.

Exercise 4.7 Let $X \in \mathscr{S}_-$, and let T be exponentially distributed with mean $1/q$, independently of X. Determine

$$\int_0^\infty \beta e^{-\beta x} \mathbb{E}_x e^{-\alpha Q_T} \mathrm{d}x$$

as $q \downarrow 0$. Interpret the result.

Exercise 4.8 Let X be a Lévy process with $\mathbb{E}X_1 < 0$. Prove that $\sup_{s\in[0,t]} X_s$ converges in distribution to $\sup_{s\in[0,\infty)} X_s$, as $t \to \infty$. Interpret this fact in the setting of the corresponding Lévy-driven queue.

Exercise 4.9 Let X correspond to $\mathbb{B}\mathrm{m}(-1,1)$ and T be exponentially distributed with mean $1/\vartheta$, independently of X. Derive $\mathbb{E}_x e^{-\alpha Q_T}$. Find $\mathbb{E}_0 Q_T$ and $\mathbb{V}\mathrm{ar}_0\, Q_T$.

Exercise 4.10 Let X correspond to $\mathbb{B}\mathrm{m}(-1,1)$, and fix a time epoch $t > 0$, and an initial workload $x \geq 0$.

(a) Argue that (4.6) entails that $\mathbb{E}_x Q_t = I(x,t) + J(x,t)$, where

$$I(x,t) := \int_{-\infty}^{(x-t)/\sqrt{t}} \int_0^{x-t-z\sqrt{t}} \frac{1}{\sqrt{2\pi}} e^{-z^2/2} \mathrm{d}y\, \mathrm{d}z,$$

$$J(x,t) := \int_{-\infty}^{(-x+t)/\sqrt{t}} \int_0^{-x+t-z\sqrt{t}} \frac{1}{\sqrt{2\pi}} e^{-z^2/2} e^{-2y} \mathrm{d}y\, \mathrm{d}z.$$

(b) Verify that

$$I(x,t) = (x-t)\Phi_{\mathrm{N}}\left(\frac{x-t}{\sqrt{t}}\right) + \sqrt{\frac{t}{2\pi}} e^{-(x-t)^2/(2t)},$$

and

$$J(x,t) = \frac{1}{2}\Phi_{\mathrm{N}}\left(\frac{-x+t}{\sqrt{t}}\right) - \frac{e^{2x}}{2}\Phi_{\mathrm{N}}\left(\frac{-x-t}{\sqrt{t}}\right).$$

(c) Check that

$$\mathbb{E}_0 Q_t = -\left(t + \frac{1}{2}\right) + (t+1)\Phi_{\mathrm{N}}\left(\sqrt{t}\right) + \sqrt{\frac{t}{2\pi}} e^{-t/2}.$$

Chapter 5
Heavy Traffic

In this chapter we study Lévy-driven queues *in heavy traffic*. More specifically, we analyze the asymptotic regime in which the drift of the input process tends to 0. Evidently, in this regime the workload explodes, but under an appropriate scaling its distribution converges to a non-degenerate limit.

We now introduce the key quantities studied in this chapter. With $(X_t)_t$ denoting a Lévy process, we consider a queue fed by the input process $(X_t^{(\varepsilon)})_t$, defined through

$$X_t^{(\varepsilon)} := \Delta(\varepsilon)\,(X_t - t\,\mathbb{E}X_1 - \varepsilon t)\,,$$

where $\Delta(\varepsilon)$ denotes an appropriately chosen *scaling function*. In the sequel the stochastic process $(Q_t^{(\varepsilon)})_t$ denotes the workload process of the Lévy-driven queue fed by $(X_t^{(\varepsilon)})_t$, whereas the corresponding stationary version is defined by the random variable $Q^{(\varepsilon)}$. Without loss of generality, we assume in this chapter that $\mathbb{E}X_1 = 0$. We consider the regime in which ε tends to 0, so that the queue under consideration is increasingly heavily loaded.

The goal of this chapter is twofold. First, depending on specific distributional properties of X_1, we identify the right scaling function $\Delta(\varepsilon)$ that provides us with a non-trivial limit of the stationary workload $Q^{(\varepsilon)}$, in the limiting regime $\varepsilon \downarrow 0$. In addition, we determine a rescaling of time, denoted by $n(\varepsilon)$, such that the transient workload $Q_{tn(\varepsilon)}^{(\varepsilon)}$ has a proper limit. Second, we explicitly find the limiting distributions of both the transient and stationary workloads, as $\varepsilon \downarrow 0$.

As it turns out, there are two regimes that should be distinguished in this chapter. To point out the difference between these two situations, which differ in terms of the scaling function $\Delta(\varepsilon)$, let us first consider a special class of Lévy input: we assume for the moment that $X \in \mathbb{S}(\alpha, \beta, 0)$, with $\alpha \in (1, 2]$. Recall that $(X_t)_t$ is self-

K. Dębicki, M. Mandjes, *Queues and Lévy Fluctuation Theory*, Universitext,
DOI 10.1007/978-3-319-20693-6_5

similar with parameter $1/\alpha$, as we observed in Section 2.3. We therefore obtain, with $\gamma := -\alpha/(\alpha-1) \le -2$,

$$\begin{aligned} Q^{(\varepsilon)} &= \Delta(\varepsilon)\sup_{t\ge 0}(X_t - \varepsilon t) = \Delta(\varepsilon)\sup_{t\ge 0}\left(X_{\varepsilon^\gamma t} - \varepsilon^{1+\gamma} t\right) \\ &\stackrel{\mathrm{d}}{=} \left(\Delta(\varepsilon)\varepsilon^{-1/(\alpha-1)}\right)\sup_{t\ge 0}(X_t - t). \end{aligned}$$

It is therefore natural to choose $\Delta(\varepsilon) = \varepsilon^{1/(\alpha-1)}$, which results in the distributional equality, for each $\varepsilon > 0$,

$$Q^{(\varepsilon)} \stackrel{\mathrm{d}}{=} \sup_{t\ge 0}(X_t - t).$$

Having found the candidate for the 'space scaling' $\Delta(\varepsilon)$, we now focus on identifying the appropriate scaling of time $n(\varepsilon)$. To this end, consider the transient workload. Assuming $Q_0^{(\varepsilon)} = x$, relying on (2.4), it is seen that picking $n(\varepsilon) = \varepsilon^\gamma$ leads to a non-degenerate limit:

$$\begin{aligned} Q^{(\varepsilon)}_{\varepsilon^\gamma t} &= \Delta(\varepsilon)\left(X_{\varepsilon^\gamma t} - \varepsilon^{\gamma+1} t\right) + \max\left\{x, -\Delta(\varepsilon)\inf_{0\le s\le \varepsilon^\gamma t}(X_s - \varepsilon s)\right\} \\ &\stackrel{\mathrm{d}}{=} \left(\Delta(\varepsilon)\varepsilon^{-1/(\alpha-1)}\right)(X_t - t) + \max\left\{x, -\left(\Delta(\varepsilon)\varepsilon^{-1/(\alpha-1)}\right)\inf_{0\le s\le t}(X_s - s)\right\}. \end{aligned}$$

The above toy example suggests that if $(X_t)_t$ is asymptotically stable with index $\alpha \in (1,2]$, then $\Delta(\varepsilon)$ should decrease polynomially to 0, roughly at the rate $1/(\alpha - 1)$, whereas time should be 'stretched' by a factor ε^γ. In particular, if $\alpha = 2$, that is, $(X_t)_t$ belongs to the domain of attraction of a normal law, then $\Delta(\varepsilon) \sim \varepsilon$ as $\varepsilon \downarrow 0$. This suggests that in cases where $\mathbb{V}\mathrm{ar}\, X_1 < \infty$, we should choose $\Delta(\varepsilon) \sim \varepsilon$ and $n(\varepsilon) \sim \varepsilon^{-2}$. We note that in this chapter we use the intuitive notation $a(\varepsilon) \sim b(\varepsilon)$ to express that $a(\varepsilon)/b(\varepsilon)$ tends to 1 as $\varepsilon \downarrow 0$.

We formalize the above observations in the next sections. The principal distinction turns out to be that if $\mathbb{V}\mathrm{ar}\, X_1 < \infty$ then the heavy-traffic limit of the workload process is that of a queue fed by Brownian motion (i.e. reflected Brownian motion), whereas if $\mathbb{V}\mathrm{ar}\, X_1 = \infty$ then we have convergence to a queue fed by an α-stable Lévy motion.

Pioneering work on queues in heavy traffic was done in the 1960s by Kingman [134–136]; Kingman's approach is in line with the one we follow in Section 5.1, in that it is based on manipulating the transform of the workload distribution (as identified in the previous chapters). More precisely, for the class of Lévy processes $(X_t)_t$ having a finite variance, we demonstrate in Section 5.1 how to find the stationary and transient heavy-traffic limiting distributions from our expressions of the Laplace transforms of $Q^{(\varepsilon)}$ and $Q_t^{(\varepsilon)}$, respectively. An alternative approach, based on the central limit theorem, was developed by e.g. Prohorov [180]. For more background, we refer to a survey by Glynn [100] and the book by Whitt [217].

The case that the variance $\mathbb{V}\mathrm{ar}\, X_1$ is *infinite* has been analyzed in detail as well. In general terms, as indicated above, in this case the heavy-traffic limit corresponds to a reflected α-stable Lévy motion (rather than a reflected Brownian motion). We focus on this case in Section 5.2, where we use an alternative (i.e. different from the one presented in Section 5.1) way of finding the limiting process, which is based on the functional central limit theorem. For further reading in this context we refer e.g. to [50, 96, 197], and again [217].

5.1 Lévy Inputs with Finite Variance

In this section we assume $\mathbb{V}\mathrm{ar}\, X_1$ to be finite. For convenience, we normalize the Lévy process such that $\mathbb{V}\mathrm{ar}\, X_1 = 1$. Realize that this normalization can be done without loss of generality; we later comment on how to translate the results to the unnormalized case $\mathbb{V}\mathrm{ar}\, X_1 \neq 1$.

The main finding of this section is that in this finite-variance case, as $\varepsilon \downarrow 0$, the workload process scales essentially as ε^{-1}. More precisely, in a steady state, picking $\Delta(\varepsilon) = \varepsilon$, it will be argued that the distribution of the steady-state workload $Q^{(\varepsilon)}$ tends to that of an exponentially distributed random variable (with mean $\frac{1}{2}$) as $\varepsilon \downarrow 0$. This finding can be intuitively understood, bearing in mind that this distribution is the steady-state distribution of a queue fed by $\mathbb{B}\mathrm{m}(-1, 1)$, in combination with the functional central limit theorem. In addition we address the corresponding *transient* behavior: stretching time by a factor ε^{-2}, we prove convergence of the transient distribution to that of reflected Brownian motion.

We subsequently treat the spectrally positive case, the spectrally negative case, and the general case. In our approach, we rely on the stationary and transient results derived in the previous chapters: we consider the transforms that we derived for $Q^{(\varepsilon)}$ and $Q^{(\varepsilon)}_{t/\varepsilon^2}$, respectively, and study their behavior in the heavy-traffic regime, that is, as $\varepsilon \downarrow 0$.

In the remainder of this chapter, E denotes an exponential random variable with mean $\frac{1}{2}$. The process $(E_t)_t$ denotes a Brownian motion $\mathbb{B}\mathrm{m}(-1, 1)$ reflected at 0; according to Eqn. (4.6),

$$\mathbb{P}(E_t \leq y \mid E_0 = x) = 1 - \Phi_{\mathrm{N}}\left(\frac{-y + x - t}{\sqrt{t}}\right) - e^{-2y}\Phi_{\mathrm{N}}\left(\frac{-y - x + t}{\sqrt{t}}\right), \qquad (5.1)$$

with, as before, $\Phi_{\mathrm{N}}(\cdot) := 1 - \Psi_{\mathrm{N}}(\cdot)$ denoting the distribution function of a standard normal random variable. Recall that E corresponds to the steady-state distribution of $\mathbb{B}\mathrm{m}(-1, 1)$ reflected at 0, which also follows when taking $t \to \infty$ in (5.1).

Spectrally positive case—We start by analyzing the stationary workload under heavy traffic, and later shift to the transient case, for $X \in \mathscr{S}_+$. Define

$$\varphi^{(\varepsilon)}(\alpha) := \log \mathbb{E}e^{-\alpha X_1^{(\varepsilon)}} = \alpha\varepsilon^2 + \varphi(\alpha\varepsilon).$$

From Thm. 3.2, we thus have

$$\mathbb{E}e^{-\alpha Q^{(\varepsilon)}} = \frac{\alpha}{\varphi^{(\varepsilon)}(\alpha)}\left(-\mathbb{E}X_1^{(\varepsilon)}\right) = \frac{\alpha\varepsilon^2}{\alpha\varepsilon^2 + \varphi(\alpha\varepsilon)}.$$

As we assume that $\mathbb{V}\mathrm{ar}\, X_1 = 1 < \infty$, and realizing that $\varphi(\alpha) = \frac{1}{2}\alpha^2 + o(\alpha^2)$, it follows that, as $\varepsilon \downarrow 0$,

$$\mathbb{E}e^{-\alpha Q^{(\varepsilon)}} = \frac{1}{1 + \frac{1}{2}\alpha + o(1)} = \frac{2}{2+\alpha} + o(1);$$

recognize the Laplace transform of an exponential random variable with mean $\frac{1}{2}$ in the right-hand side of the previous display. We obtain the following convergence in distribution; its validity is an immediate consequence of Lévy's convergence theorem, see e.g. Williams [220, Section 18.1].

Theorem 5.1 *Let $X \in \mathscr{S}_+$. Then, as $\varepsilon \downarrow 0$, $Q^{(\varepsilon)} \stackrel{\mathrm{d}}{\to} E$.*

Example 5.1 Consider the situation of compound Poisson input with exponentially distributed jobs. We choose the arrival rate $\lambda_\varepsilon = 1 - \varepsilon$, mean service time $\mu_\varepsilon^{-1} = 1$, and depletion rate 1. It is readily verified that the driving Lévy process has mean rate $(1 - \varepsilon) - 1 = -\varepsilon$. The transform of the unscaled stationary workload is, due to Thm. 3.2, $\varepsilon(1 + \alpha)/(\alpha + \varepsilon)$, which leads to $(1 + \varepsilon\alpha)/(\alpha + 1)$ when scaling the workload by ε. When taking the limit $\varepsilon \downarrow 0$, we obtain $1/(\alpha + 1)$, corresponding to an exponential distribution with mean 1.

This could have been concluded without doing any computations with transforms. For the M/M/1 queue (with arrival rate λ, the mean service time μ^{-1}, and depletion rate 1) it is known that the steady-state workload satisfies

$$\mathbb{P}(Q \le x) = 1 - \frac{\lambda}{\mu}e^{-(\lambda-\mu)x},$$

for $x \ge 0$. Plugging in the above parameters and recalling that $\Delta(\varepsilon) = \varepsilon$, we arrive at

$$\mathbb{P}(Q^{(\varepsilon)} \le x) = \mathbb{P}\left(\sup_{t\ge0}(X_t - t\,\mathbb{E}X_1 - \varepsilon t) \le \frac{x}{\Delta(\varepsilon)}\right)$$

$$= 1 - \frac{\lambda_\varepsilon}{\mu_\varepsilon}e^{-(\lambda_\varepsilon-\mu_\varepsilon)(x/\varepsilon)} \to 1 - e^{-x},$$

as $\varepsilon \downarrow 0$. ◇

Now we consider the time-dependent behavior, relying on Thm. 4.1. The first question is, how should we scale time in order to obtain a meaningful limit? In the setup of Thm. 4.1 (i.e. the result providing the transform of the transient workload in a queue with spectrally positive input) we scale the parameter ϑ by a factor $n(\varepsilon)$ (so that the exponential clock has mean $n(\varepsilon)/\vartheta$); the function $n(\varepsilon)$ we identify later on. We assume that $Q_0^{(\varepsilon)} = x$.

With $\psi^{(\varepsilon)}(\cdot)$ defined as the inverse of $\varphi^{(\varepsilon)}(\cdot)$, Thm. 4.1 yields, for $\alpha \geq 0$,

$$\mathbb{E}_x e^{-\alpha Q_{Tn(\varepsilon)}} = \frac{\vartheta/n(\varepsilon)}{\vartheta/n(\varepsilon) - \varphi^{(\varepsilon)}(\alpha)} \left(e^{-\alpha x} - \frac{\alpha}{\psi^{(\varepsilon)}(\vartheta/n(\varepsilon))} e^{-\psi^{(\varepsilon)}(\vartheta/n(\varepsilon))x} \right),$$

with T exponentially distributed with mean ϑ^{-1}. As before, we now study the behavior of this transform for $\varepsilon \downarrow 0$. To this end, we first verify that, as $\varepsilon \downarrow 0$,

$$\psi^{(\varepsilon)}(\vartheta) \approx -1 + \sqrt{1 + \frac{2\vartheta}{\varepsilon^2}},$$

so that picking $n(\varepsilon) = \varepsilon^{-2}$ leads to

$$\lim_{\varepsilon \downarrow 0} \mathbb{E}_x e^{-\alpha Q_{T/\varepsilon^2}} = \frac{\vartheta}{\vartheta - \alpha - \frac{1}{2}\alpha^2} \left(e^{-\alpha x} - \frac{\alpha}{-1 + \sqrt{1 + 2\vartheta}} e^{(1-\sqrt{1+2\vartheta})x} \right),$$

which is the Laplace transform of $\mathbb{B}\mathrm{m}(-1, 1)$ reflected at 0, after an exponential time with mean ϑ^{-1} (use Thm. 4.1). This proves the following statement.

Theorem 5.2 *Let $X \in \mathscr{S}_+$. Then, for any $t > 0$, as $\varepsilon \downarrow 0$, $Q_{t/\varepsilon^2}^{(\varepsilon)} \stackrel{\mathrm{d}}{\to} (E_t \mid E_0 = x)$.*

Spectrally negative case—We now address $X \in \mathscr{S}_-$, and we start again by considering the stationary case. We define the cumulant of $X_1^{(\varepsilon)}$ by $\Phi^{(\varepsilon)}(\cdot)$, and its right inverse by $\Psi^{(\varepsilon)}(\cdot)$. Thm. 3.3 states that $Q^{(\varepsilon)}$ obeys an exponential distribution with mean $(\Psi^{(\varepsilon)}(0))^{-1}$. As $\varepsilon \downarrow 0$, $\Psi^{(\varepsilon)}(q) \approx 1 + \sqrt{1 + 2q/\varepsilon^2}$. We find the following result.

Theorem 5.3 *Let $X \in \mathscr{S}_-$. Then, as $\varepsilon \downarrow 0$, $Q^{(\varepsilon)} \stackrel{\mathrm{d}}{\to} E$.*

In relation to the transient case, with T exponentially distributed with mean q^{-1}, Thm. 4.2 yields

$$\int_0^\infty \beta e^{-\beta x} \mathbb{E}_x e^{-\alpha Q_{T/\varepsilon^2}} \mathrm{d}x = \frac{\Psi^{(\varepsilon)}(q\varepsilon^2)}{\Psi^{(\varepsilon)}(q\varepsilon^2) + \alpha} + \frac{q\varepsilon^2}{\Phi^{(\varepsilon)}(\beta) - q\varepsilon^2} \frac{\Psi^{(\varepsilon)}(q\varepsilon^2) - \beta}{\Psi^{(\varepsilon)}(q\varepsilon^2) + \alpha} \frac{\alpha}{\alpha + \beta}.$$

As $\varepsilon \downarrow 0$, this converges to

$$\frac{1 + \sqrt{1 + 2q}}{1 + \sqrt{1 + 2q} + \alpha} + \frac{q}{\frac{1}{2}\beta^2 - \beta - q} \cdot \frac{1 + \sqrt{1 + 2q} - \beta}{1 + \sqrt{1 + 2q} + \alpha} \cdot \frac{\alpha}{\alpha + \beta}.$$

This is the triple transform corresponding to $\mathbb{B}\mathrm{m}(-1,1)$ reflected at 0, started with an initial level sampled from an exponential distribution with mean β^{-1}, after an exponentially distributed time with mean q^{-1}. We again find that the limiting marginal distribution coincides with that of reflected Brownian motion.

Theorem 5.4 *Let $X \in \mathscr{S}_-$. Then, for any $t > 0$, as $\varepsilon \downarrow 0$, $Q^{(\varepsilon)}_{t/\varepsilon^2} \xrightarrow{\mathrm{d}} (E_t \mid E_0 = x)$.*

General case—For the situation we are considering, that is, $\mathbb{V}\mathrm{ar}\, X_1 < \infty$, the results derived for $X \in \mathscr{S}_+$ and $X \in \mathscr{S}_-$ carry over to general Lévy processes. We sketch how this can be proved for the stationary case relying on the expression for the transform of the stationary workload, as given in Thm. 3.5; the result for the transient case can be established similarly (but requires a considerable additional amount of calculus).

From Thm. 3.5, we have, for $\alpha \geq 0$,

$$\begin{aligned}\mathbb{E}e^{-\alpha Q^{(\varepsilon)}} &= \exp\left(-\int_0^\infty \int_{(0,\infty)} \frac{1}{t}(1-e^{-\alpha x})\,\mathbb{P}\left(X_t^{(\varepsilon)} \in \mathrm{d}x\right)\mathrm{d}t\right)\\ &= \exp\left(-\int_0^\infty \int_{(0,\infty)} \frac{1}{t}\alpha e^{-\alpha x}\,\mathbb{P}\left(X_t^{(\varepsilon)} > x\right)\mathrm{d}x\,\mathrm{d}t\right). \qquad (5.2)\end{aligned}$$

The idea is to let $\varepsilon \downarrow 0$ in this expression. In the calculations, we need the following lemma.

Lemma 5.1 *For any $s > 0$,*

$$\int_0^\infty \frac{s}{\sqrt{2\pi x^3}} \exp\left(-\frac{1}{2}\frac{(s-x)^2}{x}\right)\mathrm{d}x = 1$$

and

$$\int_0^\infty \frac{s}{\sqrt{2\pi x^3}} \exp\left(-\frac{1}{2}\frac{(s+x)^2}{x}\right)\mathrm{d}x = e^{-2s}.$$

Proof Substituting $u := s^2/x$,

$$I := \int_0^\infty \frac{s}{\sqrt{2\pi x^3}} \exp\left(-\frac{1}{2}\frac{(s-x)^2}{x}\right)\mathrm{d}x = \int_0^\infty \frac{1}{\sqrt{2\pi u}} \exp\left(-\frac{1}{2}\frac{(s-u)^2}{u}\right)\mathrm{d}u.$$

Adding these two integrals, we obtain

$$2I = \int_0^\infty \frac{1}{\sqrt{2\pi x}}\left(\frac{s}{x}+1\right)\exp\left(-\frac{1}{2}\left(\frac{s}{\sqrt{x}}-\sqrt{x}\right)^2\right)\mathrm{d}x.$$

Now substitute $y := s/\sqrt{x} - \sqrt{x}$; it is readily verified that

$$\mathrm{d}y = -\frac{1}{2}\left(\frac{1}{\sqrt{x}}\left(\frac{s}{x}+1\right)\right)\mathrm{d}x.$$

It follows that $I = \int_{-\infty}^{\infty}(\sqrt{2\pi})^{-1}e^{-y^2/2}\mathrm{d}y = 1$. The second claim follows from the first claim, by an elementary computation. $\square$

Now let us return to our expression for $\mathbb{E}e^{-\alpha Q^{(\varepsilon)}}$. Substituting $s := t\varepsilon$, we obtain that the exponent in (5.2) reads as

$$-\int_0^\infty \int_{(0,\infty)} \frac{1}{s}\alpha e^{-\alpha x}\,\mathbb{P}\left(\frac{X_{s/\varepsilon}}{\sqrt{s/\varepsilon}} > \frac{x}{\sqrt{s\varepsilon}} + \sqrt{s\varepsilon}\right)\mathrm{d}x\,\mathrm{d}s.$$

Realize that we are considering the regime $\varepsilon \downarrow 0$, so that s/ε becomes large. The idea is that we apply the central limit theorem, and replace the random variable $X_{s/\varepsilon}/\sqrt{s/\varepsilon}$ by a standard normal random variable. We thus obtain

$$-\int_0^\infty \int_{(0,\infty)} \frac{1}{s}\alpha e^{-\alpha x}\int_0^\infty \frac{1}{\sqrt{2\pi}}\exp\left(-\frac{1}{2}\left(y+\frac{x}{\sqrt{s\varepsilon}}+\sqrt{s\varepsilon}\right)^2\right)\mathrm{d}y\,\mathrm{d}x\,\mathrm{d}s.$$

Now subsequently substituting $z := y\sqrt{s\varepsilon}$ and $v := s\varepsilon$, we find

$$-\int_0^\infty \int_{(0,\infty)} \alpha e^{-\alpha x}\int_0^\infty \frac{1}{\sqrt{2\pi v^3}}\exp\left(-\frac{1}{2}\left(\frac{(x+z+v)^2}{v}\right)\right)\mathrm{d}z\,\mathrm{d}x\,\mathrm{d}v$$

(which is independent of ε!). The next step is to interchange the order of integration; first perform the integration over v. The second claim of Lemma 5.1 implies that the above expression equals

$$\begin{aligned}
-\int_{(0,\infty)}\int_0^\infty \alpha\frac{e^{-\alpha x}}{x+z}e^{-2(x+z)}\mathrm{d}z\,\mathrm{d}x &= -\int_0^\infty\int_x^\infty \alpha e^{-\alpha x}\frac{e^{-2y}}{y}\mathrm{d}y\,\mathrm{d}x\\
&= -\int_0^\infty\int_0^y \alpha e^{-\alpha x}\frac{e^{-2y}}{y}\mathrm{d}x\,\mathrm{d}y\\
&= -\int_0^\infty \frac{e^{-2y}-e^{-(2+\alpha)y}}{y}\mathrm{d}y = \log\left(\frac{2}{2+\alpha}\right),
\end{aligned}$$

where the last step is due to 'Frullani'. We conclude that

$$\lim_{\varepsilon\downarrow 0}\mathbb{E}e^{-\alpha Q^{(\varepsilon)}} = \frac{2}{2+\alpha}.$$

This leads to the following result.

Theorem 5.5 *Let X be a general Lévy process with* $\mathbb{V}\mathrm{ar}\,X_1 = 1$. *Then, as* $\varepsilon \downarrow 0$, $Q^{(\varepsilon)} \xrightarrow{\mathrm{d}} E$.

To make the above computations into a formal proof, one step needs further justification: the replacement of $X_{s/\varepsilon}/\sqrt{s/\varepsilon}$ by a standard normal random variable. This step can be made rigorous by using detailed, explicit error bounds for the central limit theorem.

We could apply the same line of reasoning to the transient result presented in Thm. 4.3. This leads to the following transient counterpart of Thm. 5.5.

Theorem 5.6 *Let X be a general Lévy process with* $\mathbb{V}\mathrm{ar}\,X_1 = 1$. *Then, for any* $t > 0$, *as* $\varepsilon \downarrow 0$, $Q^{(\varepsilon)}_{t/\varepsilon^2} \xrightarrow{\mathrm{d}} (E_t \mid E_0 = x)$.

While we have chosen to present proofs that rely on the results derived in the previous chapters, there are various alternative ways to establish the above heavy-traffic results. We refer e.g. to Asmussen [19, Thm. X.7.1] for an elementary and straightforward proof of the discrete-time version of Thm. 5.5; it requires a minor argument to extend this result to continuous time. In [19, Prop. X.7.4] the discrete-time version of Thm. 5.6 is proved for the special case that the queue is initially empty, that is, the case that $x = 0$.

We mentioned that we assumed, without loss of generality, that $\mathbb{V}\mathrm{ar}\,X_1 = 1$. By rescaling, it is straightforward to obtain the counterparts of Thms. 5.5 and 5.6 for the case of *non*-unit variance: with $V := \mathbb{V}\mathrm{ar}\,X_1$, as $\varepsilon \downarrow 0$,

$$\frac{Q^{(\varepsilon)}}{V} \xrightarrow{\mathrm{d}} E, \qquad \frac{Q^{(\varepsilon)}_{tV/\varepsilon^2}}{V} \xrightarrow{\mathrm{d}} (E_t \mid E_0 = x). \tag{5.3}$$

5.2 Lévy Inputs in the Domain of a Stable Law

Consider now a complementary scenario, when the variance of the input process is infinite (i.e. $\mathbb{V}\mathrm{ar}\,X_1 = \infty$). More precisely, we assume that the centered Lévy process X satisfies (D_1)–(D_2), defined as follows.

Definition 5.1 We say that the condition (D_1) holds if

$$\mathbb{P}(|X_1| > x) = L(x)x^{-\alpha},$$

for $\alpha \in (1, 2)$, and a slowly varying function $L(\cdot)$, that is, $L(x)/L(tx) \to 1$ for $x \to \infty$, for any $t > 0$.

We say that the condition (D_2) holds if

$$\lim_{x\to\infty} \frac{\mathbb{P}(X_1 > x)}{\mathbb{P}(|X_1| > x)} = \frac{1+\beta}{2},$$

with $\beta \in [-1, 1]$.

The key observation is that assuming (D_1)–(D_2) is equivalent to assuming the existence of a function $d(t)$, $t > 0$, such that

$$\frac{X_t}{d(t)} \xrightarrow{\mathrm{d}} S_1^{(\alpha)}, \qquad \text{as } t \to \infty, \tag{5.4}$$

where $S_1^{(\alpha)} \stackrel{\mathrm{d}}{=} S_\alpha(1, \beta, 0)$; see e.g. Whitt [217, Thm. 4.5.1]. Put differently, X satisfies (D_1)–(D_2) if and only if it belongs to the domain of attraction of the stable law $S_\alpha(1, \beta, 0)$; for more background, see e.g. [217, Chapter IV]. Observe that the above statement entails that the tail of the complementary distribution function of X_1 decays roughly as $x^{-\alpha}$ for $\alpha \in (1, 2)$, such that the mean of X_1 exists, but the variance does not. In this sense, this case is complementary to the situation that was analyzed in Section 5.1. The goal of the remainder of this section is to establish the counterpart of the heavy-traffic results of Section 5.1 (now imposing (D_1)–(D_2) rather than $\mathbb{V}\mathrm{ar}\, X_1 < \infty$).

To simplify the proofs, and to avoid technicalities for the moment, we first assume that

$$\lim_{x \to \infty} L(x) = A > 0;$$

that is, X belongs to the *normal* domain of attraction of a stable law. We comment on the general case (i.e. not requiring $L(x) \to A$) later on.

In the above setting, the classical result describing the convergence of random sums to stable laws, e.g. [217, Thm. 4.5.2], states that

$$d(t) := \left(\frac{A}{C_{\alpha,1}}\right)^{1/\alpha} t^{1/\alpha} \tag{5.5}$$

is the right scaling function, where $C_{\alpha,1} := (1 - \alpha)/(\Gamma(2 - \alpha)\cos(\pi\alpha/2))$ (cf. Prop. 2.1).

Consider first the stationary workload $Q^{(\varepsilon)}$ of the Lévy-driven queue fed by $(X_t^{(\varepsilon)})_t$, with $X_t^{(\varepsilon)} = \Delta(\varepsilon)\,(X_t - \varepsilon t)$ (recall that throughout we assume $\mathbb{E}X_t = 0$), and analyze its behavior as $\varepsilon \downarrow 0$.

We begin by identifying the scaling function $\Delta(\varepsilon)$. Let $n(\varepsilon)$ be a positive function, to be specified later. Using that

$$Q^{(\varepsilon)} \stackrel{\mathrm{d}}{=} \sup_{t \geq 0} \Delta(\varepsilon)\,(X_t - \varepsilon t) = \sup_{t \geq 0} \Delta(\varepsilon)\,\big(X_{n(\varepsilon)t} - \varepsilon n(\varepsilon) t\big),$$

we deduce that an appropriate scaling for the convergence of $\Delta(\varepsilon) X_{n(\varepsilon)t}$ (in the heavy-traffic regime $\varepsilon \downarrow 0$) is

$$\Delta(\varepsilon) = \frac{1}{d(n(\varepsilon))}, \tag{5.6}$$

where $n(\varepsilon)$ is chosen such that $\varepsilon n(\varepsilon)\Delta(\varepsilon) \to 1$, as $\varepsilon \downarrow 0$. Hence, in view of (5.5), we should pick

$$n(\varepsilon) \sim \left(\frac{A}{C_{\alpha,1}}\right)^{1/(1-\alpha)} \varepsilon^{-\alpha/(\alpha-1)}$$

and

$$\Delta(\varepsilon) \sim \left(\frac{A}{C_{\alpha,1}}\right)^{1/(1-\alpha)} \varepsilon^{1/(\alpha-1)},$$

as $\varepsilon \downarrow 0$.

We are now in a position to find the distributional limit of $Q^{(\varepsilon)}$ as $\varepsilon \downarrow 0$, with $\Delta(\varepsilon)$ as chosen above. Whereas we used a transform-based approach in the previous section (relating to $\mathbb{V}\mathrm{ar}\, X_1 < \infty$), we now follow an alternative approach that relies on a functional central limit theorem for asymptotically stable processes.

For given $T > 0$, consider the supremum on a finite interval:

$$\sup_{t\in[0,n(\varepsilon)T]} \Delta(\varepsilon)\,(X_t - \varepsilon t) = \sup_{t\in[0,T]} \left(\frac{X_{n(\varepsilon)t}}{d(n(\varepsilon))} - \frac{\varepsilon n(\varepsilon)}{d(n(\varepsilon))}t\right).$$

Now the classical functional central limit theorem for asymptotically stable stochastic processes ensures that, with the process $(S_t^{(\alpha)})_t$ denoting an α-stable Lévy motion $\mathbb{S}(\alpha,\beta,0)$,

$$\left(\frac{X_{n(\varepsilon)t}}{d(n(\varepsilon))}\right)_{t\in[0,T]} \xrightarrow{\mathrm{d}} \left(S_t^{(\alpha)}\right)_{t\in[0,T]}, \tag{5.7}$$

weakly in the space $D[0,T]$ (with the J_1 topology); we refer to [45, 50, 96, 197, 217] for details on weak convergence in this function space.

Combining the above with the continuous mapping theorem (note that for any finite $T > 0$, the supremum over $[0,T]$ is a continuous mapping) implies that

$$\sup_{t\in[0,T]} \left(\frac{X_{n(\varepsilon)t}}{d(n(\varepsilon))} - \frac{\varepsilon n(\varepsilon)}{d(n(\varepsilon))}t\right) \xrightarrow{\mathrm{d}} \sup_{t\in[0,T]} \left(S_t^{(\alpha)} - t\right); \tag{5.8}$$

see e.g. [139, 197]. Thus we are left with extending the above convergence to the supremum over the interval $[0,\infty)$. To show this it suffices to prove that

$$\lim_{T\to\infty} \mathbb{P}\left(\sup_{t\geq T} \left(X_{n(\varepsilon)t} - \varepsilon n(\varepsilon)t\right) > 0\right) = 0, \tag{5.9}$$

uniformly in $\varepsilon > 0$.

This can be done by resorting to an idea taken from Shneer and Wachtel [197]; see also [139]. The key observation is that, by the results derived in Pruitt [181, Section 3], there exists a $C > 0$ such that for each $x, T > 0$,

$$\mathbb{P}\left(\sup_{t\in[0,T]} X_t \geq x\right) \leq CT\,\frac{\mathbb{E}(X_1^2; |X_1| \leq x)}{x^2}. \tag{5.10}$$

We combine this with the straightforward fact that

$$\mathbb{E}\left(X_1^2; |X_1| \leq x\right) \leq C_1 x^2 \mathbb{P}(|X_1| > x)$$

for sufficiently large x, and $C_1 > 0$; see e.g. [89, Prop. A.3.8]. We thus find the upper bound

$$\begin{aligned}
\mathbb{P}\left(\sup_{t\geq Tn(\varepsilon)} (X_t - \varepsilon t) > 0\right) &\leq \sum_{k=0}^{\infty} \mathbb{P}\left(\sup_{t\leq 2^{k+1}Tn(\varepsilon)} X_t > 2^k \varepsilon n(\varepsilon) T\right)\\
&\leq C_2 \sum_{k=0}^{\infty}\left(\frac{2^{k+1}Tn(\varepsilon)}{2^{2k}\varepsilon^2 n^2(\varepsilon)T^2}\right)\mathbb{E}\left(X_1^2; |X_1| \leq 2^k\varepsilon n(\varepsilon)T\right)\\
&\leq C_3\, T^{1-\alpha}\varepsilon^{-\alpha}(n(\varepsilon))^{1-\alpha}\sum_{k=0}^{\infty} 2^{k(1-\alpha)} \leq C_4\, T^{1-\alpha},
\end{aligned}$$

where we used that $\varepsilon^{-\alpha}(n(\varepsilon))^{1-\alpha} \sim C_{\alpha,1}/A$ as $\varepsilon \downarrow 0$, and where C_2, C_3, C_4 denote some positive constants. Hence the condition (5.9) holds uniformly in $\varepsilon > 0$; recall that $\alpha > 1$.

The case of a general slowly varying function $L(\cdot)$, rather than assuming that $L(x) \to A > 0$ as $x \to \infty$, follows the same line of reasoning, with the exception that

$$d(t) := \inf\left\{y : \mathbb{P}(|X_1| > y) < \frac{1}{t}\right\}$$

is the appropriate scaling [89, Prop. 2.2.13]. We refer to [139, 197], and references therein, for details.

We have thus arrived at the following result.

Theorem 5.7 *Let X be a general Lévy process satisfying (D_1)–(D_2). Then, as $\varepsilon \downarrow 0$, for $\Delta(\varepsilon) = 1/d(n(\varepsilon))$ and $n(\varepsilon)$ such that $\varepsilon n(\varepsilon)\Delta(\varepsilon) \to 1$, we have*

$$Q^{(\varepsilon)} \xrightarrow{\text{d}} \sup_{t\geq 0}\left(S_t^{(\alpha)} - t\right),$$

where $S^{(\alpha)} \in \mathbb{S}(\alpha, \beta, 0)$.

The analysis of the corresponding transient case follows from a straightforward combination of the weak convergence (5.7) with the fact that the transient solution of the Skorokhod problem is continuous with respect to the input process $(X_t)_t$ [185, Prop. D.4]. As a consequence, the following statement holds: under an appropriate scaling the workload process converges to that of a queue fed by an α-stable Lévy motion. Recall that $Q_0^{(\varepsilon)} = x$.

Theorem 5.8 *Let X be a general Lévy process satisfying (D_1)–(D_2). Then, for any $t > 0$, as $\varepsilon \downarrow 0$, for $\Delta(\varepsilon) = 1/d(n(\varepsilon))$ and $n(\varepsilon)$ such that $\varepsilon n(\varepsilon)\Delta(\varepsilon) \to 1$, we have*

$$Q_{tn(\varepsilon)}^{(\varepsilon)} \xrightarrow{\mathrm{d}} (S_t^\alpha - t) + \max\left\{x, -\inf_{0\le s\le t}(S_s^\alpha - s)\right\},$$

where $S^{(\alpha)} \in \mathbb{S}(\alpha, \beta, 0)$.

Example 5.2 Consider the case of a queue fed by compound Poisson input. More precisely, the process $(X_t)_t$ corresponds to a Poissonian arrival stream (of rate λ) of jobs with distribution function

$$\mathbb{P}(B \le x) = 1 - (x+1)^{-\delta},$$

for $x \ge 0$ and $\delta \in (1, 2)$. To ensure stability, the queue is emptied at a constant rate r that is larger than

$$\lambda\,\mathbb{E}B = \frac{\lambda}{\delta(\delta-1)}.$$

Observe that $\mathbb{E}B < \infty$, whereas $\mathbb{V}\mathrm{ar}\,B = \infty$. Considering (D_1)–(D_2), it can be verified that we have to pick $\alpha = \delta$ and $\beta = 1$ (note that the process is spectrally positive). We conclude from Thm. 5.7 that $Q^{(\varepsilon)}$ converges to the stationary workload in a queue fed by $(S_t^{(\delta)} - t)_t$, with $S^{(\delta)} \in \mathbb{S}(\delta, 1, 0)$, irrespective of the values of λ and r (as long as $r > \lambda\,\mathbb{E}B$). ◇

Exercises

Exercise 5.1 Prove (5.3) from Thms. 5.5 and 5.6.

Exercise 5.2 In this exercise we prove Eqn. (4.6), which plays a crucial role in Chapter 5. We rely on elementary arguments to do this; an alternative, martingale-based proof can be found in e.g. [19, Thm. XIII.4.3] for the special case $Q_0 = x = 0$.

For ease we start with the case $d = -1$, $\sigma = 1$, and $t = 1$, and later translate that into the setting of Eqn. (4.6). Define $X_t := B_t - t$, with $(B_t)_t$ standard Brownian motion.

(a) Prove that, for $x \geq 0$,

$$\mathbb{P}\left(\sup_{s\in[0,1]} B_s > x \,\middle|\, B_1 = 0 \right) = e^{-2x^2}.$$

Hint: First argue that

$$\mathbb{P}\left(\sup_{s\in[0,1]} B_s > x \,\middle|\, B_1 = 0 \right) = \mathbb{P}\left(\sup_{s\geq 0} B_{(s+1)^{-1}} > x \,\middle|\, B_1 = 0 \right)$$
$$= \mathbb{P}\left(\sup_{s\geq 0} \frac{1}{1+s} B_{s+1} > x \,\middle|\, B_1 = 0 \right).$$

Then use $\mathbb{P}(\sup_{s>0}(B_s - as) > x) = \exp(-2ax)$ for $a > 0$. (For an alternative derivation, see [184].)

(b) Prove that

$$\mathbb{P}\left(\sup_{s\in[0,1]} (B_s - s) > x \,\middle|\, B_1 = y \right) = e^{-2x(x+1-y)}$$

for $x \geq \max\{0, y-1\}$, and 1 otherwise.

(c) We first assume the workload process is in stationarity at time 0. Denote the density of a standard normal random variable by $\phi_N(\cdot) := \Phi_N'(\cdot)$. Show that

$$\mathbb{P}(Q_0 \leq x, Q_1 \leq y) = \int_{-\infty}^{y+1} \mathbb{P}\Big(\sup_{s\leq 0} (-X_s) \leq x, \sup_{t\leq 1} (X_1 - X_t)$$
$$\leq y \mid B_1 = z \Big) \phi_N(z) \mathrm{d}z.$$

(d) Prove that

$$\mathbb{P}\left(\sup_{s\leq 0} (-X_s) \leq x, \sup_{t\leq 1} (X_1 - X_t) \leq y \,\middle|\, B_1 = z \right)$$
$$= \mathbb{P}\left(\sup_{s\leq 0} (-X_s) \leq \min\{x, y+1-z\} \right) \mathbb{P}\left(\sup_{t\in[0,1]} (-X_t) \leq y+1-z \,\middle|\, B_1 = z \right).$$

(e) Use the above results to prove that

$$\mathbb{P}(Q_0 > x, Q_1 > y) = e^{-2y}\Phi_N(-x+y-1) + e^{-2x}\Phi_N(x-y-1) +$$
$$e^{-2(x+y)}\Phi_N(-x-y+1) - \Phi_N(-x-y-1)$$

(this requires quite a bit of calculus!), and

$$\mathbb{P}(Q_0 \in \mathrm{d}x, Q_1 > y) = 2e^{-2x}\Phi_{\mathrm{N}}(x - y - 1)\mathrm{d}x + 2e^{-2(x+y)}\Phi_{\mathrm{N}}(-x - y + 1)\mathrm{d}x.$$

(f) Conclude that

$$\mathbb{P}(Q_1 \le y \mid Q_0 = x) = 1 - \Phi_{\mathrm{N}}(x - y - 1) - e^{-2y}\Phi_{\mathrm{N}}(-x - y + 1).$$

(g) Derive Eqn. (4.6).

Exercise 5.3 Let $X_t^{(\varepsilon)} := X_t^{(1)} + X_t^{(2)} - \varepsilon t$, where it is assumed that $X^{(1)} \in \mathbb{S}(\alpha_1, \beta_1, 0)$ and $X^{(2)} \in \mathbb{S}(\alpha_2, \beta_2, 0)$ are mutually independent, with $\alpha_1, \alpha_2 \in (1, 2)$.

(a) Find the heavy-traffic scaling function, that is, a function $\Delta(\varepsilon)$, such that $Q^{(\varepsilon)}$ converges in distribution to a non-degenerate limit, as $\varepsilon \downarrow 0$.
(b) Assume now that $X^{(2)}$ is a standard Brownian motion, independent of $X^{(1)}$. Find the heavy-traffic scaling function, that is, a function $\Delta(\varepsilon)$, such that $Q^{(\varepsilon)}$ converges in distribution to a non-degenerate limit, as $\varepsilon \downarrow 0$.

Exercise 5.4 Let $X_t^{(\varepsilon)} = X_t^{(1)} + X_t^{(2)}$, where $X^{(1)} \in \mathbb{S}(\alpha, \beta, 0)$ and $X^{(2)}$ is an independent centered Lévy process with $\mathbb{V}\mathrm{ar}\, X_1^{(2)} < \infty$.

(a) Prove that X belongs to the domain of attraction of the stable law, and find its parameters.
(b) Find the heavy-traffic scaling $\Delta(\varepsilon)$ such that $Q^{(\varepsilon)}$ converges in distribution to a non-degenerate limit as $\varepsilon \downarrow 0$.

Exercise 5.5 Suppose that $X \in \mathbb{S}(\alpha, \beta, 0)$ for $\alpha \in (1, 2)$ and $\beta \in (-1, 1)$. Prove that

$$\frac{X_t}{t^{1/\alpha}} \xrightarrow{\mathrm{d}} S_1^{(\alpha)}, \text{ as } t \to \infty,$$

where $S_1^{(\alpha)} \stackrel{\mathrm{d}}{=} S_\alpha(1, \beta, 0)$.

Exercise 5.6 Suppose that a centered Lévy process $(X_t)_t$ satisfies (D_1)–(D_2) with $\alpha > 2$ and $\beta \in (-1, 1)$. Find the domain of attraction of X.

Exercise 5.7 Let $(X_t)_t$ be a centered Lévy process such that, for $\alpha \in (1, 2)$,

$$\lim_{x\to\infty} \frac{\mathbb{P}(|X_1| > x)}{x^{-\alpha}} = 1.$$

(a) Suppose that $\mathbb{P}(X_1 > x) = o(\mathbb{P}(X_1 < -x))$ as $x \to \infty$. Find the distribution of $\lim_{\varepsilon\downarrow 0} Q^{(\varepsilon)}$ under an appropriate heavy-traffic parameterization.
(b) Suppose that $\mathbb{P}(X_1 < -x) = o(\mathbb{P}(X_1 > x))$ as $x \to \infty$. Find the distribution of $\lim_{\varepsilon\downarrow 0} Q^{(\varepsilon)}$ under an appropriate heavy-traffic parameterization.

Chapter 6
Busy Period

Besides the (stationary and transient) workload distribution, as analyzed in the previous chapters, a primary object of study in queueing theory is the so-called *busy period*. In this chapter we analyze the busy period in a Lévy-driven queue, in that we characterize the time it takes for the queue to drain given that it starts off in the queue's stationary distribution. In the sequel, we let τ denote the busy-period duration:

$$\tau := \inf\{t \geq 0 : Q_t = 0\},$$

where it is assumed that Q_0 obeys the stationary workload distribution.

As for the stationary and transient workloads, all results are in terms of Laplace transforms. Again we will focus on spectrally positive input, spectrally negative input, and general (i.e. spectrally two-sided) input. The last section, Section 6.4, is about a metric that directly relates to the busy period: the minimum workload attained by the Lévy-driven queue over an interval of given length, where we again start from the stationary workload at time 0.

However, we start with a general result, reflecting the duality between hitting times and running maxima, which will be applied in the spectrally negative case and the general case. Recall the definition of the *first passage time*:

$$\tau(x) := \inf\{t \geq 0 : X_t < -x\}.$$

As before, let $X'_t := -X_t$, and let T be an exponential random variable with mean q^{-1}. Then the following result holds for any Lévy process X.

Lemma 6.1 *For* $q \geq 0$, $\beta > 0$,

$$\int_0^\infty e^{-\beta x} \mathbb{E} e^{-q\tau(x)} \mathrm{d}x = \frac{1}{\beta}\left(1 - \mathbb{E} e^{-\beta \bar{X}'_T}\right). \tag{6.1}$$

K. Dębicki, M. Mandjes, *Queues and Lévy Fluctuation Theory*, Universitext,
DOI 10.1007/978-3-319-20693-6_6

Proof A (by now) standard integration-by-parts argument yields

$$\int_0^\infty e^{-\beta x}\mathbb{E}e^{-q\tau(x)}\mathrm{d}x = \int_0^\infty e^{-\beta x}\int_0^\infty qe^{-qt}\mathbb{P}(\tau(x)\le t)\mathrm{d}t\,\mathrm{d}x. \tag{6.2}$$

Observe that

$$\{\tau(x)\le t\} = \left\{\sup_{0\le s\le t} X'_s > x\right\} = \{\bar{X}'_t > x\}.$$

Also changing the order of integration, we thus obtain that (6.2) can be rewritten as

$$\int_0^\infty qe^{-qt}\int_0^\infty e^{-\beta x}\mathbb{P}(\bar{X}'_t > x)\mathrm{d}x\,\mathrm{d}t.$$

Again integration by parts yields that this equals

$$\int_0^\infty qe^{-qt}\frac{1}{\beta}\left(1-\int_0^\infty e^{-\beta x}\mathbb{P}\left(\bar{X}'_t\in\mathrm{d}x\right)\right)\mathrm{d}t,$$

which can be interpreted as the right-hand side of (6.1). □

Remark 6.1 In the case that the stability condition is *not* fulfilled, the first passage time $\tau(x)$ is defective. It is readily checked that the following version of Lemma 6.1 still applies:

$$\int_0^\infty e^{-\beta x}\left(\mathbb{E}e^{-q\tau(x)}1_{\{\tau(x)<\infty\}}\right)\mathrm{d}x = \frac{1}{\beta}\left(1-\mathbb{E}e^{-\beta\bar{X}'_T}\right). \tag{6.3}$$

This, evidently, also yields an expression for the transform of $\mathbb{P}(\tau(x)<\infty)$ (by putting $q=0$). ◇

6.1 Spectrally Positive Case

When introducing the inverse Gaussian process in Section 2.2, it was observed that for $X\in\mathbb{B}\mathrm{m}(d,\sigma^2)$, the process $e^{-\varphi(\alpha)t}\,e^{-\alpha X_t}$ is a mean-1 martingale; it is readily verified that this property carries over to any $X\in\mathscr{S}_+$ with $\mathbb{E}X_1<0$. Now apply 'optional sampling' with respect to the stopping time $\tau(x)$; realize that due to the fact that there are no negative jumps, $X_{\tau(x)}=-x$. It follows that

$$1 = e^{\alpha x}\,\mathbb{E}e^{-\varphi(\alpha)\tau(x)},$$

or, equivalently, $\mathbb{E}e^{-\vartheta\tau(x)} = e^{-\psi(\vartheta)x}$. This leads to the interesting observation that $(\tau(x))_{x\geq 0}$ is an increasing Lévy process with Laplace exponent $-\psi(\vartheta)$. This is formalized in the following property, which is useful when studying the busy period of a Lévy-driven queue with spectrally positive input. As an aside, it is noted that Lemma 6.2 also holds for $\mathbb{E}X_1 \geq 0$; in the case $\mathbb{E}X_1 > 0$ the random variable $\tau(x)$ is defective.

Lemma 6.2 *Let* $X \in \mathscr{S}_+$, *and* $\mathbb{E}X_1 < 0$. *For* $\vartheta \geq 0$, $x > 0$,

$$\mathbb{E}e^{-\vartheta\tau(x)} = e^{-\psi(\vartheta)x}.$$

With Lemma 6.2 characterizing the time it takes before the buffer idles starting at level x, we are now in a position to find an expression for the Laplace transform of the busy period τ (with the workload at time 0 distributed according to its stationary distribution).

Proposition 6.1 *Let* $X \in \mathscr{S}_+$. *For* $\vartheta \geq 0$,

$$\mathbb{E}e^{-\vartheta\tau} = \frac{\psi(\vartheta)\varphi'(0)}{\vartheta}.$$

Proof By virtue of Lemma 6.2,

$$\mathbb{E}e^{-\vartheta\tau} = \int_0^\infty \mathbb{E}e^{-\vartheta\tau(x)}\mathbb{P}(Q_0 \in \mathrm{d}x) = \int_0^\infty e^{-\psi(\vartheta)x}\mathbb{P}(Q_0 \in \mathrm{d}x) = \mathbb{E}e^{-\psi(\vartheta)Q_0}.$$

The statement follows by applying Thm. 3.2. □

In the special case $X \in \mathbb{CP}(r, \lambda, b(\cdot))$, the notion of a busy period starting at 0 is well defined. More precisely, such a busy period starts with a job arriving in an empty queue, and ends at the first epoch that the workload attains 0 again. We denote the associated random variable by τ^0; let $\pi(\vartheta) := \mathbb{E}e^{-\vartheta\tau^0}$ be the corresponding Laplace–Stieltjes transform, which is known to satisfy the fixed-point equation $\pi(\vartheta) = b(\vartheta + \lambda - \lambda\pi(\vartheta))$, after having renormalized time such that $r = 1$. This fixed-point equation can be obtained by the following standard argument.

First observe that the duration of the busy period does not depend on the specific service policy, as long as it is *work conserving*; for instance, the busy period when serving on a first-come-first-served basis coincides with the busy period when serving on a last-come-first-served basis. Now consider the following work-conserving service mechanism. After the arrival of the first job, the server starts processing that job. In the case that the next job arrives before the service of the first has been completed, the server starts serving that second job, and in fact the entire 'sub-busy period' associated with the second job, and only after having finished this does it resume serving the first job. This policy is continued until the first job has been fully served, and the busy period ends. Observe that the busy period has, by virtue of the above construction, a *self-similar* nature, in the sense that, with $\tau_1^0, \tau_2^0, \ldots$ being i.i.d. copies of τ_0 (independent of the service requirement B corresponding to the first job), and N_t a Poisson process with rate λ (independent of B and $\tau_1^0, \tau_2^0, \ldots$),

$$\tau^0 \stackrel{\mathrm{d}}{=} B + \sum_{k=1}^{N_{\tau^0}} \tau_k^0$$

(where $\sum_{k=1}^{0} \tau_k^0 := 0$). As a consequence, when conditioning on $B = t$, and realizing that the number of sub-busy periods has a Poisson distribution with mean λt,

$$\pi(\vartheta) = \int_0^\infty e^{-\vartheta t} \sum_{k=0}^{\infty} e^{-\lambda t} \frac{(\lambda t)^k}{k!} (\pi(\vartheta))^k \, \mathbb{P}(B \in \mathrm{d}t).$$

Elementary calculus now yields $\pi(\vartheta) = b(\vartheta + \lambda - \lambda\pi(\vartheta))$.

Now recall that the Laplace exponent is $\varphi(\alpha) = \alpha - \lambda + \lambda b(\alpha)$. Therefore

$$0 = b(\vartheta + \lambda - \lambda\pi(\vartheta)) - \pi(\vartheta) = \frac{1}{\lambda}\varphi(\vartheta + \lambda - \lambda\pi(\vartheta)) - \frac{\vartheta}{\lambda},$$

and hence $\varphi(\vartheta + \lambda - \lambda\pi(\vartheta)) = \vartheta$. Apply $\psi(\cdot)$ to both sides, and we obtain the following result.

Proposition 6.2 *Let $X \in \mathbb{C}\mathrm{P}(1, \lambda, b(\cdot))$. For $\vartheta \geq 0$,*

$$\pi(\vartheta) = \frac{\lambda + \vartheta}{\lambda} - \frac{1}{\lambda}\psi(\vartheta).$$

Let us return to the setting of general $X \in \mathscr{S}_+$ (i.e. not just compound Poisson). In fact, more refined results than Prop. 6.1 can be found; see e.g. Mandjes et al. [158]. Consider for instance

$$\begin{aligned} L(\vartheta; \alpha, \bar{\alpha}) &:= \int_0^\infty e^{-\vartheta t} \mathbb{E}\left[e^{-\alpha Q_0 - \bar{\alpha} Q_t} 1_{\{\tau > t\}}\right] \mathrm{d}t \\ &= \int_0^\infty e^{-(\alpha+\bar{\alpha})x} \mathbb{E}\left[\int_0^{\tau(x)} e^{-\bar{\alpha} X_t - \vartheta t} \mathrm{d}t\right] \mathbb{P}(Q_0 \in \mathrm{d}x); \end{aligned}$$

here it is used that for $t \leq \tau$ it holds that $Q_t = Q_0 + X_t$. Now observe that for any Lévy process $(Z_t)_t$ and δ for which the expressions are well defined, we have that

$$M_s := e^{-\delta Z_s} - 1 - \left[\log \mathbb{E} e^{-\delta Z_1}\right] \cdot \int_0^s e^{-\delta Z_t} \mathrm{d}t \tag{6.4}$$

is a martingale. Now pick $\delta = \bar{\alpha}$ and $Z_t = X_t + (\vartheta/\bar{\alpha})t$, and use 'optional sampling' to obtain

$$\mathbb{E}\left[\int_0^{\tau(x)} e^{-\bar{\alpha}X_t - \vartheta t}\mathrm{d}t\right] = \frac{1 - e^{(\bar{\alpha}-\psi(\vartheta))x}}{\vartheta - \varphi(\bar{\alpha})};$$

here it is used that $X_{\tau(x)} = -x$. Combining the above, we end up with

$$L(\vartheta;\alpha,\bar{\alpha}) = \frac{\varphi'(0)}{\vartheta - \varphi(\bar{\alpha})}\left(\frac{\alpha + \bar{\alpha}}{\varphi(\alpha + \bar{\alpha})} - \frac{\alpha + \psi(\vartheta)}{\varphi(\alpha + \psi(\vartheta))}\right).$$

It is actually also possible to compute the *joint* transform of the stationary workload and residual busy period:

$$\begin{aligned}\mathbb{E}e^{-\alpha Q - \vartheta\tau} &= \int_0^\infty e^{-\alpha x}\mathbb{E}e^{-\vartheta\tau(x)}\mathbb{P}(Q \in \mathrm{d}x)\\ &= \int_0^\infty e^{-\alpha x}e^{-\psi(\vartheta)x}\mathbb{P}(Q \in \mathrm{d}x) = \kappa(\alpha + \psi(\vartheta)),\end{aligned}$$

with $\kappa(\cdot)$ as given in Thm. 3.2.

6.2 Spectrally Negative Case

The following lemma, which is crucial in the analysis of the busy period distribution for $X \in \mathscr{S}_-$, presents the double transform of $\tau(x)$. It is a direct consequence of Lemma 6.1 and Thm. 3.4: because of Lemma 6.1,

$$\int_0^\infty e^{-\beta x}\mathbb{E}e^{-q\tau(x)}\mathrm{d}x = \frac{1}{\beta}\left(1 - \mathbb{E}e^{-\beta(\bar{X}_T - X_T)}\right),$$

and because of Thm. 3.4, with T being exponentially distributed with mean q^{-1} (as usual independently of X),

$$\mathbb{E}e^{-\beta(\bar{X}_T - X_T)} = \frac{\bar{k}(q,\beta)}{\bar{k}(q,0)} = \frac{q}{\Psi(q)}\frac{\Psi(q) - \beta}{q - \Phi(\beta)},$$

using the explicit expression for $\bar{k}(\cdot,\cdot)$ for $X \in \mathscr{S}_-$; see Section 3.3.

Lemma 6.3 *Let $X \in \mathscr{S}_-$, and $\mathbb{E}X_1 < 0$. For $q \geq 0$, $x > 0$, $\beta > 0$,*

$$\int_0^\infty e^{-\beta x}\mathbb{E}e^{-q\tau(x)}\mathrm{d}x = \frac{1}{\beta}\left(1 - \frac{q}{\Psi(q)}\frac{\Psi(q) - \beta}{q - \Phi(\beta)}\right).$$

Now recall from Thm. 3.3 that Q_0 is exponentially distributed with mean β_0^{-1}. As a result, Lemma 6.3 yields

$$\mathbb{E}e^{-q\tau} = \int_0^\infty \beta_0 e^{-\beta_0 x}\mathbb{E}e^{-q\tau(x)}\mathrm{d}x = 1 - \frac{q}{\Psi(q)}\frac{\Psi(q)-\beta_0}{q-\Phi(\beta_0)}.$$

Using that $\Phi(\beta_0) = 0$ and $\beta_0 = \Psi(0)$, we find the following result.

Proposition 6.3 *Let $X \in \mathscr{S}_-$. For $q \geq 0$,*

$$\mathbb{E}e^{-q\tau} = \frac{\Psi(0)}{\Psi(q)}.$$

Similarly to what we did above for $X \in \mathscr{S}_+$, we can find more detailed results for $X \in \mathscr{S}_-$. To this end, we first state and prove a lemma, known as the *second factorization identity*, which can be found in e.g. Kyprianou [146, p. 176]; it is a slight extension of Eqn. (6.3). Importantly, it does not require the underlying Lévy process to be spectrally one sided. Realize that $x + X_{\tau(x)} \leq 0$. In the lemma below, we assume T to have an exponential distribution with mean q^{-1}, which is, as usual, independent of anything else.

Lemma 6.4 *For $q, \bar{q} \geq 0$, $\beta > 0$,*

$$\int_0^\infty e^{-\beta x}\mathbb{E}\left(e^{-q\tau(x)+\bar{q}(x+X_{\tau(x)})}1_{\{\tau(x)<\infty\}}\right)\mathrm{d}x = \frac{1}{\beta-\bar{q}}\left(1-\frac{\mathbb{E}e^{-\beta\bar{X}'_T}}{\mathbb{E}e^{-\bar{q}\bar{X}'_T}}\right).$$

Proof We follow the proof of [146, Exercise 6.7], which is in line with [7, 66]. Essentially due to the memoryless property of the exponential distribution, we have

$$\mathbb{E}\left(e^{-\bar{q}\bar{X}'_T}1_{\{\bar{X}'_T>x\}}\right) = \mathbb{E}\left(e^{-\bar{q}\bar{X}'_T}1_{\{\tau(x)<T\}}\right) = \mathbb{E}\left(e^{\bar{q}X_{\tau(x)}}1_{\{\tau(x)<T\}}\right)\mathbb{E}\left(e^{-\bar{q}\bar{X}'_T}\right);$$

the reasoning behind these equations immediately becomes clear when drawing a picture (the 'memoryless argument' is due to the underlying Markovian structure). In addition,

$$\begin{aligned}\mathbb{E}\left(e^{\bar{q}X_{\tau(x)}}1_{\{\tau(x)<T\}}\right) &= \int_0^\infty e^{-qs}\int_0^\infty qe^{-q(t-s)}\,\mathbb{E}\left(1_{\{s<t\}}e^{\bar{q}X_s}\right)\mathrm{d}t\,\mathbb{P}(\tau(x)\in\mathrm{d}s)\\ &= \mathbb{E}\left(e^{-q\tau(x)+\bar{q}X_{\tau(x)}}1_{\{\tau(x)<\infty\}}\right).\end{aligned}$$

Combining the above, it is verified that it is left to prove that

$$\int_0^\infty (\beta-\bar{q})e^{-(\beta-\bar{q})x}\mathbb{E}\left(e^{-\bar{q}\bar{X}'_T}1_{\{\bar{X}'_T>x\}}\right)\mathrm{d}x = \mathbb{E}\left(e^{-\bar{q}\bar{X}'_T}\right) - \mathbb{E}\left(e^{-\beta\bar{X}'_T}\right).$$

This equality follows by writing the left-hand side of the previous display as

$$\int_0^\infty \int_x^\infty (\beta - \bar{q}) e^{-(\beta-\bar{q})x} e^{-\bar{q}u} \mathrm{d}\mathbb{P}(\bar{X}_T' \in \mathrm{d}u)\, \mathrm{d}x,$$

and interchanging the order of the integration. □

Now observe that

$$\begin{aligned} L(q;\beta,\bar\beta) &:= \int_0^\infty e^{-qt}\mathbb{E}\left[e^{-\beta Q_0-\bar\beta Q_t};\tau>t\right]\mathrm{d}t \\ &= \int_0^\infty \beta_0 e^{-(\beta+\beta_0+\bar\beta)x}\mathbb{E}\left[\int_0^{\tau(x)} e^{-\bar\beta X_t-qt}\mathrm{d}t\right]\mathrm{d}x, \end{aligned}$$

using that Q_0 is exponentially distributed with mean $1/\beta_0$. As in the spectrally positive case,

$$\mathbb{E}\left[\int_0^{\tau(x)} e^{-\bar\beta X_t-qt}\mathrm{d}t\right] = \frac{1-\mathbb{E}(e^{-\bar\beta X_{\tau(x)}-q\tau(x)})}{q-\Phi(-\bar\beta)}.$$

Applying the second factorization identity, as derived in Lemma 6.4, we obtain that

$$L(q;\beta,\bar\beta) = \frac{\beta_0}{\beta+\beta_0+\bar\beta}\frac{1}{q-\Phi(-\bar\beta)}\frac{\mathbb{E}e^{-(\beta+\beta_0)\bar{X}_T'}}{\mathbb{E}e^{\bar\beta\bar{X}_T'}},$$

which, by virtue of Thm. 4.1 ($X' \in \mathscr{S}_+!$; take $x=0$), leads to

$$L(q;\beta,\bar\beta) = \frac{\beta_0}{\beta+\beta_0+\bar\beta}\frac{\Psi(q)-\beta-\beta_0}{q-\Phi(\beta+\beta_0)}\frac{1}{\Psi(q)+\bar\beta}.$$

The joint transform of the stationary workload and busy period follows directly from Lemma 6.3 and Thm. 3.3:

$$\mathbb{E}e^{-\beta Q-q\tau} = \beta_0\int_0^\infty e^{-(\beta+\beta_0)x}\mathbb{E}e^{-q\tau(x)}\mathrm{d}x = \frac{\beta_0}{\beta+\beta_0}\left(1-\frac{q}{\Psi(q)}\frac{\Psi(q)-\beta-\beta_0}{q-\Phi(\beta+\beta_0)}\right).$$

6.3 Spectrally Two-Sided Case

In this section we do not assume the driving Lévy process to be necessarily spectrally one sided. It turns out that in this case the results on the busy period are less explicit than in the one-sided case, that is, in terms of the Wiener–Hopf factors. We detail two approaches: one is based on Lemma 6.1, while the other exploits a duality property between the busy period and the epoch in which the driving Lévy process attains its all-time maximum.

Approach using 'Pecherskii–Rogozin'—The following result is usually attributed to Pecherskii and Rogozin [174]; see also [146, Exercise 6.7(ii)]. It follows immediately from Lemma 6.1 and Thm. 3.4. The function $\bar{k}(\cdot,\cdot)$ is defined in (3.11).

Lemma 6.5 *Let X be a general Lévy process. For $q \geq 0$, $\beta > 0$,*

$$\int_0^\infty e^{-\beta x}\mathbb{E}e^{-q\tau(x)}\mathrm{d}x = \frac{1}{\beta}\left(1 - \frac{\bar{k}(q,\beta)}{\bar{k}(q,0)}\right).$$

When inverting the above double transform with respect to β (which could be done numerically), we obtain, for any $x \geq 0$, $\mathbb{E}e^{-q\tau(x)}$. The Laplace transform of the busy period is then, evidently, obtained by computing

$$\int_0^\infty \mathbb{E}e^{-q\tau(x)}\mathbb{P}(Q_0 \in \mathrm{d}x),$$

where the distribution of Q_0 follows from Thm. 3.5.

Approach using duality—This approach uses a sequence of elementary arguments, e.g. from renewal theory. It is first observed that the busy period τ equivalently reads $\tau(Q)$ with Q the stationary workload, and $\tau(x)$ the duration of the busy period given that the initial workload is x. Then it is noticed that this random quantity is distributed as the *age* $\tau^{\leftarrow}$ of the same busy period, which is defined as

$$\tau^{\leftarrow} := -\inf\{s \leq 0 : \forall r \in [s,0] : X_s - X_r \geq 0\};$$

this distributional equality follows directly from standard properties for forward and backward recurrence times; see e.g. Asmussen [19, Section V.3]. It takes a little thought to realize that time-reversibility arguments entail that

$$\tau^{\leftarrow} \stackrel{\mathrm{d}}{=} \sup\{s \geq 0 : X_s = \bar{X}\} = G,$$

with $\bar{X} := \lim_{T\to\infty} \bar{X}_T$ and $G := \lim_{T\to\infty} G_T$, where $\bar{X}_T$ and G_T are as used in the context of Thm. 3.4. It now follows that, as an immediate consequence of Thm. 3.4,

$$\mathbb{E}\,e^{-\vartheta\tau} = \mathbb{E}\,e^{-\vartheta G} = \frac{k(\vartheta,0)}{k(0,0)}.$$

6.4 Infimum Over Given Time Interval

In this section we consider the distribution of the random variable $\underline{Q}_t := \inf_{s\in[0,t]} Q_s$, assuming the workload is in stationarity at time 0. In the first part of this section, which is based on Dębicki et al. [74], we find explicit expressions for Laplace transforms for the spectrally one-sided situation. We then consider the spectrally two-sided case, and find expressions in terms of the Wiener–Hopf factors, relying on the techniques developed in Section 4.3 (and in particular a representation in the spirit of the one given in Remark 4.4).

Spectrally one-sided case—Observe, for $u \geq 0$, that the event $\{\underline{Q}_t > u\}$ corresponds to $\{Q_0 + \inf_{s\in[0,t]} X_s > u\}$. Hence

$$\int_0^\infty e^{-\vartheta t} \int_0^\infty e^{-\alpha u} \mathbb{P}(\underline{Q}_t > u) \mathrm{d}u \, \mathrm{d}t$$

$$= \int_0^\infty e^{-\vartheta t} \int_0^\infty e^{-\alpha u} \int_u^\infty \mathbb{P}\left(\inf_{s\in[0,t]} X_s > u - q\right) \mathbb{P}(Q_0 \in \mathrm{d}q) \mathrm{d}u \, \mathrm{d}t$$

$$= \int_0^\infty \int_0^q e^{-\alpha u} \int_0^\infty e^{-\vartheta t} \mathbb{P}(\tau(q-u) > t) \mathrm{d}t \, \mathrm{d}u \, \mathbb{P}(Q_0 \in \mathrm{d}q).$$

The inner integral is the transform of the tail probability $\mathbb{P}(\tau(q-u) > t)$, so that integration by parts yields

$$\int_0^\infty \int_0^q e^{-\alpha u} \frac{1}{\vartheta} \left(1 - \mathbb{E}e^{-\vartheta \tau(q-u)}\right) \mathrm{d}u \, \mathbb{P}(Q_0 \in \mathrm{d}q). \tag{6.5}$$

Now we have to distinguish between $X \in \mathscr{S}_+$ and $X \in \mathscr{S}_-$. In the former case we can use Lemma 6.2 to evaluate the inner integral; then we have to perform a bit of straightforward calculus, in combination with Thm. 3.2. We obtain the following result, with T exponentially distributed with mean ϑ^{-1}.

Proposition 6.4 *Let $X \in \mathscr{S}_+$. For $\alpha, \vartheta \geq 0$,*

$$\int_0^\infty e^{-\alpha u} \mathbb{P}(\underline{Q}_T > u) \mathrm{d}u = \int_0^\infty \vartheta e^{-\vartheta t} \int_0^\infty e^{-\alpha u} \mathbb{P}(\underline{Q}_t > u) \mathrm{d}u \, \mathrm{d}t$$

$$= \left(\frac{1}{\alpha} - \frac{\varphi'(0)}{\varphi(\alpha)}\right) - \frac{\varphi'(0)}{\alpha - \psi(\vartheta)} \left(\frac{\psi(\vartheta)}{\vartheta} - \frac{\alpha}{\varphi(\alpha)}\right).$$

In the latter case, that is, $X \in \mathscr{S}_-$, we recall from Thm. 3.3 that Q_0 has an exponential distribution with parameter $\beta_0 = \Psi(0)$. Interchanging the order of integration in (6.5), and applying Lemma 6.3, we obtain the following result. Here T is exponentially distributed with mean q^{-1}.

Proposition 6.5 *Let $X \in \mathscr{S}_-$. For $\beta, q \geq 0$,*

$$\int_0^\infty e^{-\beta u} \mathbb{P}(\underline{Q}_T > u) \mathrm{d}u = \int_0^\infty q e^{-qt} \int_0^\infty e^{-\beta u} \mathbb{P}(\underline{Q}_t > u) \mathrm{d}u \, \mathrm{d}t$$

$$= \frac{1}{\beta + \Psi(0)} \frac{\Psi(q) - \Psi(0)}{\Psi(q)}.$$

Example 6.1 The special case of $X \in \mathbb{B}\mathrm{m}(d, \sigma^2)$ can be solved explicitly. To simplify the notation assume that $d = -1, \sigma^2 = 1$. Then, using that Q_0 is exponentially distributed with mean $\frac{1}{2}$ (see Example 3.1), for each $u, t > 0$, with $X'_t = -X_t$,

$$\begin{aligned}\mathbb{P}(\underline{Q}_t > u) &= \mathbb{P}(Q_0 + \inf_{s\in[0,t]} X_s > u) \\ &= \int_u^\infty \mathbb{P}\left(\inf_{s\in[0,t]} X_s > u - x\right) 2e^{-2x}\mathrm{d}x \\ &= 2e^{-2u} \int_0^\infty \mathbb{P}\left(\sup_{s\in[0,t]} X'_s < y\right) e^{-2y}\mathrm{d}y = e^{-2u}\mathbb{E}e^{-2\bar{X}'_t}.\end{aligned}$$

Now, by (4.6), we can explicitly find $\mathbb{E}e^{-2\bar{X}'_t}$, which finally leads to

$$\mathbb{P}(\underline{Q}_t > u) = e^{-2u}\left(2(1+t)\Psi_N(\sqrt{t}) - \sqrt{\frac{2t}{\pi}}e^{-t/2}\right),$$

with $\Psi_N(\cdot)$, as before, the complementary distribution function of a standard normal random variable. ◇

Spectrally two-sided case—To analyze the two-sided case, we first recall that, with $\underline{X}_T := \inf_{s\in[0,T]} X_s$,

$$Q_T := X_T + \max\{Q_0, -\underline{X}_T\} = (X_T - \underline{X}_T) + (\underline{X}_T + \max\{Q_0, -\underline{X}_T\}). \qquad (6.6)$$

Let T be exponentially distributed with mean $1/\vartheta$. For the moment we do not specify the distribution of Q_0; later in our exposition we assume that it is distributed as the stationary workload Q (and hence also as the all-time supremum $\bar{X}$).

Observe that, due to Thm. 3.4, (i) these two terms in the right-hand side of (6.6) are independent, and (ii) the first of these is distributed as $\bar{X}_T$. As a consequence,

$$\mathbb{E}e^{-\alpha Q_T} = \frac{k(\vartheta,\alpha)}{k(\vartheta,0)}\left(\mathbb{E}\left[e^{-\alpha(Q_0+\underline{X}_T)}1_{\{Q_0+\underline{X}_T>0\}}\right] + \mathbb{P}(Q_0 + \underline{X}_T \le 0)\right). \qquad (6.7)$$

If we were to assume that Q_0 is sampled from an exponential distribution with mean $1/\beta$, we would recover the result featuring in Remark 4.4. The idea now, however, is to let Q_0 follow the steady-state workload distribution, as characterized through Thm. 3.5. As starting in stationarity implies that the workload still obeys the stationary distribution after an exponential time, we have that the left-hand side of (6.7) equals $\mathbb{E}e^{-\alpha\bar{X}} = k(0,\alpha)/k(0,0)$. In addition, it is easily seen that

$$\mathbb{E}\left[e^{-\alpha(Q_0+\underline{X}_T)}1_{\{Q_0+\underline{X}_T>0\}}\right] + \mathbb{P}(Q_0 + \underline{X}_T \le 0) = \mathbb{E}e^{-\alpha\underline{Q}_T}.$$

Upon combining these two properties, it thus follows that

$$\mathbb{E}e^{-\alpha \underline{Q}_T} = \frac{k(\vartheta,0)}{k(0,0)}\frac{k(0,\alpha)}{k(\vartheta,\alpha)}.$$

Proposition 6.6 *Let X be a general Lévy process, and let T be exponentially distributed with mean $1/\vartheta$, independently of X. For $\alpha \geq 0$,*

$$\int_0^\infty e^{-\alpha u}\mathbb{P}(\underline{Q}_T > u)\mathrm{d}u = \frac{1}{\alpha}\left(1 - \frac{k(\vartheta,0)}{k(0,0)}\frac{k(0,\alpha)}{k(\vartheta,\alpha)}\right).$$

It is readily checked that the results for the spectrally one-sided case are in agreement with Prop. 6.6.

From the above steps some more refined results can be derived, as follows. To this end, first observe that from the two ways to characterize the busy-period distribution that were presented in Section 6.3, for any $t > 0$, with $\bar{X}$ and $\underline{X}_t$ independent,

$$\mathbb{P}(Q_0 + \underline{X}_t \leq 0) = \mathbb{P}(\bar{X} + \underline{X}_t \leq 0) = \mathbb{P}(\tau \leq t) = \mathbb{P}(G \leq t).$$

It thus follows that

$$\mathbb{P}(Q_0 + \underline{X}_T \leq 0) = \mathbb{P}(G \leq T) = \mathbb{E}e^{-\vartheta G} = \frac{k(\vartheta,0)}{k(0,0)},$$

and as an immediate consequence,

$$\mathbb{E}\left[e^{-\alpha \underline{Q}_T}1_{\{\underline{Q}_T>0\}}\right] = \mathbb{E}\left[e^{-\alpha(Q_0+\underline{X}_T)}1_{\{Q_0+\underline{X}_T>0\}}\right] = \frac{k(\vartheta,0)}{k(0,0)}\left(\frac{k(0,\alpha)}{k(\vartheta,\alpha)} - 1\right).$$

This leads to the identity

$$\mathbb{E}\left(e^{-\alpha \underline{Q}_T} \mid \underline{Q}_T > 0\right) = \frac{k(\vartheta,0)}{k(\vartheta,\alpha)}\frac{k(0,\alpha) - k(\vartheta,\alpha)}{k(0,0) - k(\vartheta,0)}.$$

Having found a characterization of the distribution of $\underline{Q}_T$ conditional on $\underline{Q}_T > 0$, it is remarked that we can also find the transform of Q_T given that $\underline{Q}_T > 0$. This is done as follows. The two terms in the right-hand side of (6.7) correspond to the transforms of Q_T in the scenarios that $\underline{Q}_T > 0$ and $\underline{Q}_T = 0$. We therefore have

$$\mathbb{E}\left[e^{-\alpha Q_T}1_{\{Q_0+\underline{X}_T>0\}}\right] = \frac{k(\vartheta,\alpha)}{k(\vartheta,0)}\mathbb{E}\left[e^{-\alpha(Q_0+\underline{X}_T)}1_{\{Q_0+\underline{X}_T>0\}}\right],$$

and hence

$$\mathbb{E}\left(e^{-\alpha Q_T} \mid \underline{Q}_T > 0\right) = \frac{k(0,\alpha) - k(\vartheta,\alpha)}{k(0,0) - k(\vartheta,0)}.$$

We also note that we have found the remarkable identity

$$\mathbb{E}\left(e^{-\alpha Q_T} \mid \underline{Q}_T > 0\right) = \mathbb{E}\left(e^{-\alpha \underline{Q}_T} \mid \underline{Q}_T > 0\right) \mathbb{E}e^{-\alpha Q}.$$

We now turn to behavior related to the queue conditional on being non-empty on the entire interval $[0, \infty)$. With $\underline{Q}$ the all-time infimum of the workload during the time interval $[0, \infty)$, then by letting $\vartheta \downarrow 0$ and by applying L'Hôpital's rule, we obtain

$$\lim_{t\to\infty} \mathbb{E}\left(e^{-\alpha \underline{Q}_t} \mid \underline{Q}_t > 0\right) = \mathbb{E}\left(e^{-\alpha \underline{Q}} \mid \underline{Q} > 0\right) = \frac{k'(0,\alpha)}{k(0,\alpha)} \frac{k(0,0)}{k'(0,0)},$$

where the derivation is with respect to the first argument of $k(\cdot,\cdot)$. Likewise,

$$\lim_{t\to\infty} \mathbb{E}\left(e^{-\alpha Q_t} \mid \underline{Q}_t > 0\right) = \frac{k'(0,\alpha)}{k'(0,0)}.$$

There is a connection between these results, where one conditions on the rare event of the workload process never hitting 0, and the topic of *quasi-stationarity*; see e.g. [158] and references therein.

Exercises

Exercise 6.1 In the case $X \in \mathbb{CP}(r, \lambda, b(\cdot))$, the notion of a busy period starting at 0 is well defined. We denote this random variable by τ^0; let $\pi(\vartheta) := \mathbb{E}e^{-\vartheta\tau^0}$ be the corresponding Laplace–Stieltjes transform. Renormalizing time such that $r = 1$, give a detailed proof of the fact that $\pi(\vartheta)$ is the unique solution of the fixed-point equation

$$\pi(\vartheta) = b(\vartheta + \lambda - \lambda\pi(\vartheta)).$$

Exercise 6.2 Prove that $(M_t)_t$, defined in (6.4), is a martingale.

Exercise 6.3 Compute $\mathbb{E}e^{-\vartheta\tau}$ for X corresponding to $\mathbb{B}\mathrm{m}(-1, 1)$, both by using the result for $X \in \mathscr{S}_+$ and the result for $X \in \mathscr{S}_-$.

Exercise 6.4 This exercise is on the mean of the busy period τ, with the workload starting in stationarity.

(a) Prove that

$$\mathbb{E}\tau = \frac{\varphi''(0)}{2(\varphi'(0))^2}$$

for $X \in \mathscr{S}_+$.

Hint: Realize that

$$\varphi'(\psi(\vartheta))\psi'(\vartheta) = 1, \qquad \varphi''(\psi(\vartheta))(\psi'(\vartheta))^2 + \varphi'(\psi(\vartheta))\psi''(\vartheta) = 0.$$

(b) Prove that $\mathbb{E}\tau = -(\Phi'(0)\beta_0)^{-1}$ for $X \in \mathscr{S}_-$.

Exercise 6.5 Show that Eqn. (4.14) is in line with Lemma 6.3.

Exercise 6.6 As $\mathbb{B}\mathrm{m}(d,\sigma^2)$ (with $d < 0$) is in $\mathscr{S}_+$ as well as $\mathscr{S}_-$, both Lemmas 6.2 and 6.3 can be used to identify $\mathbb{E}e^{-\vartheta\tau(x)}$. Show that both lemmas lead to the same result.

Exercise 6.7 Let X correspond to $\mathbb{B}\mathrm{m}(-1,1)$ and $(Q_t)_t$ be the stationary workload process.

(a) Prove that

$$\mathbb{P}\left(\inf_{s\in[0,t]} Q_s > u\right) = e^{-2u}\,\mathbb{E}\exp\left(-2\sup_{s\in[0,t]}(-X_s)\right).$$

(b) Use this to show that

$$\mathbb{P}\left(\inf_{s\in[0,t]} Q_s > u\right) = e^{-2u}\left(2(1+t)\Psi_{\mathrm{N}}(\sqrt{t}) - \sqrt{\frac{2t}{\pi}}\exp\left(-\frac{t}{2}\right)\right).$$

(c) Find an explicit function $f(\cdot)$ such that

$$\frac{\mathbb{P}(\inf_{s\in[0,t]} Q_s > u)}{f(t)} \to 1$$

as $t \to \infty$.

Exercise 6.8 Let X correspond to $\mathbb{B}\mathrm{m}(-1,1)$. Suppose that $Q_0 = x > u > 0$.

(a) Determine $\mathbb{P}(\inf_{s\in[0,t]} Q_s > u)$.

(b) Find an explicit function $f(\cdot)$ such that

$$\frac{\mathbb{P}(\inf_{s\in[0,t]} Q_s > u)}{f(t)} \to 1$$

as $t \to \infty$.

Exercise 6.9 In this exercise we study the distribution of $(\underline{Q} \mid \underline{Q} > 0)$ for spectrally one-sided case.

(a) For $X \in \mathscr{S}_+$ determine

$$\mathbb{E}\left(e^{-\alpha \underline{Q}} \mid \underline{Q} > 0\right).$$

Show that this conditional distribution corresponds to the residual lifetime distribution associated with the steady-state workload (to be denoted by $Q^{\rm res}$).
Hint: Show that

$$\mathbb{E}\left(e^{-\alpha \underline{Q}} \mid \underline{Q} > 0\right) = \frac{1 - \mathbb{E}e^{-\alpha Q}}{\alpha\, \mathbb{E}Q} = \mathbb{E}e^{-\alpha Q^{\rm res}},$$

with Q denoting the stationary workload.

(b) Show that $(\underline{Q} \mid \underline{Q} > 0)$ has an exponential distribution with mean $1/\beta_0$ for $X \in \mathscr{S}_-$.

(c) For $X \in \mathscr{S}_+$ show that

$$\lim_{t\to\infty} \mathbb{E}\left(e^{-\alpha Q_t} \mid \underline{Q}_t > 0\right) = \frac{\zeta(\alpha)}{\zeta(0)},$$

where

$$\zeta(\alpha) := \frac{\varphi'(0)}{\varphi(\alpha)}\left(1 - \frac{\alpha\varphi'(0)}{\varphi(\alpha)}\right) = \frac{\alpha\varphi'(0)}{\varphi(\alpha)} \cdot \frac{1}{\alpha}\left(1 - \frac{\alpha\varphi'(0)}{\varphi(\alpha)}\right).$$

Conclude that when $t \to \infty$, the distribution of $(Q_t \mid \underline{Q}_t > 0)$ converges to that of the sum of two independent random variables, where the first of these is distributed as the stationary workload Q, and the second as the residual stationary workload $Q^{\rm res}$.

(d) Show that for $X \in \mathscr{S}_-$, the distribution of $(Q_t \mid \underline{Q}_t > 0)$ converges (as $t \to \infty$) to that of an Erlang(2) random variable, where each of the phases has mean $1/\beta_0$, that is,

$$\lim_{t\to\infty} \mathbb{E}\left(e^{-\beta Q_t} \mid \underline{Q}_t > 0\right) = \left(\frac{\Psi(0)}{\Psi(0) + \beta}\right)^2.$$

Exercise 6.10 In this exercise we analyze $(\underline{Q} \mid \underline{Q} > 0)$ for $X \in \mathscr{S}_+$. The special feature is that Q_0 is now not sampled from the stationary workload distribution, but rather we assume that $Q_0 = x > 0$. Let T have an exponential distribution with mean $1/\vartheta$, independent of the driving Lévy process.

(a) Use the identity

$$\mathbb{E}_x e^{-\alpha Q_T} = \mathbb{E}_x\left(e^{-\alpha Q_T} 1_{\{\underline{Q}_T > 0\}}\right) + \mathbb{P}(\tau(x) \le T)\mathbb{E}_0 e^{-\alpha Q_T}, \tag{6.8}$$

in conjunction with Thm. 4.1, to show that

$$\mathbb{E}_x\left(e^{-\alpha \underline{Q}_T} \mid \underline{Q}_T > 0\right) = \frac{\psi(\vartheta)}{\psi(\vartheta) - \alpha} \frac{e^{-\alpha x} - e^{-\psi(\vartheta)x}}{1 - e^{-\psi(\vartheta)x}}.$$

Hint: Realize that

$$\mathbb{E}_x\left(e^{-\alpha Q_T} 1_{\{\underline{Q}_T > 0\}}\right) = \mathbb{E}_x\left(e^{-\alpha \underline{Q}_T} 1_{\{\underline{Q}_T > 0\}}\right) \mathbb{E}_0 e^{-\alpha \bar{X}_T}.$$

(b) Show that when $t \to \infty$, the distribution of $(Q_t \mid \underline{Q}_t > 0)$ converges to that of a uniform random variable on the interval $(0, x]$.

Remark The case of $X \in \mathscr{S}_-$ and $Q_0 = x > 0$ will be dealt with in Exercise 11.3.

Chapter 7
Workload Correlation Function

Where we analyzed the busy period in the previous chapter, in this chapter we study a second transience-related metric: the correlation function of the workload process. Assuming the Lévy-driven queue is in stationarity at time 0, we concentrate, for $t \geq 0$, on the function

$$r(t) := \mathbb{C}\mathrm{orr}(Q_0, Q_t) := \frac{\mathbb{C}\mathrm{ov}(Q_0, Q_t)}{\sqrt{\mathbb{V}\mathrm{ar}\, Q_0 \cdot \mathbb{V}\mathrm{ar}\, Q_t}} = \frac{\mathbb{E}(Q_0 Q_t) - (\mathbb{E} Q_0)^2}{\mathbb{V}\mathrm{ar}\, Q_0}$$

(note that $\mathbb{E} Q_0 = \mathbb{E} Q_t$ and $\mathbb{V}\mathrm{ar}\, Q_0 = \mathbb{V}\mathrm{ar}\, Q_t$ due to the stationarity). This function offers us insight into the 'memory' of the workload process: to what extent does the value of Q_0 provide us with information on the value of Q_t? Knowledge of the workload correlation is helpful if we are asked to determine a threshold T such that for $t \geq T$ the workloads Q_0 and Q_t can be safely assumed independent (in the sense that the correlation is negligibly small, i.e. below some given level $\varepsilon > 0$).

In this chapter we first explicitly compute, for the case of spectrally one-sided input, the Laplace transform $\hat{r}(\cdot)$ corresponding to the correlation function $r(\cdot)$, in terms of the model primitives; this we do intensively relying on our results for the transient workload, as stated in Thms. 4.1 and 4.2. Then we show how these transforms can be used to prove a set of structural properties; more specifically, relying on the theory of completely monotone functions, it is shown that $r(\cdot)$ is positive, decreasing, and convex.

7.1 Spectrally Positive Case: Transform

In this section we determine the transform of the workload correlation function for $X \in \mathscr{S}_+$. As mentioned above, we assume that the workload process is in steady state at time 0, which in this case means that Q_0 obeys the distribution featuring in

K. Dębicki, M. Mandjes, *Queues and Lévy Fluctuation Theory*, Universitext,
DOI 10.1007/978-3-319-20693-6_7

Thm. 3.2. Let T have an exponential distribution with mean $1/\vartheta$. First realize that

$$\mathbb{E}(e^{-\alpha Q_T} \mid Q_0 = x) = \int_0^\infty \vartheta e^{-\vartheta t}\mathbb{E}(e^{-\alpha Q_t} \mid Q_0 = x)\mathrm{d}t.$$

By differentiation with respect to α and subsequently letting $\alpha \downarrow 0$, we obtain, by applying Thm. 4.1, that

$$\int_0^\infty \vartheta e^{-\vartheta t}\mathbb{E}(Q_t \mid Q_0 = x)\mathrm{d}t = -\frac{\varphi'(0)}{\vartheta} + x + \frac{e^{-\psi(\vartheta)x}}{\psi(\vartheta)}. \tag{7.1}$$

Concentrate on the Laplace transform $\gamma(\vartheta)$ of $\mathbb{C}\mathrm{ov}(Q_0, Q_t)$. Straightforward calculus reveals that

$$\begin{aligned}\gamma(\vartheta) &:= \int_0^\infty \mathbb{C}\mathrm{ov}(Q_0, Q_t)e^{-\vartheta t}\mathrm{d}t = \int_0^\infty (\mathbb{E}(Q_0 Q_t) - \mu^2)e^{-\vartheta t}\mathrm{d}t \\ &= \int_0^\infty \int_0^\infty x \cdot \mathbb{E}(Q_t \mid Q_0 = x) \cdot e^{-\vartheta t}\mathrm{d}\mathbb{P}(Q_0 \leq x)\mathrm{d}t - \frac{\mu^2}{\vartheta};\end{aligned}$$

we use the notation $\mu = \mathbb{E}\, Q_0$. By invoking (7.1) we find that the expression in the previous display equals

$$\begin{aligned}&\int_0^\infty \frac{x}{\vartheta}\left(-\frac{\varphi'(0)}{\vartheta} + x + \frac{e^{-\psi(\vartheta)x}}{\psi(\vartheta)}\right)\mathrm{d}\mathbb{P}(Q_0 \leq x) - \frac{\mu^2}{\vartheta} \\ &= -\frac{\mu\varphi'(0)}{\vartheta^2} + \frac{v}{\vartheta} + \frac{1}{\vartheta\psi(\vartheta)}\mathbb{E}(Q_0 e^{-\psi(\vartheta)Q_0}),\end{aligned} \tag{7.2}$$

with v, as defined in (3.6), the variance $\mathbb{V}\mathrm{ar}\, Q_0$. From Thm. 3.2 we obtain by differentiating

$$\mathbb{E}(Q_0 e^{-\alpha Q_0}) = \varphi'(0)\left(-\frac{1}{\varphi(\alpha)} + \alpha\frac{\varphi'(\alpha)}{(\varphi(\alpha))^2}\right).$$

Inserting this relation, in addition to (3.5), into Eqn. (7.2), we obtain the Laplace transform of $\mathbb{C}\mathrm{ov}(Q_0, Q_t)$:

$$\gamma(\vartheta) = -\frac{\varphi''(0)}{2\vartheta^2} + \frac{v}{\vartheta} + \frac{\varphi'(0)}{\vartheta^2}\left(\frac{1}{\vartheta\psi'(\vartheta)} - \frac{1}{\psi(\vartheta)}\right).$$

This trivially provides us with the Laplace transform of $\mathbb{C}\mathrm{orr}(Q_0, Q_t)$ as well. It is stated in the following theorem, which is due to Es-Saghouani and Mandjes [90]; when specializing to compound Poisson input, we find [39, Eqn. (6.2)] again.

Theorem 7.1 *Let $X \in \mathscr{S}_+$. For $\vartheta \geq 0$, and v as in (3.6),*

$$\hat{r}(\vartheta) := \int_0^\infty r(t)\, e^{-\vartheta t} \mathrm{d}t$$

$$= \frac{\gamma(\vartheta)}{v} = \frac{1}{\vartheta} - \frac{\varphi''(0)}{2v\vartheta^2} + \frac{\varphi'(0)}{v\vartheta^2}\left(\frac{1}{\vartheta\psi'(\vartheta)} - \frac{1}{\psi(\vartheta)}\right). \tag{7.3}$$

Remark 7.1 Using Thm. 3.2, it is readily verified that the result in Thm. 7.1 can be simplified to

$$\hat{r}(\vartheta) = \frac{1}{\vartheta} - \frac{1}{v}\left(\frac{\varphi''(0)}{2\vartheta^2} + \frac{\kappa'(\psi(\vartheta))}{\vartheta\psi(\vartheta)}\right),$$

with $\kappa(\alpha)$, as before, denoting $\mathbb{E}e^{-\alpha Q}$. ◇

Example 7.1 Suppose $X \in \mathbb{B}\mathrm{m}(-1, 1)$. Then the Laplace exponent of $(X_t)_t$ is given by $\varphi(\alpha) = \alpha + \frac{1}{2}\alpha^2$, and its inverse is $\psi(\vartheta) = -1 + \sqrt{1 + 2\vartheta}$. Thm. 7.1 yields that the Laplace transform of $r(\cdot)$ is given by

$$\hat{r}(\vartheta) = \frac{1}{\vartheta} - \frac{2}{\vartheta^2} + \frac{2}{\vartheta^3}\left(\sqrt{1+2\vartheta} - 1\right).$$

It turns out to be possible to explicitly invert $\hat{r}(\cdot)$:

$$r(t) = 2(1 - 2t - t^2)\left(1 - \Phi_{\mathrm{N}}(\sqrt{t})\right) + 2\sqrt{t}(1 + t)\phi_{\mathrm{N}}(\sqrt{t}), \tag{7.4}$$

with $\Phi_{\mathrm{N}}(\cdot)$ (respectively, $\phi_{\mathrm{N}}(\cdot)$) the standard normal distribution (respectively, standard normal density). It is remarked that Eqn. (7.4) is in agreement with the results in [1] and [157, Section 12.1]. ◇

7.2 Spectrally Negative Case: Transform

The analysis of the spectrally negative case is similar to that of the spectrally positive case. The derivation of the transform of the workload correlation function relies on the facts (i) that we have the double transform of Q_t through Thm. 4.2, and (ii) that Q_0 is exponentially distributed (as we know from Thm. 3.3). Now observe that, defining by T an exponentially distributed random variable with mean q^{-1},

$$\int_0^\infty q e^{-qt}\mathbb{E}(Q_0 Q_t)\mathrm{d}t = \int_0^\infty \beta_0 x e^{-\beta_0 x}\mathbb{E}_x Q_T\, \mathrm{d}x \tag{7.5}$$

$$= \lim_{\alpha\downarrow 0}\frac{\mathrm{d}}{\mathrm{d}\alpha}\left[\beta \cdot \frac{\mathrm{d}}{\mathrm{d}\beta}\int_0^\infty e^{-\beta x}\mathbb{E}_x e^{-\alpha Q_T}\mathrm{d}x\bigg|_{\beta=\beta_0}\right].$$

Upon combining the explicit expression in Thm. 4.2 with (7.5), and recalling that we have in the spectrally negative case that $\mathbb{V}\mathrm{ar}\, Q_0 = v = 1/\beta_0^2$, we eventually find, after considerable calculus, the following result. For computational details we refer to Glynn and Mandjes [101].

Theorem 7.2 *Let $X \in \mathscr{S}_-$. For $q \geq 0$,*

$$\hat{r}(q) := \int_0^\infty r(t)\, e^{-qt}\mathrm{d}t = \frac{1}{q} + \frac{\beta_0^2}{q^2}\Phi'(\beta_0)\left(\frac{1}{\Psi(q)} - \frac{1}{\beta_0}\right).$$

The following corollary follows from applying L'Hôpital's rule twice. It implies that in the spectrally negative case the workload process is necessarily short-range dependent, that is, $\int_0^\infty r(t)\mathrm{d}t < \infty$. Use that $\Psi'(0)\Phi'(\beta_0) = 1$ and $\Phi''(\beta_0) + (\Phi'(\beta_0))^3\Psi''(0) = 0$, which follow from repeated differentiation of the relation $\Phi(\Psi(q)) = q$.

Corollary 7.1 *Let $X \in \mathscr{S}_-$. Then*

$$\int_0^\infty r(t)\mathrm{d}t = \frac{1}{\beta_0\Phi'(\beta_0)} + \frac{\Phi''(\beta_0)}{2(\Phi'(\beta_0))^2} < \infty.$$

In the spectrally positive case the workload process is not necessarily short-range dependent; we return to this issue in Section 7.3.

7.3 Spectrally Positive Case: Structural Results

Relying on the theory of *completely monotone functions* [41], and in particular on the fact that completely monotone functions can be regarded as Laplace transforms of non-negative random variables, various structural properties of $r(\cdot)$ can be proved. In this section we do so for $X \in \mathscr{S}_+$; the next section deals with $X \in \mathscr{S}_-$. More specifically, we show that $r(\cdot)$ is positive, decreasing, and convex.

Before presenting the main result of this section (which is Prop. 7.1) we introduce the concept of complete monotonicity. A function $f(\alpha)$ on $[0, \infty)$ is said to be completely monotone if for all $n \in \mathbb{N}$, $\alpha \geq 0$,

$$(-1)^n \frac{\mathrm{d}^n}{\mathrm{d}\alpha^n} f(\alpha) \geq 0.$$

We write $f(\alpha) \in \mathscr{C}$. Bernstein [41] proved that there is equivalence between $f(\alpha)$ being completely monotone, and the possibility of writing $f(\alpha)$ as a Laplace transform of a non-negative random variable (up to a multiplicative constant); for more background on completely monotone functions, see Feller [92, pp. 439–442]. More precisely, it holds that a function $f(\alpha)$ on $[0, \infty)$ is the Laplace transform of a non-negative random variable if and only if (i) $f(\alpha) \in \mathscr{C}$, and (ii) $f(0) = 1$.

The concept of complete monotonicity is easy to work with, as a consequence of the fact that one can use a set of practical ‘generation rules’; the proof of the following lemma is standard, and can be found in e.g. [90].

Lemma 7.1 *The following properties apply.*

(1) *$\mathscr{C}$ is closed under addition: if $f(\alpha) \in \mathscr{C}$ and $g(\alpha) \in \mathscr{C}$, then $f(\alpha) + g(\alpha) \in \mathscr{C}$. This can be extended: if $f_x(\alpha) \in \mathscr{C}$ for $x \in \Xi$, then $\int_{x\in\Xi} f_x(\alpha)\mu(\mathrm{d}x) \in \mathscr{C}$ for any measure $\mu(\cdot)$.*
(2) *$\mathscr{C}$ is closed under multiplication: if $f(\alpha) \in \mathscr{C}$ and $g(\alpha) \in \mathscr{C}$, then $f(\alpha)g(\alpha) \in \mathscr{C}$.*
(3) *Properties of composite $\mathscr{C}$ functions: if $f(\alpha) \in \mathscr{C}$ and $g(\alpha) \geq 0$ with $g'(\alpha) \in \mathscr{C}$, then $f(g(\alpha)) \in \mathscr{C}$.*
(4) *Let $U(\alpha)$ be non-decreasing on $[0, \infty)$, and $U(0) = 0$, $u := \lim_{\alpha\to\infty} U(\alpha) < \infty$, and*

$$f(\alpha) := \int_{[0,\infty)} e^{-\alpha x} \mathrm{d}U(x);$$

clearly $f(\alpha) \in \mathscr{C}$ and $u = f(0)$. Then also $g(\alpha) := \alpha^{-1} \cdot (f(0) - f(\alpha)) \in \mathscr{C}$.
(5) *$\mathscr{C}$ is closed under differentiation: if $f(\alpha) \in \mathscr{C}$, then $-f'(\alpha) \in \mathscr{C}$.*

We now state the main result of this section; see [90]. The version for compound Poisson input only is due to [172].

Proposition 7.1 *Let $X \in \mathscr{S}_+$. Then $r(\cdot)$ is positive, decreasing, and convex.*

To prove this result, we first observe that it is readily seen that the Laplace exponent $\upsilon(\cdot)$ of an *increasing* Lévy process $(Y_t)_{t\geq 0}$ is necessarily of the form

$$\upsilon(\alpha) = -\alpha d + \int_0^\infty e^{-\alpha x} \Pi_\upsilon(\mathrm{d}x),$$

for $d \geq 0$ and spectral measure $\Pi_\upsilon(\cdot)$; importantly, there is *no* Brownian component. Then Lemma 7.1(1) directly implies the following result.

Lemma 7.2 *Let $(Y_t)_{t\geq 0}$ be an increasing Lévy process, with Laplace exponent $\upsilon(\alpha)$. Then $-\upsilon'(\alpha) \in \mathscr{C}$.*

Suppose that $X \in \mathscr{S}_+$ with Laplace exponent $\varphi(\alpha)$. We now make the following observations.

(A) $\psi'(\vartheta) \in \mathscr{C}$. This is because $-\psi(\cdot)$ is the Laplace exponent of an *increasing* Lévy process, as follows from Lemma 6.2, in conjunction with Lemma 7.2.
(B) If $f(\alpha) \in \mathscr{C}$, then so is

$$\frac{f(0) - f(\alpha) + \alpha f'(\alpha)}{\alpha^2}.$$

This is a consequence of consecutively applying Lemma 7.1(4) and 7.1(5).

(C) Recalling that

$$\varphi(\alpha) = -\alpha d + \frac{1}{2}\alpha^2\sigma^2 + \int_{(0,\infty)} (e^{-\alpha x} - 1 + \alpha x 1_{(0,1)})\Pi(\mathrm{d}x),$$

we obtain

$$\frac{\alpha\varphi'(\alpha) - \varphi(\alpha)}{\alpha^2} = \frac{1}{2}\sigma^2 + \frac{1}{\alpha^2}\int_{(0,\infty)} (1 - e^{-\alpha x} - \alpha x e^{-\alpha x})\Pi_\varphi(\mathrm{d}x),$$

which is in $\mathscr{C}$, as follows from the fact that any positive constant is in $\mathscr{C}$, together with claim (B) above, and Lemma 7.1(1).

Lemma 7.3 *Define* $\xi(\vartheta)$ *by*

$$\xi(\vartheta) := \frac{1}{\mu}\left(\frac{1}{\vartheta\psi'(\vartheta)} - \frac{1}{\psi(\vartheta)}\right); \tag{7.6}$$

then $\xi(\vartheta)$ *is the Laplace transform of some (non-negative) random variable.*

Proof To prove Lemma 7.3 we first factorize

$$\frac{1}{\vartheta\psi'(\vartheta)} - \frac{1}{\psi(\vartheta)} = \eta_1(\vartheta)\eta_2(\vartheta),$$

with

$$\eta_1(\vartheta) := \frac{\psi(\vartheta)}{\vartheta}, \qquad \eta_2(\vartheta) := \frac{1}{\psi(\vartheta)\psi'(\vartheta)} - \frac{\vartheta}{(\psi(\vartheta))^2}.$$

Because of Thm. 3.2, $\alpha/\varphi(\alpha) \in \mathscr{C}$; now applying Lemma 7.1(3), in conjunction with claim (A) above, we obtain that $\eta_1(\vartheta) \in \mathscr{C}$.

To show that also $\eta_2(\vartheta) \in \mathscr{C}$, we first recall from claim (C) above that $(\alpha\varphi'(\alpha) - \varphi(\alpha))/\alpha^2 \in \mathscr{C}$. Again applying Lemma 7.1(3), in conjunction with claim (A), it follows that $\eta_2(\vartheta) \in \mathscr{C}$.

As both $\eta_1(\vartheta)$ and $\eta_2(\vartheta)$ are in $\mathscr{C}$, Lemma 7.1(2) yields that $\xi(\vartheta) \in \mathscr{C}$. Applying 'L'Hôpital' twice, and using that $\psi''(0)(\varphi'(0))^3 = -\varphi''(0)$, it is readily verified that $\xi(0) = 1$. Now Lemma 7.3 follows from Bernstein's result. □

We are now ready to prove Prop. 7.1. The proof demonstrates how the concept of complete monotone functions facilitates elegant proofs of structural results.

Let $\hat{r}^{(1)}(\vartheta)$ and $\hat{r}^{(2)}(\vartheta)$ be the Laplace transforms of $r'(t)$ and $r''(t)$, respectively:

$$\hat{r}^{(1)}(\vartheta) := \int_0^\infty r'(t)\, e^{-\vartheta t}\mathrm{d}t = -\frac{\varphi''(0)}{2v\vartheta}(1 - \xi(\vartheta)), \tag{7.7}$$

$$\hat{r}^{(2)}(\vartheta) := \int_0^\infty r''(t)\, e^{-\vartheta t}\mathrm{d}t = \frac{\varphi''(0)}{2v}\xi(\vartheta), \tag{7.8}$$

for $\vartheta \geq 0$. Here the properties that $r(0) = 1$ and

$$r'(0) = \lim_{\varepsilon \downarrow 0} \frac{\mathbb{E}(Q_0 Q_\varepsilon) - \mathbb{E}(Q_0^2)}{\varepsilon \mathbb{V}\mathrm{ar}\, Q_0} = \lim_{\varepsilon \downarrow 0} \frac{\mathbb{E}(Q_0 X_\varepsilon)}{\varepsilon \mathbb{V}\mathrm{ar}\, Q_0} = -\frac{\varphi''(0)}{2v},$$

in conjunction with integration by parts, are used.

Hence, convexity of $r(\cdot)$ follows from the expression for $\hat{r}^{(2)}(\vartheta)$ in (7.8); it is concluded from Lemma 7.3 that $\hat{r}^{(2)}(\vartheta) \in \mathscr{C}$, so that $r''(t)$ is non-negative (for $t \geq 0$). The monotonicity of $r(\cdot)$ follows from the expression for $\hat{r}^{(1)}(\vartheta)$ in Eqn. (7.7), by applying Lemma 7.1(4) to $\hat{r}^{(2)}(\vartheta) \in \mathscr{C}$; we find that $-\hat{r}^{(1)}(\vartheta)$ is in $\mathscr{C}$, implying that $r'(t) \leq 0$ (for $t \geq 0$). Then it is easily verified that applying Lemma 7.1(4) to $-\hat{r}^{(1)}(\vartheta) \in \mathscr{C}$, in conjunction with Eqn. (7.3), implies $\hat{r}(\vartheta) \in \mathscr{C}$, and hence $r(t) \geq 0$ (for $t \geq 0$). We have thus proved that $r(\cdot)$ is positive, decreasing, and convex.

In Cor. 7.1 we observed that for $X \in \mathscr{S}_-$ the workload process is short-range dependent. This statement is not valid for $X \in \mathscr{S}_+$; only if $\varphi^{(4)}(0)$ is well defined is $r(t)$ then integrable and

$$\int_0^\infty r(t)\mathrm{d}t = \frac{1}{8v}\frac{\varphi^{(4)}(0)}{\varphi'(0)^2} - \frac{5}{12v}\frac{\varphi''(0)\varphi^{(3)}(0)}{\varphi'(0)^3} + \frac{1}{4v}\frac{\varphi''(0)^3}{\varphi'(0)^4};$$

see [90].

7.4 Spectrally Negative Case: Structural Results

The proof for $X \in \mathscr{S}_-$ works quite similarly to the one for $X \in \mathscr{S}_+$ (but is considerably easier). This result is due to Glynn and Mandjes [101].

Proposition 7.2 *Let $X \in \mathscr{S}_-$. Then $r(\cdot)$ is positive, decreasing, and convex.*

Proof As mentioned, we mimic the proof of the spectrally positive case, as originally developed in [90]. Using integration by parts, we find that

$$\hat{r}^{(1)}(q) := \int_0^\infty r'(t)e^{-qt}\mathrm{d}t = \frac{\beta_0^2}{q}\Phi'(\beta_0)\left(\frac{1}{\Psi(q)} - \frac{1}{\beta_0}\right),$$

which also entails that $r'(0) = \beta_0 \Phi'(\beta_0)$. Analogously,

$$\hat{r}^{(2)}(q) := \int_0^\infty r''(t)e^{-qt}\mathrm{d}t$$
$$= -r'(0) + \beta_0^2\Phi'(\beta_0)\left(\frac{1}{\Psi(q)} - \frac{1}{\beta_0}\right) = \beta_0^2\frac{\Phi'(\beta_0)}{\Psi(q)}. \quad (7.9)$$

Prop. 6.3 shows that $\Psi(0)/\Psi(q) \in \mathscr{C}$. We conclude from (7.9) that $\hat{r}^{(2)}(q)$ is in $\mathscr{C}$, and hence $r''(\cdot)$ is positive, that is, $r(\cdot)$ is convex.

We know from Lemma 7.1(4) that $f(q) \in \mathscr{C}$ implies that, with $g(q)$ defined as $(f(0)-f(q))/q$, also $g(q) \in \mathscr{C}$. Taking $f(q) = \hat{r}^{(2)}(q)$, we obtain that $-\hat{r}^{(1)}(q) \in \mathscr{C}$, and hence $r'(\cdot)$ is negative, that is, $r(\cdot)$ is decreasing. Applying the same procedure again, we find that $\hat{r}(q) \in \mathscr{C}$, and hence $r(\cdot)$ is positive. □

Exercises

Exercise 7.1 This exercise is on the workload correlation function $r(t)$ for X corresponding to $\mathbb{B}\mathrm{m}(-1,1)$.

(a) Verify Eqn. (7.4).

(b) Find an explicit function $f(\cdot)$ so that $r(t)/f(t) \to 1$ as $t \to \infty$. Use the relation

$$\lim_{x\to\infty} \frac{1-\Phi_{\mathrm{N}}(x)}{\phi_{\mathrm{N}}(x)} \cdot \left(\frac{1}{x} - \frac{1}{x^3} + \frac{3}{x^5} - \frac{15}{x^7}\right) = 1,$$

with $\Phi_{\mathrm{N}}(\cdot)$ and $\phi_{\mathrm{N}}(\cdot)$ the cumulative distribution function and the density, respectively, of a standard normal random variable.

Exercise 7.2 Verify Thm. 7.2 from Eqn. (7.5).

Chapter 8
Stationary Workload Asymptotics

Virtually all results presented so far have been in terms of transforms. These in principle uniquely characterize the entity under study, and with numerical inversion they can be evaluated up to a great level of precision, but in specific cases one would prefer closed-form expressions. However, if one is willing to settle for a less ambitious goal, then such explicit results *can* be achieved: when focusing on just *asymptotics* of the quantity of interest, various limit results can be established. In this chapter the objective is to characterize the tail asymptotics of the *steady-state workload*, that is, we analyze $\mathbb{P}(Q > u)$ for u large. More specifically, our goal is to obtain the *exact asymptotics* of $\mathbb{P}(Q > u)$, that is, we wish to identify an explicitly given function $f(\cdot)$ such that, as $u \to \infty$, $\mathbb{P}(Q > u)/f(u) \to 1$.

When analyzing stationary workload asymptotics, it turns out that the subdivision into the two spectrally one-sided cases and the case with jumps in both directions (that we have repeatedly come across in the previous chapters) is less relevant. Crucial when characterizing these tail probabilities is a subdivision along the lines of the 'heaviness' of the upper tail of the driving Lévy process X. The asymptotics for cases when this tail is 'light' are intrinsically different from those corresponding to 'heavy-tailed' scenarios (and there turns out to be an intermediate regime too).

Before proceeding with the analysis of the stationary workload distribution we include a short technical note. Recall that, due to Eqn. (2.5), we have the identity

$$\mathbb{P}(Q > u) = \mathbb{P}(\exists t \geq 0 : X_t > u) = \mathbb{P}(\sigma(u) < \infty),$$

where $\sigma(u)$ is defined as the first passage time of level u, that is, $\inf\{t \geq 0 : X_t > u\}$.

K. Dębicki, M. Mandjes, *Queues and Lévy Fluctuation Theory*, Universitext,
DOI 10.1007/978-3-319-20693-6_8

8.1 Light-Tailed Regime

We denote by $\mathscr{L}$ the class of Lévy processes such that there exists an $\omega > 0$ such that $\mathbb{E}e^{\omega X_1} = 1$ and $\mathbb{E}X_1 e^{\omega X_1} < \infty$; in the literature this case is often referred to as the 'Cramér case' or the 'light-tailed case'. Notice that a necessary condition for $X \in \mathscr{L}$ is that all moments of X_1 are finite (this is not a sufficient condition though: think of a compound Poisson input process with jobs that have a Weibull or lognormal distribution).

For ease we start our analysis by considering $X \in \mathbb{C}\mathrm{P}(r, \lambda, b(\cdot)) \cap \mathscr{L}$, for which the asymptotics of $\mathbb{P}(Q > u)$ for u large can be determined in very explicit terms relying on the concept of *change of measure*. Then we focus on spectrally positive input; there the crucial observation is that in this light-tailed regime the Laplace exponent $\varphi(\alpha)$ is well defined not only for $\alpha \geq 0$ but also for a range of negative values. The last part of the section presents the asymptotics for general $X \in \mathscr{L}$; we briefly sketch the proof of this result.

Compound Poisson case—As was mentioned above, we first assume $X \in \mathbb{C}\mathrm{P}(r, \lambda, b(\cdot))$, where we set, without loss of generality, $r = 1$. We assume that $\varrho = \lambda\,\mathbb{E}B/r = \lambda\,\mathbb{E}B < 1$ to ensure stability. Then let ω solve the equation

$$\varphi(-\omega) = -\omega - \lambda + \lambda b(-\omega) = 0;$$

due to the convexity of $\varphi(\cdot)$ in combination with the fact that $\varphi'(0) = -\mathbb{E}X_1 > 0$, we conclude that this ω is positive.

Referring to the original probability measure as $\mathbb{P}$, we introduce an alternative measure $\mathbb{Q}$ that is characterized as $\mathbb{C}\mathrm{P}(1, \lambda + \omega, \bar{b}(\cdot))$, where the Laplace transform $\bar{b}(\alpha)$ is given by $b(\alpha - \omega)/b(-\omega)$. The underlying random variable that corresponds to the job size under $\mathbb{Q}$ can be thought of as a random variable whose density is

$$\mathbb{P}(B \in \mathrm{d}x)\,\frac{e^{\omega x}}{\mathbb{E}\,e^{\omega B}} = \mathbb{P}(B \in \mathrm{d}x)\,\frac{e^{\omega x}}{b(-\omega)}$$

(where it is readily verified that this function is positive and integrates to 1, as required). As a consequence, the Laplace exponent of the driving Lévy process under $\mathbb{Q}$ is

$$\alpha - (\lambda + \omega) + (\lambda + \omega)\frac{b(\alpha - \omega)}{b(-\omega)}.$$

But we know that $-\omega - \lambda + \lambda b(-\omega) = 0$, so that the Laplace exponent under $\mathbb{Q}$ can be rewritten as

$$\alpha - (\lambda + \omega) + \lambda b(\alpha - \omega) = -(\omega - \alpha) - \lambda + \lambda b(\alpha - \omega) = \varphi(\alpha - \omega).$$

Recall that the Laplace exponent was $\varphi(\alpha)$ under $\mathbb{P}$; the above computation shows that under $\mathbb{Q}$ this Laplace exponent is shifted by ω (recall this is a positive number!) to the right.

This procedure to generate an alternative probability model is often referred to as *exponential twisting*; it owes its name to the relation

$$\mathbb{Q}(X_t \in \mathrm{d}x) = e^{\omega x}\,\mathbb{P}(X_t \in \mathrm{d}x),$$

which holds because

$$\int_{-\infty}^{\infty} e^{-\alpha x}\mathbb{Q}(X_t \in \mathrm{d}x) = e^{\varphi(\alpha-\omega)t} = \int_{-\infty}^{\infty} e^{(\omega-\alpha)x}\,\mathbb{P}(X_t \in \mathrm{d}x)$$

for all α for which these expressions are well defined.

We now check whether the queue is stable under the new measure $\mathbb{Q}$. To this end, first note that the corresponding load can be expressed as $(\lambda+\omega)\,\mathbb{E}_{\mathbb{Q}}B$. From the definition of ω and the convexity of $\mathbb{E}e^{\omega X_1}$, it can be concluded (in self-evident notation) that

$$(\lambda+\omega)\,\mathbb{E}_{\mathbb{Q}}B = (\lambda+\omega)\left(-\frac{b'(-\omega)}{b(-\omega)}\right) = -\lambda b'(-\omega) =: \varrho_{\mathbb{Q}} > 1, \tag{8.1}$$

so that under $\mathbb{Q}$ the queue is *unstable*. In other words, under $\mathbb{Q}$ we have that $\sigma(u) < \infty$ almost surely, for any $u > 0$.

A change-of-measure argument yields that

$$\mathbb{P}(Q > u) = \mathbb{E}_{\mathbb{P}}1_{\{\sigma(u)<\infty\}} = \mathbb{E}_{\mathbb{Q}}\left(\frac{f^{(\mathbb{P})}_{X_{\sigma(u)}}(X_{\sigma(u)})}{f^{(\mathbb{Q})}_{X_{\sigma(u)}}(X_{\sigma(u)})}1_{\{\sigma(u)<\infty\}}\right), \tag{8.2}$$

with $f^{(\mathbb{P})}_{X_t}(\cdot)$ and $f^{(\mathbb{Q})}_{X_t}(\cdot)$ denoting the densities of X_t under the original and alternative measures, respectively; see e.g. Asmussen [19, Thm. XIII.3.2]. This identity will appear, in various forms, several times in this monograph. If under $\mathbb{Q}$ the event of overflow is more likely than under $\mathbb{P}$, the above equality states that this is compensated for by suitably small values of the 'likelihood ratio' $f^{(\mathbb{P})}_{X_{\sigma(u)}}(X_{\sigma(u)})/f^{(\mathbb{Q})}_{X_{\sigma(u)}}(X_{\sigma(u)})$.

But now realize that $\mathbb{Q}$ is constructed (by exponential twisting) such that

$$f^{(\mathbb{Q})}_{X_t}(x) = f^{(\mathbb{P})}_{X_t}(x)\frac{e^{\omega x}}{\mathbb{E}e^{\omega X_t}} = f^{(\mathbb{P})}_{X_t}(x)e^{\omega x}.$$

Using that $\sigma(u) < \infty$ almost surely, we thus find the powerful identity

$$\mathbb{P}(Q > u) = \mathbb{E}_{\mathbb{Q}}e^{-\omega X_{\sigma(u)}}. \tag{8.3}$$

Our goal now is to use this result (which holds for any $u > 0$) to obtain the *exact asymptotics* of $\mathbb{P}(Q > u)$.

To this end, the next step is to observe that $X_{\sigma(u)} = u + R_u$, where R_u is the *overshoot* over level u. Let L_n be the nth *ladder height*, that is, the difference between the nth and $(n-1)$st record value; these random variables are positive and i.i.d., and, due to (8.1), non-defective (use that under $\mathbb{Q}$ the load is larger than 1!). Renewal theory (see e.g. [19, Section V.4]) now yields that the overshoot R_u converges to a limiting random variable R, whose distribution is given through

$$\mathbb{Q}(R \leq v) = \frac{1}{\mathbb{E}_{\mathbb{Q}}L}\int_0^v (1 - \mathbb{Q}(L \leq y))\mathrm{d}y,$$

with the random variable L corresponding to a single ladder height. Due to the definition of the new measure $\mathbb{Q}$, we have

$$\mathbb{Q}(L \in \mathrm{d}y) = e^{\omega y}\,\mathbb{P}(L \in \mathrm{d}y) = e^{\omega y}\lambda\mathbb{P}(B > y)\mathrm{d}y; \tag{8.4}$$

it is an exercise to verify that it follows from the definition of ω that this density indeed integrates to 1. Upon combining the above findings, we obtain that, as $u \to \infty$,

$$\mathbb{P}(Q > u)e^{\omega u} \to \frac{1}{\mathbb{E}_{\mathbb{Q}}L}\int_0^\infty e^{-\omega y}(1 - \mathbb{Q}(L \leq y))\mathrm{d}y.$$

Further, with straightforward calculus we can evaluate the constant in the left-hand side of the previous display, as follows. Integration by parts yields

$$\int_0^\infty e^{-\omega y}(1 - \mathbb{Q}(L \leq y))\mathrm{d}y = \frac{1}{\omega}\left(1 - \int_0^\infty e^{-\omega y}\mathbb{Q}(L \in \mathrm{d}y)\right). \tag{8.5}$$

Inserting relation (8.4) then reduces (8.5) to $(1 - \varrho)/\omega$. Again by integration by parts,

$$\mathbb{E}_{\mathbb{Q}}L = \frac{\lambda}{\omega^2}(1 - b(-\omega)) - \frac{\lambda}{\omega}b'(-\omega). \tag{8.6}$$

Recalling that ω solves $-\omega - \lambda + \lambda b(-\omega) = 0$, and using the definition of $\varrho_{\mathbb{Q}}$, we arrive at the following result, known as the *Cramér–Lundberg asymptotics*. It states that, in the case of compound Poisson input, in the light-tailed regime $\mathbb{P}(Q > u)$ decays essentially exponentially.

Theorem 8.1 *Let $X \in \mathbb{CP}(1, \lambda, b(\cdot)) \cap \mathscr{L}$. Then, as $u \to \infty$,*

$$\mathbb{P}(Q > u)e^{\omega u} \to \frac{1 - \varrho}{\varrho_{\mathbb{Q}} - 1}.$$

In passing, we also proved that, for all $u \geq 0$, $\mathbb{P}(Q > u) \leq e^{-\omega u}$ (use the identity (8.3), write $X_{\sigma(u)} = u + R_u$, and realize that $R_u \geq 0$). In [44, Remark 2] and [72] it is argued that this uniform bound applies for all $X \in \mathscr{L}$, that is, not just for compound Poisson; the proof in [72] relies on a change-of-measure argument.

Corollary 8.1 *Let $X \in \mathscr{L}$. Then, for any $u > 0$, $\mathbb{P}(Q > u) \leq e^{-\omega u}$.*

Spectrally positive case—A next step is to consider asymptotics for more general Lévy processes in $\mathscr{L}$: is it for instance possible to extend Thm. 8.1 to $\mathscr{S}_+ \cap \mathscr{L}$? To this end, realize that we have the Laplace–Stieltjes transform of Q, that is, $\alpha\varphi'(0)/\varphi(\alpha)$; see Thm. 3.2. Then one idea is to exploit this transform to obtain the tail asymptotics. An approach to doing so is through application of the so-called *Heaviside principle*, as advocated in e.g. Abate and Whitt [3]. We now point out how 'Heaviside' works, with a focus on the asymptotics of complementary distribution functions. We restrict ourselves to special cases, that is, the case that the transform of the complementary distribution has a pole, and the one that it has a branching point.

Recipe 8.1 ('Heaviside') Let $\zeta(\cdot)$ be the Laplace transform of the complementary distribution function of a (non-negative) random variable Z:

$$\zeta(\alpha) := \int_0^\infty e^{-\alpha x}\, \mathbb{P}(Z > x)\mathrm{d}x.$$

(i) Suppose there is an $\omega > 0$ such that, for some $A > 0$, $\zeta(\alpha) \sim A(\alpha + \omega)^{-1}$, as $\alpha \downarrow -\omega$. Then,

$$\mathbb{P}(Z > u) \sim Ae^{-\omega u},$$

as $u \to \infty$.

(ii) Suppose there is an $\omega > 0$ such that, for some $A > 0$ and (irrelevant) B, $\zeta(\alpha) \sim A\sqrt{\alpha + \omega} + B$, as $\alpha \downarrow -\omega$. Then,

$$\mathbb{P}(Z > u) \sim \frac{Ae^{-\omega u}}{u\sqrt{u}\,\Gamma(-\frac{1}{2})},$$

as $u \to \infty$. ◇

It is noted, however, that the Heaviside principle, although well established in the literature and frequently used [65], lacks full mathematical rigor. This is why we have called this technique a 'recipe'; in [3] the term 'operational principle' is used.

Now let us use 'Heaviside' to find the tail asymptotics of $\mathbb{P}(Q > u)$. To this end, first note that an elementary integration-by-parts argument yields

$$\int_0^\infty e^{-\alpha x}\mathbb{P}(Q > x)\mathrm{d}x = \frac{\mathbb{P}(Q > 0)}{\alpha} - \frac{\varphi'(0)}{\varphi(\alpha)}.$$

Now observe that when $X \in \mathscr{L}$, $\varphi(\cdot)$ has a zero in $-\omega$, and

$$\lim_{\alpha\downarrow-\omega}(\alpha+\omega)\int_0^\infty e^{-\alpha x}\mathbb{P}(Q > x)\mathrm{d}x = \frac{\varphi'(0)}{-\varphi'(-\omega)} > 0;$$

note that we assumed that the denominator of the last expression is finite (due to the requirement $\mathbb{E}X_1 e^{\omega X_1} < \infty$). Now the Heaviside principle says that, as $u \to \infty$,

$$\mathbb{P}(Q > u)e^{\omega u} \to \frac{\varphi'(0)}{-\varphi'(-\omega)};$$

it is readily checked that for the compound Poisson case this expression agrees with that of Thm. 8.1.

General case—The most general (rigorously proven) result is due to Bertoin and Doney [44]: there, tail asymptotics for $\mathbb{P}(Q > u)$ are derived for the full class $\mathscr{L}$. These are of the form $Ce^{-\omega u}$, where ω solves $\mathbb{E}e^{\omega X_1} = 1$, but with some rather involved expression for C. A nice alternative proof of this result, relying on an embedding approach, was given in [82]. Here we present the compact proof that was derived in Asghari et al. [15], and that heavily rests on the identity (8.3). It uses Lemma 6.4, which can be rephrased as

$$\begin{aligned}\int_0^\infty e^{-\beta x}\mathbb{E}\left(e^{-q\sigma(x)-\bar{q}(X_{\sigma(x)}-x)}1_{\{\sigma(x)<\infty\}}\right)\mathrm{d}x &= \frac{1}{\beta-\bar{q}}\left(1-\frac{\mathbb{E}e^{-\beta\bar{X}_T}}{\mathbb{E}e^{-\bar{q}\bar{X}_T}}\right)\\ &= \frac{1}{\beta-\bar{q}}\left(1-\frac{k(q,\beta)}{k(q,\bar{q})}\right). \quad (8.7)\end{aligned}$$

In Eqn. (8.3) we found $\mathbb{P}(Q > u) = \mathbb{E}_{\mathbb{Q}}e^{-\omega X_{\sigma(u)}}$; recall that $\mathbb{Q}$ is the exponentially twisted version of $\mathbb{P}$, with parameter ω, under which the Lévy process $(X_t)_t$ has a positive drift. Now observe that, with the overshoot $R_u := X_{\sigma(u)} - u$,

$$\mathbb{E}_{\mathbb{Q}}e^{-\omega X_{\sigma(u)}} = e^{-\omega u}\,\mathbb{E}_{\mathbb{Q}}e^{-\omega(X_{\sigma(u)}-u)} = e^{-\omega u}\,\mathbb{E}_{\mathbb{Q}}e^{-\omega R_u}.$$

Then, relying on (8.7), in self-evident notation we have that

$$\lim_{u\to\infty}\mathbb{E}_{\mathbb{Q}}e^{-\omega R_u} = \lim_{\beta\downarrow 0}\frac{\beta}{\beta-\omega}\left(1-\frac{k_{\mathbb{Q}}(0,\beta)}{k_{\mathbb{Q}}(0,\omega)}\right). \quad (8.8)$$

From

$$k_{\mathbb{Q}}(0,\beta) = \exp\left(-\int_0^\infty \int_{(0,\infty)} \frac{1}{t}\left(e^{-t} - e^{-\beta x}\right) e^{\omega x}\mathbb{P}(X_t \in \mathrm{d}x)\,\mathrm{d}t\right)$$

$$= \exp\left(-\int_0^\infty \int_{(0,\infty)} \frac{1}{t}\left(\left(e^{-t} - e^{-(\beta-\omega)x}\right) - \left(e^{-t} - e^{-t+\omega x}\right)\right)\mathbb{P}(X_t \in \mathrm{d}x)\,\mathrm{d}t\right)$$

$$= \frac{k(0,\beta-\omega)}{k(1,-\omega)},$$

it is immediate that

$$\frac{k_{\mathbb{Q}}(0,\beta)}{k_{\mathbb{Q}}(0,\omega)} = \frac{k(0,\beta-\omega)}{k(0,0)}.$$

As argued in Kyprianou [146, p. 188], $\ell(0,\beta-\omega) := 1/k(0,\beta-\omega) \to 0$ as $\beta \downarrow 0$. It follows that (8.8) equals

$$\frac{1}{\omega k(0,0)} \lim_{\beta\downarrow 0} \frac{\beta}{\ell(0,\beta-\omega)} = \frac{1}{\omega k(0,0)} \frac{1}{\ell'(0,-\omega)},$$

where the derivative $\ell'(0,-\omega)$ is to be understood as the partial derivative with respect to the second argument. The final result is given in the theorem below; a detailed version of the above sketch-of-proof is found in [146, Section VII.2]. It is checked that the result is in agreement with the asymptotics for the light-tailed spectrally positive case.

Theorem 8.2 *Let $X \in \mathscr{L}$. Then, as $u \to \infty$,*

$$\mathbb{P}(Q > u)e^{\omega u} \to \frac{1}{\omega k(0,0)} \frac{1}{\ell'(0,-\omega)}.$$

8.2 Intermediate Regime

We now focus on a second regime, in which the upper tail of X_1 is still essentially exponential, but the equation $\mathbb{E}e^{\omega X_1} = 1$ now lacks a proper positive solution.

We define

$$\omega := \sup\{\delta \geq 0 : \mathbb{E}e^{\delta X_1} < \infty\}.$$

We say that $X \in \mathscr{I}$ if $\omega \in (0,\infty)$ and $\mathbb{E}e^{\omega X_1} < 1$; this basically means that at $\delta = \omega$, the moment generating function $\mathbb{E}e^{\delta X_1}$ jumps from a value strictly smaller than 1 to ∞.

Interestingly, again the change-of-measure technique can be used to find a uniform upper bound. Defining $M(\delta) := \mathbb{E}e^{\delta X_1}$, the new measure $\mathbb{Q}(\vartheta)$ is such that the Lévy process under $\mathbb{Q}(\vartheta)$ has moment generating function

$$\mathbb{E}_{\mathbb{Q}(\vartheta)}e^{\delta X_1} = \frac{M(\delta+\vartheta)}{M(\vartheta)}.$$

As before, for all $\vartheta < \omega$, we obtain the inequality

$$\mathbb{P}(Q > u) = \mathbb{E}_{\mathbb{Q}(u)}\left(e^{-\vartheta X_{\sigma(u)}} \cdot (M(\vartheta))^{\sigma_u}\right) \leq e^{-\vartheta u},$$

which leads to the following uniform bound.

Corollary 8.2 *Let $X \in \mathscr{I}$. Then $\mathbb{P}(Q > u) \leq e^{-\omega u}$.*

The following exact asymptotics were derived in Dieker [82]; see also [137]. Remarkably, they show that for $X \in \mathscr{I}$ the tail distribution of Q is asymptotically proportional to that of X_1. We do not include a proof here.

Proposition 8.1 *Let $X \in \mathscr{I}$. Then, as $u \to \infty$,*

$$\frac{\mathbb{P}(Q > u)}{\mathbb{P}(X_1 > u)} \to \frac{\mathbb{E}e^{\omega Q}}{M(\omega)\log M(\omega)}.$$

8.3 Heavy-Tailed Regime

In this section we consider Lévy processes for which $\mathbb{E}e^{\delta X_1} = \infty$ for all $\delta > 0$. An important subclass of these processes is the class of regularly varying Lévy processes $\mathscr{R}$.

Considering the class of compound Poisson inputs, regular variation refers to the tail of the distribution of the jobs: it is assumed that for an index α and all $y > 0$, as $x \to \infty$,

$$\frac{\mathbb{P}(B > yx)}{\mathbb{P}(B > x)} \to y^{\alpha}.$$

We begin with a heuristic argument that leads to the correct asymptotics. Recall that, due to (3.2),

$$\mathbb{P}(Q > u) = \mathbb{P}\left(\sum_{i=1}^{N} B_i^{\rm res} > u\right),$$

where $B_1^{\rm res}, B_2^{\rm res}, \ldots$ are i.i.d. samples from the residual lifetime distribution of B and $\mathbb{P}(N = n) = (1-\rho)\,\rho^n$, with $\rho = \lambda\mathbb{E}B/r$. A known property of distributions with a regularly varying tail is that

$$\mathbb{P}\left(\sum_{i=1}^{n} B_i^{\rm res} > u\right) \sim n\mathbb{P}\left(B^{\rm res} > u\right)$$

for each n, as $u \to \infty$. Under the proviso that we can interchange the sum (i.e. $\sum_{n=1}^{\infty}\cdots$) with the limit (i.e. $\lim_{u\to\infty}\cdots$), we have

$$\mathbb{P}(Q > u) = \sum_{n=1}^{\infty} \mathbb{P}\left(\sum_{i=1}^{n} B_i^{\rm res} > u\right)(1-\rho)\,\rho^n$$

$$\sim (1-\rho)\,\mathbb{P}\left(B^{\rm res} > u\right)\sum_{n=1}^{\infty} n\rho^n = \frac{\varrho}{1-\varrho}\mathbb{P}\left(B^{\rm res} > u\right),$$

as $u \to \infty$.

Now we sketch an approach to formally finding the tail asymptotics of $\mathbb{P}(Q > u)$ for u large and $X \in \mathbb{CP}(r, \lambda, b(\cdot)) \cap \mathscr{R}$, following a recipe outlined in Zwart [223, pp. 36–39]. This recipe is based on the insight that in these heavy-tailed scenarios a large workload is (with overwhelming probability) due to *a single big job*. The approach therefore consists of a lower bound, in which the probability of this most likely scenario is evaluated, and an upper bound in which it is shown that the contributions of other scenarios (e.g. no big job, multiple big jobs) can be neglected. Here we demonstrate how the lower bound is derived; the upper bound is more involved and left out.

We apply the usual time normalization, in that we assume $r = 1$ (without loss of generality). We consider $\mathbb{CP}(1, \lambda, b(\cdot))$, and we denote, as earlier, $\varrho := \lambda\,\mathbb{E}B$. First it is noted that due to the law of large numbers, we can find (for any $\delta, \varepsilon > 0$) a number $t_{\delta,\varepsilon}$ such that for all $t \geq t_{\delta,\varepsilon}$,

$$\mathbb{P}(X_t > (\varrho - 1 - \varepsilon)t) > 1 - \delta.$$

It is noted that a sufficient condition for the workload Q_0 to exceed u is that a job of size at least $u + (1-\varrho)t + \varepsilon t$ arrived at time $-t$, and that the amount of work that arrived between $-t$ and 0 is at least $(\varrho - \varepsilon)t$; notice that the former event is rare, as opposed to the latter. We thus obtain

$$\mathbb{P}(Q > u) \geq \int_{t_{\delta,\varepsilon}}^{\infty} \lambda\mathbb{P}(B > u + (1-\varrho)t + \varepsilon t)\mathbb{P}(-X_{-t} > (\varrho - 1 - \varepsilon)t)\mathrm{d}t$$

$$\geq (1-\delta)\int_{t_{\delta,\varepsilon}}^{\infty} \lambda\mathbb{P}(B > u + (1-\varrho)t + \varepsilon t)\mathrm{d}t$$

$$= (1-\delta)\frac{\varrho}{1-\varrho+\varepsilon}\mathbb{P}(B^{\rm res} > u + t_{\delta,\varepsilon}) \sim \frac{(1-\delta)\varrho}{1-\varrho+\varepsilon}\mathbb{P}(B^{\rm res} > u),$$

where the last step is an immediate consequence of the definition of regular variation. Now let $\delta, \varepsilon \downarrow 0$, and the lower bound follows.

As pointed out above, the upper bound considers the contributions of other scenarios, and proves that they can be neglected relative to $\mathbb{P}(B^{\rm res} > u)$. This is in general quite a cumbersome procedure. After having established this upper bound, the following theorem is obtained. The result dates back e.g. to [49, 62], but there entirely different proof techniques were used.

Theorem 8.3 *Let $X \in \mathbb{CP}(1, \lambda, b(\cdot)) \cap \mathscr{R}$. Then, as $u \to \infty$,*

$$\mathbb{P}(Q > u) \sim \frac{\varrho}{1-\varrho}\mathbb{P}(B^{\rm res} > u).$$

There is an alternative approach though, that is helpful if the Laplace–Stieltjes transform is available: Tauberian inversion. To explain this method, we first define the following notion.

Definition 8.1 We say that $f(x) \in \mathscr{R}_\delta(n, \eta)$, with $\delta \in (n, n+1)$, for $x \downarrow 0$, if

$$f(x) = \sum_{i=0}^{n} \frac{f^{(i)}(0)}{i!}x^i + \eta x^\delta L(1/x), \quad x \downarrow 0,$$

for a slowly varying function $L(\cdot)$, that is, $L(x)/L(tx) \to 1$ for $x \to \infty$, for any $t > 0$.

The Tauberian theorem in [47, Thm. 8.1.6] states that the property that $\mathbb{E}e^{-\alpha X} \in \mathscr{R}_\delta(n, \eta)$ as $\alpha \downarrow 0$, for some non-negative random variable X, is equivalent to

$$\mathbb{P}(X > u) \sim \frac{(-1)^{n-1}}{\Gamma(1-\delta)} \cdot \eta \cdot u^{-\delta} L(u),$$

as $u \to \infty$; see also [46].

Suppose now that $X \in \mathscr{S}_+$ and $\varphi(\alpha) \in \mathscr{R}_\nu(n, \eta)$; we then write $X \in \mathscr{S}_+ \cap \mathscr{R}$. It is readily checked that

$$\mathbb{E}e^{-\alpha Q} = \frac{\alpha\varphi'(0)}{\varphi(\alpha)} \in \mathscr{R}_{\nu-1}\left(n-1, \frac{\eta}{\varphi'(0)}\right).$$

Then the Tauberian theorem yields the following result.

Theorem 8.4 *Let $X \in \mathscr{S}_+ \cap \mathscr{R}$, with $\varphi(\alpha) \in \mathscr{R}_\nu(n, \eta)$. Then, as $u \to \infty$,*

$$\mathbb{P}(Q > u) \sim \frac{(-1)^n}{\Gamma(2-\nu)} \cdot \left(\frac{\eta}{\varphi'(0)}\right) u^{1-\nu} L(u).$$

Example 8.1 Consider $X \in \mathbb{CP}(1, \lambda, b(\cdot))$. Suppose $\mathbb{P}(B > x) \sim x^{-\delta}L(x)$. From $\varphi(\alpha) = \alpha + \lambda b(\alpha) - \lambda$, it follows that $\varphi(\alpha) \in \mathscr{R}_\delta(n, \lambda\Gamma(1-\delta)(-1)^{n-1})$ by applying 'Tauber'. Then the above theorem confirms the result presented in Thm. 8.3. ◇

It should be realized that the class $\mathscr{R}$ is a subset of the class of all Lévy processes for which $\mathbb{E}e^{\delta X_1} = \infty$ for all $\delta > 0$. Now define the (broader) class of heavy-tailed (or *subexponential*) Lévy processes $\mathscr{H}$, as follows.

To this end, we first introduce the notion of subexponential distribution functions, following the terminology of [82]. With $D(\cdot)$ being a distribution function on $[0, \infty)$ and $D^{\star 2}$ the twofold convolution of D, we say that D is subexponential if

$$1 - D^{\star 2}(x) \sim 2(1 - D(x))$$

as $x \to \infty$. For a measure $\mu(\cdot)$ we say that it is subexponential if the following two conditions are fulfilled: (i) $\mu([1, \infty)) < \infty$, and (ii) $\mu([1, \cdot])/\mu([1, \infty))$ is subexponential. Then, for the spectral measure $\Pi(\cdot)$ associated with the Lévy process $(X_t)_t$, define

$$\bar{\Pi}((x, \infty)) := \int_x^\infty \Pi((y, \infty))\mathrm{d}y.$$

We say that $X \in \mathscr{H}$ if $\bar{\Pi}(\cdot)$ is subexponential.

The following result is found in Asmussen [18]; a version also containing local asymptotics was first presented in [82]. Realize that $-\mathbb{E}X_1$ is a positive number.

Theorem 8.5 *Let $X \in \mathscr{H}$. Then, as $u \to \infty$,*

$$\mathbb{P}(Q > u) \sim \frac{1}{-\mathbb{E}X_1} \int_u^\infty \mathbb{P}(X_1 > x)\mathrm{d}x.$$

It is straightforward to check that the class of α-stable Lévy motions belongs to $\mathscr{H}$. The following result, attributed to Port [178], is an immediate consequence of Thm. 8.5, Prop. 2.1, and Karamata's theorem [47, Section 1.6]; recall that we assumed $m < 0$.

Proposition 8.2 *Let $X \in \mathbb{S}(\alpha, \beta, m)$, with $\alpha \in (1, 2)$, $\beta \in (-1, 1]$. Then, as $u \to \infty$,*

$$\mathbb{P}(Q > u) \sim \frac{1}{(-m)} \int_u^\infty x^{-\alpha} C_{\alpha,1}\left(\frac{1+\beta}{2}\right)\mathrm{d}x \sim \frac{1}{(-m)}\frac{1}{\alpha - 1}u^{-\alpha+1}C_{\alpha,1}\left(\frac{1+\beta}{2}\right).$$

It is noted that there is a seeming incompatibility between the above asymptotics and the corresponding result in [178] (which is Prop. 3.7 in [96]), but it is a matter of (straightforward but tedious) calculus to verify that both expressions match.

The case of Lévy input that is an aggregate of α-stable Lévy motion and compound Poisson with regularly varying jobs was considered in Furrer [96]; in that setting it turns out that the heaviest tail essentially dominates the asymptotics, as could be expected.

Exercises

Exercise 8.1 Prove that $\varrho_{\mathbb{Q}} > 1$ in (8.1).

Exercise 8.2 This exercise is on the Cramér–Lundberg asymptotics.

(a) Let X correspond to $\mathbb{B}\mathrm{m}(-1, 1)$. Determine the asymptotics of $\mathbb{P}(Q > u)$.
(b) Let X correspond to $\mathbb{C}\mathrm{P}(1, \lambda, b(\cdot))$, with B exponentially distributed with mean μ^{-1}. Determine the asymptotics of $\mathbb{P}(Q > u)$.

Exercise 8.3 Check Eqns. (8.5) and (8.6), and show that these equations imply the limiting constant appearing in the statement of Thm. 8.1.

Exercise 8.4 Define

$$b(\alpha) := \frac{3}{2} + \alpha - \sqrt{\left(\frac{3}{2} + \alpha\right)^2 - 2}$$

for $\alpha > \alpha_0 := -\frac{3}{2} + \sqrt{2}$ and ∞ otherwise.

(*Remark*: Here, $b(\cdot)$ is the Laplace transform of a busy period in an M/M/1 queue with arrival rate $\frac{1}{2}$ and service rate 1; see e.g. [19, Prop. III.8.10].)

Let X correspond to $\mathbb{C}\mathrm{P}(r, \lambda, b(\cdot))$.

(a) Verify that the queue's stability condition is $r/\lambda > 2$.
(b) Determine $\mathbb{E}e^{\omega X_1} = e^{\varphi(-\omega)}$.
(c) Verify that $X \in \mathscr{L}$ if $\varphi(\alpha_0) \geq 0$ (in addition to $r/\lambda > 2$), that is,

$$\frac{r}{\lambda} \leq \frac{\sqrt{2} - 1}{\frac{3}{2} - \sqrt{2}},$$

and that $X \in \mathscr{I}$ otherwise. In both cases compute the corresponding asymptotics of $\mathbb{P}(Q > u)$.

Hint: Verify that when $X \in \mathscr{L}$,

$$\omega = \frac{\lambda r - 2\lambda^2}{r^2 + 2\lambda r}.$$

Exercise 8.5 Verify Example 8.1.

Exercise 8.6 In the case $X \in \mathbb{S}(\alpha, \beta, m)$, where $\alpha \in (1,2)$, $\beta \in (-1,1]$, and $m < 0$, [96, Prop. 3.7] gives that

$$\mathbb{P}(Q > u) \sim \frac{1}{(-m)} \frac{1}{\alpha - 1} u^{-\alpha+1} A(\alpha, \beta),$$

with

$$A(\alpha, \beta) := \frac{\Gamma(1+\alpha)}{\pi\alpha} \sqrt{1 + \beta^2 \tan^2\left(\frac{\pi\alpha}{2}\right)} \times \sin\left(\frac{\pi\alpha}{2} + \arctan\left(\beta \tan\left(\frac{\pi\alpha}{2}\right)\right)\right)$$

(realize that β needs to be replaced by $-\beta$ in [96, Prop. 3.7] to make Furrer's setting compatible with ours!). Prove that this is consistent with Prop. 8.2.

Exercise 8.7 Let $X \in \mathbb{S}(\alpha_1, \beta_1, m_1)$ and $Y \in \mathbb{S}(\alpha_2, \beta_2, m_2)$ be independent, with $\alpha_1, \alpha_2 \in (1,2]$ and $m_1, m_2 < 0$. Consider $Z_t = X_t + Y_t$.

(a) Assume that $\alpha_1 \geq \alpha_2$ and $\beta_2 > -1$. Find the asymptotics of the stationary workload distribution for a queue driven by the Lévy process Z.
(b) Assume that $\alpha_1 < \alpha_2$. Find the asymptotics of the stationary workload distribution for a queue driven by the Lévy process Z.

Exercise 8.8 Let $Y^{(1)}, Y^{(2)}, \ldots$ be i.i.d. copies of $Y \in \mathbb{S}(\alpha, \beta, -m)$, with $\alpha \in (1,2)$ and $m > 0$. Let $Q^{(N)}$ be the stationary workload of a queue driven by $X_t^{(N)} := \sum_{i=1}^{N} Y_t^{(i)}$. Find the exact asymptotics of

$$\mathbb{P}\left(Q^{(N)} > NB\right),$$

as $N \to \infty$, with $B > 0$.

Chapter 9
Transient Asymptotics

In this chapter we discuss asymptotics that relate to transient properties of the workload. We start in Section 9.1 by considering the asymptotics of the transient workload distribution (for given initial value x and time t); in line with what we found in Chapter 8, various scenarios are possible, depending on the shape of the upper tail of X_1.

Then in Section 9.2 we focus on 'joint exceedance probabilities' of the type

$$\mathbb{P}\left(Q_0 > pu, Q_{T(u)} > qu\right),$$

for u large, $p, q > 0$, and various shapes of the function $T(u)$. It is assumed throughout that the workload is in stationarity at time 0. There are various regimes, depending on the specific values of p and q, with appealing intuitive interpretations.

In the third section we focus on the tail distribution of the busy period $p(t) := \mathbb{P}(\tau > t)$ for t large; as it turns out, similar techniques can also be used to characterize the asymptotic behavior of the correlation function $r(t)$ for t large. This chapter is concluded in Section 9.4 by analyzing the minimum value attained over an interval of length $T(u)$, for u large and various shapes of $T(u)$.

9.1 Transient Workload Asymptotics

In this section we concentrate on the asymptotic properties of

$$\mathbb{P}_x(Q_t > u) := \mathbb{P}(Q_t > u \mid Q_0 = x), \tag{9.1}$$

as $u \to \infty$, for given $x, t \geq 0$.

K. Dębicki, M. Mandjes, *Queues and Lévy Fluctuation Theory*, Universitext,
DOI 10.1007/978-3-319-20693-6_9

Before we proceed with the analysis of (9.1) for u large, let us first make a useful observation. It follows straightforwardly from representation (2.4) that, with $Q_0 = x$,

$$\mathbb{P}(\bar{X}_t > u) \leq \mathbb{P}(Q_t > u) \leq \mathbb{P}(\bar{X}_t > u - x), \tag{9.2}$$

for each $x, t \geq 0$, with as before $\bar{X}_t := \sup_{s\in[0,t]} X_s$; to see this, observe that we have the trivial inequality $a \leq \max\{a, b\} \leq a + b$, and the distributional equality

$$X_t - \inf_{0\leq s\leq t} X_s \stackrel{\mathrm{d}}{=} \sup_{0\leq s\leq t} X_s = \bar{X}_t.$$

The above bounds indicate that the asymptotic behavior of $\bar{X}_t$ plays the key role in the analysis of (9.1). The following lemma, due to Willekens [219], will be intensively used in this section.

Lemma 9.1 *Let $(X_t)_t$ be a Lévy process. Then, for any $0 < u_0 < u$ we have*

$$\mathbb{P}\left(\sup_{s\in[0,t]} X_s > u\right)\mathbb{P}\left(\inf_{s\in[0,t]} X_s > -u_0\right) \leq \mathbb{P}(X_t > u - u_0).$$

Proof Recall the stopping time $\sigma(u) = \inf\{t \geq 0 : X_t \geq u\}$, as introduced in Section 3.2. We have

$$\begin{aligned}\mathbb{P}(\sigma(u) \leq t) &\leq \mathbb{P}(X_t > u - u_0) + \mathbb{P}(\sigma(u) \leq t, X_t \leq u - u_0)\\ &\leq \mathbb{P}(X_t > u - u_0) + \mathbb{P}\left(\sigma(u) \leq t, \inf_{s\in[\sigma(u),\sigma(u)+t]} (X_s - X_{\sigma(u)}) \leq -u_0\right)\\ &= \mathbb{P}(X_t > u - u_0) + \mathbb{P}(\sigma(u) \leq t)\mathbb{P}\left(\inf_{s\in[0,t]} X_s \leq -u_0\right),\end{aligned}$$

by the strong Markov property. As a consequence,

$$\mathbb{P}(\sigma(u) \leq t)\mathbb{P}\left(\inf_{s\in[0,t]} X_s > -u_0\right) \leq \mathbb{P}(X_t > u - u_0),$$

which combined with the fact that

$$\mathbb{P}\left(\sup_{s\in[0,t]} X_s > u\right) \leq \mathbb{P}(\sigma(u) \leq t),$$

completes the proof. □

The combination of Lemma 9.1 with inequality (9.2) enables us to obtain the asymptotics of (9.1) for a wide class of Lévy processes. We distinguish between a number of specific scenarios.

Long-tailed case—We start by analyzing the transient probability (9.1) for X satisfying the property

$$\lim_{u\to\infty} \frac{\mathbb{P}(X_t > u - y)}{\mathbb{P}(X_t > u)} = 1, \tag{9.3}$$

for each $y > 0$, that is, we assume that X_t has a *long-tailed* distribution function. We refer to Foss et al. [93] for properties of this family of distribution functions.

Upon combining (9.2) with Lemma 9.1, we obtain the inequality

$$\frac{\mathbb{P}_x(Q_t > u)}{\mathbb{P}(X_t > u)} \leq \frac{\mathbb{P}(\bar{X}_t > u - x)}{\mathbb{P}(X_t > u)} \leq \left(\mathbb{P}\left(\inf_{s\in[0,t]} X_s \leq -u_0\right)\right)^{-1} \frac{\mathbb{P}(X_t > u - x)}{\mathbb{P}(X_t > u)},$$

for $u - x > u_0 > 0$ such that $\mathbb{P}(\inf_{s\in[0,t]} X_s \leq -u_0) > 0$. Using the assumption (9.3) and letting $u_0 \to \infty$, it follows that

$$\limsup_{u\to\infty} \frac{\mathbb{P}_x(Q_t > u)}{\mathbb{P}(X_t > u)} \leq 1.$$

The above, together with the fact that by (9.2),

$$\frac{\mathbb{P}_x(Q_t > u)}{\mathbb{P}(X_t > u)} \geq 1,$$

justifies the following result.

Theorem 9.1 *Suppose that X_t satisfies (9.3). Then, as $u \to \infty$,*

$$\mathbb{P}_x(Q_t > u) \sim \mathbb{P}(X_t > u).$$

Example 9.1 Consider the case of $X \in \mathbb{S}(\alpha, \beta, -r)$ with $\alpha \in (1, 2)$, $\beta \in (-1, 1]$. Following Thm. 9.1 and Prop. 2.1,

$$\mathbb{P}_x(Q_t > u) \sim \frac{1-\alpha}{\Gamma(2-\alpha)\cos(\pi\alpha/2)} \left(\frac{1+\beta}{2}\right) tu^{-\alpha}.$$

We note that the obtained asymptotic behavior is insensitive with respect to the initial workload $Q_0 = x$ and the shift parameter r. ◇

Weibullian-tailed case—Now we address the case that also covers light-tailed marginals of $(X_t)_t$. Suppose that there exist numbers $A > 0$ and $\gamma > 0$, such that

$$\lim_{u\to\infty} \frac{\log \mathbb{P}(X_t > u)}{u^\gamma} = -A, \tag{9.4}$$

that is, X_t has an (asymptotically) *Weibullian* distribution function. Following the same line of reasoning as used in the case of long-tailed Lévy processes, we arrive at the following result, now for logarithmic asymptotics.

Theorem 9.2 *Suppose that X_t satisfies (9.4). Then, as $u \to \infty$,*

$$\lim_{u\to\infty} \frac{\log \mathbb{P}_x(Q_t > u)}{\log \mathbb{P}(X_t > u)} = 1.$$

Example 9.2 Let $X \in \mathbb{S}(\alpha, -1, -r)$ with $\alpha \in (1, 2)$. It follows that

$$\lim_{u\to\infty} \frac{\log \mathbb{P}_x(Q_t > u)}{u^{\alpha/(\alpha-1)}} = -t^{-1/(\alpha-1)} \frac{\alpha - 1}{\alpha^{\alpha/(\alpha-1)}} \left(\cos\left(\frac{\pi(2-\alpha)}{2}\right)\right)^{1/(\alpha-1)},$$

by combining Thm. 9.1 with Prop. 2.2 (independently of the values $Q_0 = x$ and r). ◇

Example 9.3 Suppose that $X_t = J_t - rt$, where $J \in \mathbb{G}(\gamma, \beta)$. Then

$$\lim_{u\to\infty} \frac{\log \mathbb{P}_x(Q_t > u)}{u} = -\gamma.$$

Observe that these logarithmic asymptotics do not depend on either $Q_0 = x$, or r, or time t. ◇

9.2 Joint Transient Distribution

In Dębicki et al. [72] the focus is on probabilities of the type

$$\mathbb{P}(Q_0 > pu, Q_{T(u)} > qu),$$

for $p, q > 0$. We summarize the main findings of this paper, and provide an intuitive justification of these results.

Reference [72] first identifies conditions under which the probability of interest is essentially dominated by the 'most demanding event', in the sense that it is asymptotically equivalent to $\mathbb{P}(Q > \max\{p, q\}u)$ for u large; here, as before, Q denotes the steady-state workload. These conditions turn out to reduce to $T(u)$ being sublinear (i.e. $T(u)/u \to 0$ as $u \to \infty$).

This condition makes sense, as it means that the time epochs 0 and $T(u)$ are 'close', relative to the workload levels to be achieved (i.e. pu and qu). Informally, for the scenario that $p > q$ it means that if at time 0 level pu is exceeded, then with high chance also qu is exceeded at time $T(u)$; a similar reasoning applies when $q > p$.

Then a second condition is derived under which the probability of interest 'decouples', in that it is asymptotically equivalent to the product of the corresponding marginal probabilities, $\mathbb{P}(Q > pu)\mathbb{P}(Q > qu)$, for u large (meaning that the ratio of this product of probabilities to the joint probability tends to 1 as $u \to \infty$). In this condition a crucial role is played by the random quantity Q^D, for $D > \mathbb{E}X_1$, which is distributed as $\sup_{t \geq 0}(X_t - Dt)$; as a result Q^D resembles the original queue Q but with the drain rate adapted by D, due to (2.5). Then the decoupling condition is that for all $\eta > 0$, $D > \mathbb{E}X_1$, we should have that

$$\lim_{u \to \infty} \frac{\mathbb{P}(Q^D > \eta T(u))}{\mathbb{P}(Q > pu)\mathbb{P}(Q > qu)} = 0.$$

For various types of input considered in the literature this 'decoupling condition' reduces to requiring that $T(u)$ is superlinear (i.e. $T(u)/u \to \infty$ as $u \to \infty$). This is the case for instance if the tails of Q and Q^D decay exponentially, as is verified easily.

This decoupling property can be heuristically justified as follows. If $T(u)$ grows fast (faster than the buffer levels pu and qu, to be reached at times 0 and $T(u)$, respectively), it is conceivable that time epochs 0 and $T(u)$ are contained in different busy periods. This means that the event of reaching qu at $T(u)$ 'does not benefit' from the positive correlation of the fact that pu was reached at 0, and hence the events are essentially independent.

However, for $X \in \mathscr{R}$, requiring that $T(u)/u \to \infty$ is *not* sufficient for decoupling. In this case it is seen that, due to the fact that the tails of Q and Q^D decay in a regularly varying fashion, the 'decoupling condition' reduces to $T(u)/u^2 \to \infty$. The rationale behind the fact that we have decoupling only for $T(u)$ increasing superquadratically is that for $T(u)$ increasing subquadratically, with overwhelming probability it suffices to have a *single* big jump to cause overflow both over pu at time 0, and over qu at time $T(u)$; 'decoupling', on the contrary, would correspond to a scenario with *two* big jumps. These findings imply that for $X \in \mathscr{R}$ there is a third regime, that is, $T(u)$ increasing superlinearly but subquadratically; [72] also identifies the asymptotics for this case.

In [72] special attention is paid to the case $T(u) = Ru$ for some $R > 0$; for $X \in \mathscr{L}$, intuitively appealing asymptotics are derived, intensively relying on sample-path large-deviations results [69]. The regimes obtained can be interpreted in terms of most likely paths to overflow. If R is small (i.e. fulfilling an explicit criterion in terms of p, q, and the characteristics of the driving Lévy process $(X_t)_t$), then one has asymptotics of the type $\mathbb{P}(Q > \max\{p, q\}u)$. If this condition does not apply, two cases are possible: for large R the most likely scenario is that the buffer fills up to level pu, then drains, remains empty for a while, and starts building up a relatively short time before Ru, to reach level qu at Ru (in this case the asymptotics look like $\mathbb{P}(Q > pu)\mathbb{P}(Q > qu)$), whereas for moderate R the buffer remains (most likely) non-empty between 0 and R.

We thus obtain the following structural result. Let $\omega > 0$ solve, as before, $\mathbb{E}e^{\omega X_1} = 1$. Then there are (uniquely characterized) thresholds $\bar{R}$ and $\check{R}$ such that for all R smaller than $\bar{R}$, as $u \to \infty$,

$$\frac{1}{u}\log \mathbb{P}(Q_0 > pu, Q_{Ru} > qu) \to -\max\{p, q\}\omega;$$

for R between $\bar{R}$ and $\check{R}$,

$$\frac{1}{u}\log \mathbb{P}(Q_0 > pu, Q_{Ru} > qu) \to -p\omega - R \cdot \sup_\delta \left(\delta\left(\frac{q-p}{R}\right) - \log \mathbb{E}e^{\delta X_1}\right);$$

and for R larger than $\check{R}$,

$$\frac{1}{u}\log \mathbb{P}(Q_0 > pu, Q_{Ru} > qu) \to -(p+q)\omega.$$

Summarizing, in the first scenario the 'most demanding' event dominates the asymptotics, in the last scenario the events are essentially independent, while in the intermediate scenario the (scaled) most likely path attains p at time 0, then follows a straight line with slope $(q-p)/R$ during R time units, to reach q at time R.

9.3 Busy Period and Correlation Function

Relying on the Heaviside recipe (i.e. Recipe 8.1), we now study the tail distribution function of the busy period $p(t) := \mathbb{P}(\tau > t)$, as well as the correlation function $r(t)$. We show the computations for $p(t)$, but note that the computations for $r(t)$ work very similarly, and are provided in detail in Es-Saghouani and Mandjes [90] and Glynn and Mandjes [101].

Let us first consider the light-tailed case. For $X \in \mathscr{L}$ we have that $\mathbb{E}e^{\omega X_1} = 1$ for some $\omega > 0$, which implies that $\mathbb{E}e^{sX_1}$ has a minimizer for s in the interval between 0 and ω. This observation will be used several times later on.

As usual, we start by considering the spectrally positive case. As before, we assume that the equation $\varphi(\alpha) = 0$ has a negative root. Observe that then (obviously) Prop. 6.1 holds for any positive ϑ, but that we can consider the *analytic continuation* up to the branching point $\vartheta^\star < 0$ of $\psi(\cdot)$; in the sequel let $\zeta < 0$ denote the minimizer of $\varphi(\cdot)$, so that $\varphi(\zeta) = \vartheta^\star < 0$ (where it is noticed that $v_\varphi := \varphi''(\zeta) > 0$).

Then the idea is to write, for $\vartheta \downarrow \vartheta^\star$, that $\psi(\vartheta) - \zeta \sim \sqrt{2/v_\varphi} \cdot \sqrt{\vartheta - \vartheta^\star}$. Hence, around $\vartheta^\star$, we have that, for some (irrelevant) constant κ,

$$\int_0^\infty e^{-\vartheta t}\mathbb{P}(\tau > t)\mathrm{d}t = \frac{1}{\vartheta} - \varphi'(0)\frac{\psi(\vartheta)}{\vartheta^2} \sim \kappa + A_\varphi\sqrt{\vartheta - \vartheta^\star},$$

where

$$A_\varphi := -\frac{\varphi'(0)}{(\vartheta^\star)^2}\sqrt{\frac{2}{v_\varphi}} < 0,$$

and hence, applying 'Heaviside', we estimate the tail distribution of the busy period by

$$\mathbb{P}(\tau > t) \sim \frac{A_\varphi}{\Gamma(-\frac{1}{2})} \cdot \frac{e^{\vartheta^\star t}}{t\sqrt{t}}. \tag{9.5}$$

We now turn to the spectrally negative case. Prop. 6.3 holds for any positive q, but we can consider the analytic continuation up to the branching point $q^\star < 0$ of $\Psi(\cdot)$. Let $\zeta > 0$ denote the minimizer of $\Phi(\cdot)$, so that $\Phi(\zeta) = q^\star < 0$. Similarly to the spectrally positive case, we obtain, with $v_\Phi := \Phi''(\zeta) > 0$ and κ being some (irrelevant) number,

$$\int_0^\infty e^{-qt}\mathbb{P}(\tau > t)\mathrm{d}t = \frac{1}{q}\left(1 - \frac{\Psi(0)}{\Psi(q)}\right) \sim \kappa + A_\Phi\sqrt{q - q^\star},$$

where

$$A_\Phi := \frac{\Psi(0)}{q^\star\zeta^2}\sqrt{\frac{2}{v_\Phi}} < 0,$$

and hence 'Heaviside' estimates the tail of the busy-period distribution as

$$\mathbb{P}(\tau > t) \sim \frac{A_\Phi}{\Gamma(-\frac{1}{2})} \cdot \frac{e^{q^\star t}}{t\sqrt{t}}. \tag{9.6}$$

For related results on the light-tailed case, we refer e.g. to [65, 173]; importantly, the asymptotic shape $Ae^{\vartheta t}/(t\sqrt{t})$, for $\vartheta < 0$ and $A > 0$, appears also if $X \in \mathscr{L}$ without necessarily being spectrally one sided.

We now consider the heavy-tailed case. In the case $X \in \mathscr{S}_+ \cap \mathscr{R}$, the following 'Tauberian lemma' is useful [90].

Lemma 9.2 *Suppose $\varphi'(\alpha) \in \mathscr{R}_{\nu-1}(n-1,\eta)$. Then the following three statements hold, for $\alpha \downarrow 0$ or $\vartheta \downarrow 0$:*

(A) $\varphi(\alpha) \in \mathscr{R}_\nu(n, \eta/\nu)$;
(B) $\psi(\vartheta) \in \mathscr{R}_\nu(n, \bar\eta)$, *with* $\bar\eta := -\eta\nu^{-1}(\varphi'(0))^{-\nu-1}$;
(C) $\psi'(\vartheta) \in \mathscr{R}_{\nu-1}(n-1, \bar\eta\nu)$.

For the heavy-tailed case with compound Poisson input, we refer for results on τ^0 (the busy period starting at 0; see Section 6.1) e.g. to [26, 78]. The main intuition is that a single big job causes a long busy period. It is found that if $X \in \mathbb{CP}(1, \lambda, b(\cdot)) \cap \mathscr{R}$, then

$$\mathbb{P}(\tau^0 > t) \sim \frac{1}{1-\varrho}\,\mathbb{P}(B > t(1-\varrho)) \tag{9.7}$$

as $t \to \infty$; as usual $\varrho := \lambda\,\mathbb{E}B$. The validity of (9.7) can be seen as follows. First recall that $\mathbb{E}e^{-\vartheta\tau^0} = (\lambda + \vartheta - \psi(\vartheta))/\lambda$. Suppose that $\varphi(\alpha) \in \mathscr{R}_\nu(n, \eta)$, that is, $b(\alpha) \in \mathscr{R}_\nu(n, \eta/\lambda)$. In view of the above lemma,

$$\psi(\vartheta) \in \mathscr{R}_\nu(n, \check\eta), \quad \text{with } \check\eta := -\frac{\eta}{(\varphi'(0))^{\nu+1}},$$

and as a consequence, $\mathbb{E}e^{-\vartheta\tau^0} \in \mathscr{R}_\nu(n, -\check\eta/\lambda)$. It follows from 'Tauber' that

$$\mathbb{P}(\tau^0 > t) \sim \frac{(-1)^{n-1}}{\Gamma(1-\nu)}\frac{\eta/\lambda}{(\varphi'(0))^{\nu+1}} t^{-\nu} L(t),$$

but also

$$\mathbb{P}(B > t) \sim \frac{(-1)^{n-1}}{\Gamma(1-\nu)}(\eta/\lambda)\, t^{-\nu} L(t).$$

Relation (9.7) now follows by noting that $\varphi'(0) = 1 - \varrho$.

Now consider the situation that $X \in \mathscr{S}_+ \cap \mathscr{R}$, in which $\varphi(\alpha) \in \mathscr{R}_\nu(n, \eta)$. Our objective now is to find the tail asymptotics $\mathbb{P}(\tau > t)$. Again, $\psi(\vartheta) \in \mathscr{R}_\nu(n, \check\eta)$. Straightforward calculations yield that

$$\mathbb{E}e^{-\vartheta\tau} \in \mathscr{R}_{\nu-1}(n-1, \check\eta\varphi'(0)),$$

such that 'Tauber' eventually gives

$$\mathbb{P}(\tau > t) \sim \frac{(-1)^n}{\Gamma(2-\nu)}\check\eta\, t^{1-\nu} L(t).$$

For results on the asymptotics of the correlation function $r(t)$ in the case $X \in \mathscr{S}_+ \cap \mathscr{R}$, we refer to [90, Section 5]. Under the assumption that $\varphi'(\alpha) \in \mathscr{R}_{\nu-1}(n-1, \eta)$, with Lemma 9.2, we can characterize the behavior of $\psi(\vartheta)$ and $\psi'(\vartheta)$ for ϑ small, so that Thm. 7.1 yields $\hat{r}(\vartheta)$ for ϑ small (this requires a substantial amount of algebra!). Then, again with 'Tauber', we can identify the asymptotics of $r(t)$ for t large. The case of $X \in \mathscr{S}_-$ is covered in [101].

9.4 Infimum over Given Time Interval

In this section we analyze the tail probabilities related to the minimum value attained by the workload process, as introduced in Section 6.4. More precisely, we consider the lowest value attained over a period of length $T(u)$, given the workload is in steady state at time 0, and we do so for various shapes of $T(u)$. We focus on the behavior of $\mathbb{P}(\underline{Q}_{T(u)} > u)$ for u large, recalling that $\underline{Q}_t$ is defined as the minimum of the stationary workload process in the interval $[0, t]$.

In Dębicki et al. [74] it is proved that for $X \in \mathscr{L}$ the following logarithmic asymptotics hold:

$$\log \mathbb{P}\left(\underline{Q}_{T(u)} > u\right) \sim -\omega u + \vartheta^\star T(u),$$

as $u \to \infty$; here, $\omega > 0$ and $\vartheta^\star < 0$ are as defined before (i.e. $\omega > 0$ solves $\mathbb{E}e^{sX_1} = 1$ and $\vartheta^\star < 0$ is the minimum attained by $\log \mathbb{E}e^{sX_1}$). We now sketch the proof of this result.

Observe that the probability of interest is, for any given $\varepsilon > 0$, bounded from below by

$$\mathbb{P}\left(Q_0 > u + \varepsilon T(u)\right) \mathbb{P}\left(\inf_{s \in [0, T(u)]} X_s > -\varepsilon T(u)\right),$$

where we used that Q_0 is independent of $(X_t)_{t \geq 0}$. Now the former factor reads, for ε small, as the probability of the steady-state workload exceeding u (which roughly looks like $e^{-\omega u}$), whereas the latter factor reads as the probability of a busy period lasting at least $T(u)$ (which roughly looks like $e^{\vartheta^\star T(u)}$). The lower bound follows after taking decay rates and sending ε to 0. In the upper bound it is proved that, relative to this scenario, all other scenarios can be ignored.

The above result entails that for $X \in \mathscr{L}$ the probability $\mathbb{P}(\underline{Q}_{T(u)} > u)$ decays roughly as $\exp(-\omega u)$ if $T(u) = o(u)$, and as $\exp(\vartheta^\star T(u))$ if $u = o(T(u))$. From an intuitive standpoint this makes sense: in the former case it is crucial that the workload reaches u at time 0, and then with high probability it remains above u during the 'short' interval $[0, T(u)]$, whereas in the latter case remaining above u during the 'long' interval $[0, T(u)]$ is the most demanding requirement.

Reference [74] also covers the case of $X \in \mathscr{R}$. The main result is that, under a mild regularity condition, the following exact asymptotics hold:

$$\mathbb{P}\left(\underline{Q}_{T(u)} > u\right) \sim \mathbb{P}\,(Q > u + T(u)) \;+\; T(u)\,\mathbb{P}(X_1 > u + T(u)),$$

as $u \to \infty$. It is easily verified that if $T(u) = o(u)$, then these asymptotics look like those of $\mathbb{P}(Q > u)$ (see Thm. 8.5); if $u = o(T(u))$, then they look like those of $T(u)\,\mathbb{P}(X_1 > T(u))$.

Exercises

Exercise 9.1 Let X correspond to $\mathbb{B}\mathrm{m}(-1, 1)$. Determine, for $p, q, R > 0$,

$$\lim_{u\to\infty} \frac{1}{u} \log \mathbb{P}(Q_0 > pu, Q_{Ru} > qu).$$

Determine the thresholds $\bar{R}$ and $\check{R}$ as well.

Exercise 9.2 Determine the asymptotics of $\mathbb{P}(\tau > t)$ for $X \in \mathbb{CP}(1, \lambda, b(\cdot)) \cap \mathscr{R}$ using Tauberian inversion.

Exercise 9.3 Let X correspond to $\mathbb{B}\mathrm{m}(-1, 1)$. Determine the asymptotics of $\mathbb{P}(\tau > t)$.

Exercise 9.4 Let X correspond to $\mathbb{CP}(1, \lambda, b(\cdot))$, where B is an exponentially distributed random variable with mean μ^{-1}. Determine the asymptotics of $\mathbb{P}(\tau > t)$.

Exercise 9.5 Let $X \in \mathscr{S}_+ \cap \mathscr{R}$.

(a) Assume $\varphi(\alpha) \in \mathscr{R}_\nu(n, \eta)$, and fix an $x > 0$. Apply 'Tauber' to prove that

$$\mathbb{P}(\tau(x) > t) \sim \frac{(-1)^n}{\Gamma(1-\nu)}\,\check{\eta} x\, t^{-\nu} L(t), \quad \text{with} \quad \check{\eta} := -\frac{\eta}{(\varphi'(0))^{\nu+1}},$$

as $t \to \infty$.

(b) Let $X \in \mathbb{CP}(1, \lambda, b(\cdot))$ with $b(\alpha) \in \mathscr{R}_\nu(n, \eta/\lambda)$. Show that the result proved under (a) is consistent with Eqn. (9.7) by considering the asymptotics of

$$\mathbb{P}(\tau^0 > t) = \int_0^\infty \mathbb{P}(\tau(x) > t)\mathbb{P}(B \in \mathrm{d}x).$$

Exercise 9.6 Let $X \in \mathscr{S}_+$.

(a) Show that for any $\vartheta \geq 0$,

$$\mathbb{P}(\tau > t) \leq e^{-\vartheta t}\,\mathbb{E}e^{\vartheta \tau}.$$

(b) Show that the smallest value of ϑ for which

$$\limsup_{t\to\infty} \frac{1}{t} \log \mathbb{P}(\tau > t) \leq \vartheta$$

holds is equal to $\vartheta^{\star}$.

(c) Prove the following exponential bound on the tail distribution of τ:

$$\mathbb{P}(\tau > t) \leq \varphi'(0) \frac{\zeta}{\vartheta^{\star}} e^{\vartheta^{\star} t}.$$

Exercise 9.7 Determine the asymptotics of $r(t)$ for $X \in \mathscr{S}_{+} \cap \mathscr{L}$ and for $X \in \mathscr{S}_{-}$ using the Heaviside approach. Check the findings with the results from [90, 101].

Exercise 9.8 Let X correspond to $\mathbb{B}\mathrm{m}(-1, 1)$. Determine the logarithmic asymptotics of $\mathbb{P}(\underline{Q}_{\sqrt{u}} > u)$ and $\mathbb{P}(\underline{Q}_{u\sqrt{u}} > u)$.

Exercise 9.9 Let X correspond to $X \in \mathbb{S}(\alpha, \beta, m)$ with $\alpha \in (0, 2]$, $\beta \in (-1, 1]$, and $m < 0$. Determine the exact asymptotics of $\mathbb{P}(\underline{Q}_{\sqrt{u}} > u)$ and $\mathbb{P}(\underline{Q}_{u\sqrt{u}} > u)$.

Chapter 10
Simulation of Lévy-Driven Queues

This chapter focuses on the use of stochastic simulation when analyzing Lévy-driven queues. In the first part, it is explained how (transient and stationary) Lévy-driven queues can be simulated. In the second part, the focus is on efficiently estimating rare-event probabilities using importance sampling: we subsequently treat fast simulation of the tail probabilities related to the stationary workload and the busy period. In the last part of this chapter attention is paid to estimating the workload correlation function $r(t)$ for t large.

10.1 Simulation of Lévy-Driven Queues

There are various accessible texts that describe how Lévy processes can be simulated; we refer in particular to Asmussen and Glynn [24, Chapter XII] and Cont and Tankov [63, Chapter VI]. The difficulty lies in dealing with the 'small jumps' (for processes with infinitely many jumps in a finite amount of time, such as the gamma process); various techniques have been proposed to incorporate these.

In Section 3.3 it was pointed out how a general Lévy process (with possibly 'small jumps') can be approximated by the sum of a Brownian motion and a compound Poisson process (with drift), which we could write as $\mathbb{B}\mathrm{m}(d, \sigma^2) + \mathbb{C}\mathrm{P}(r, \lambda, b(\cdot))$—evidently, without loss of generality we can take $r = 0$, by incorporating the 'full drift' in d. Therefore we concentrate in this section on pointing out how to simulate a queue fed by such a process. It it stressed that the approximation proposed in Section 3.3 becomes more accurate when $\varepsilon \downarrow 0$, but this also increases the arrival rate of the compound Poisson process. In other words, there is an evident trade-off between the accuracy of the approximating process and the simulation effort needed.

Simulation of the transient workload—Suppose that $Q_0 = x$, and that we wish to simulate Q_t for some $t > 0$. We assume that we know how to simulate random

K. Dębicki, M. Mandjes, *Queues and Lévy Fluctuation Theory*, Universitext,
DOI 10.1007/978-3-319-20693-6_10

variables that are uniformly distributed on [0, 1] (to be denoted by U in the sequel), and that we are also capable of sampling from the distribution of the random variable B, that is, the distribution characterized by the Laplace transform $b(\cdot)$ (see e.g. [24, Chapter II]).

We first point out how Q_t could be simulated if X consists of just a Brownian term $\mathbb{B}\mathrm{m}(d,\sigma^2)$ (i.e. no compound Poisson term); see e.g. [63, p. 177] and [101]. Recall that, by (2.4),

$$Q_t = X_t + \max\left\{x, -\inf_{0\le s\le t} X_s\right\}.$$

The idea is to first simulate X_t from a normal distribution with mean dt and variance $\sigma^2 t$; say it attains the value z. It is then observed that, with $(W_t)_t$ denoting a standard Brownian motion, using standard calculation rules for Brownian motion, it holds that

$$\mathbb{P}\left(-\inf_{0\le s\le t} X_s \le x \,\middle|\, X_t = z\right) = \mathbb{P}\left(\forall s \in [0,t] : W_s \le \frac{x+ds}{\sigma} \,\middle|\, W_t = \frac{dt-z}{\sigma}\right).$$

Relying on standard results for the Brownian bridge, this equals

$$1 - \exp\left(-\frac{2x}{\sigma^2 t}(x+z)\right);$$

observe that this expression does not depend on d (why?). Then it is readily verified that

$$Y_z(\sigma,t) := \left(-\inf_{0\le s\le t} X_s \le x \,\middle|\, X_t = z\right) \stackrel{\mathrm{d}}{=} -\frac{z}{2} + \frac{1}{2}\sqrt{z^2 - 2\sigma^2 t \log U}.$$

This gives us a way to sample $-\inf_{0\le s\le t} X_s$, conditional on the value of X_t. As a result, we have found an efficient way to sample Q_t.

Now return to the setting of X corresponding to $\mathbb{B}\mathrm{m}(d,\sigma^2) + \mathbb{C}\mathrm{P}(0,\lambda,b(\cdot))$. The idea is to iteratively simulate the workload at the jump epochs of the Poisson process. In the following pseudocode $E(\lambda)$ stands for a sample from the exponential distribution with mean λ^{-1}, and $N(d,\sigma^2)$ for a sample from the normal distribution with mean d and variance σ^2; B refers to a sample from the distribution of the job sizes in the compound Poisson process. The pseudocode generates an exact sample of Q_t in the case that X corresponds to the sum of a Brownian motion and a compound Poisson process.

Pseudocode 10.1 Input: $Q_0 = x$ and t; $\mathbb{B}\mathrm{m}(d,\sigma^2)$; $\mathbb{C}\mathrm{P}(0,\lambda,b(\cdot))$. Output: Q_t.

```
T := 0; Q = x;
while T < t do
    s := E(λ); T := T + s;
    if T < t then
```

```
            z := N(ds, σ²s); Q := z + max{Q, Y_z(σ, s)}; Q := max{Q + B, 0};
        else
            r := s − (T − t); z = N(dr, σ²r); Q := z + max{Q, Y_z(σ, r)};
        end (of 'if');
    end (of 'while'); return Q_t := Q.
```

Simulation of the steady-state workload—We recall that, due to the distributional identity (2.5), this amounts to simulating the all-time maximum of the Lévy process X. Let us approximate the all-time maximum with the maximum of X up until N jumps of the compound Poisson process, which we denote by $\bar{Q}_N$. Then N is chosen sufficiently large, such that we are guaranteed to obtain an 'almost exact' sample from the stationary workload Q; after describing the simulation algorithm, we point out how to select N. We propose the following pseudocode; it uses the property that the maximum is attained either immediately after the arrival of a job, or between jobs (where the process locally behaves as a Brownian motion).

Pseudocode 10.2 Input: N; $\mathbb{B}\mathrm{m}(d,\sigma^2)$; $\mathbb{C}\mathrm{P}(0,\lambda,b(\cdot))$. Output: $\bar{Q}_N$.

```
    n := 0; Q = 0; X := 0;
    while n ≤ N do
        n := n + 1; s := E(λ); z = N(ds, σ²s); m := Y_z(σ, s);
        if X + m > Q then
            Q := X + m;
        end (of 'if');
        X := X + z + B;
        if X > Q then
            Q := X;
        end (of 'if');
    end (of 'while'); return Q.
```

The issue that remains to be addressed is how big N should be chosen to ensure that the 'truncation error' is negligible. Here we describe a procedure to determine N for the case $X \in \mathscr{L}$.

Let T_i be the ith job arrival as generated by the compound Poisson process, and let the time interval $I_i = (T_i, T_{i+1}]$. Evidently, applying the union bound we can bound the error from above by

$$\mathbb{P}\left(\sup_{t\geq T_N} X_t > u\right) \leq \sum_{n=N}^{\infty} \mathbb{P}\left(\sup_{t\in I_n} X_t > u\right) \leq \sum_{n=N}^{\infty} \mathbb{P}\left(\sup_{t\in I_n} X_t > 0\right).$$

Observe that $\sup_{t\in I_n} X_t$ is, in distribution, equal to X_{T_n}, increased by an (independently sampled) random variable V representing the maximum in I_1. Now construct the random variable V as follows. Let S be exponentially distributed with mean λ^{-1}, and B be a sample from the job size distribution. Let $G(s)$ have a normal distribution with mean ds and variance $\sigma^2 s$. Then, assuming all samples are drawn independently,

$$V := \max\{Y_{G(S)}(\sigma, S), G(S) + B\}.$$

By applying the Markov inequality—in this form often referred to as the Chernoff bound [77, p. 93]—we have for any $\vartheta > 0$,

$$\mathbb{P}(X_{T_n} + V > 0) \leq \mathbb{E}e^{\vartheta X_{T_n}}\,\mathbb{E}e^{\vartheta V} = \left(\beta(-\vartheta) \cdot \frac{\lambda}{\lambda + d\vartheta + \frac{1}{2}\sigma^2\vartheta^2}\right)^n \mathbb{E}e^{\vartheta V}.$$

To make the upper bound tight, we pick the $\vartheta > 0$ that minimizes the expression $\beta(-\vartheta)/(\lambda + d\vartheta + \frac{1}{2}\sigma^2\vartheta^2)$; say that the minimum is attained at $\check{\vartheta}$. Notice that the value of the minimum, say $\check{m}$, is smaller than 1 due to the requirement that $\mathbb{E}B = \lambda\mathbb{E}X_1 + d$ be smaller than 0.

Collecting the above findings, we arrive at

$$\mathbb{P}\left(\sup_{t \geq X_N} X_t > u\right) \leq \sum_{n=N}^{\infty} \check{m}^n \cdot \mathbb{E}e^{\check{\vartheta} V} = \frac{\check{m}^N}{1 - \check{m}} \cdot \mathbb{E}e^{\check{\vartheta} V}.$$

We conclude that by picking N larger than

$$\frac{\log\varepsilon + \log(1 - \check{m}) - \log\mathbb{E}e^{\check{\vartheta} V}}{\log\check{m}},$$

it is ensured that the error made is below ε.

10.2 Estimation of Workload Asymptotics

Suppose we wish to estimate $\mathbb{P}(Q > u)$ by simulation. It is well known (see e.g. Mandjes [157, Section 8.2]) that the number of simulation runs needed to obtain an estimate with a given predefined precision (expressed in terms of the ratio of the width of the confidence interval and the estimate) is inversely proportional to the probability to be estimated. This insight follows from the following reasoning; we focus on the regime in which the probability of interest is *small*.

Suppose we perform n independent trials, and I_i is the indicator function for the event under consideration happening in run i, for $i = 1, \ldots, n$. Evidently, $\hat{p}_n := n^{-1}\sum_{i=1}^n I_i$ is an unbiased estimator of the target probability, say, p. The variance of this estimator is $n^{-1}p(1-p) \approx n^{-1}p$. Suppose that we continue simulating until the width of the confidence interval is below a fraction f of the probability p, and that our confidence level is α. As a result the confidence interval is of the form $(\hat{p}_n - t_\alpha\sqrt{\hat{p}_n/n}, \hat{p}_n + t_\alpha\sqrt{\hat{p}_n/n})$. This yields the rule of thumb that n must be sufficiently large such that

$$n \geq \frac{t_\alpha^2}{f^2 p}.$$

The above condition shows that the number of runs is indeed inversely proportional to p. In the light-tailed situation at hand this means that this number grows roughly exponentially in u, and as a result simulation experiments may take prohibitively long. Our objective now is to devise techniques that have the potential to speed up the simulation procedure. Focusing on $X \in \mathscr{L}$, ideas that date back to Siegmund [198] can be applied to estimate $\mathbb{P}(Q > u)$ highly efficiently.

As before, let ω solve $\mathbb{E}e^{\omega X_1} = 1$. The idea is now not to perform the simulation under the original measure $\mathbb{P}$, corresponding to the characteristic triplet (d, σ^2, Π), but rather under an alternative measure $\mathbb{Q}$ under which the event of interest occurs more frequently. After weighing the simulation output with an appropriate likelihood ratio, unbiasedness is recovered. This procedure is commonly referred to [24, pp. 127–128] as *importance sampling.*

This $\mathbb{Q}$ is an *exponentially twisted* version of $\mathbb{P}$, in the way it was constructed in Section 8.1. More concretely, the measure $\mathbb{Q}$ is such that, in self-evident notation, for all δ,

$$\mathbb{E}_{\mathbb{Q}} e^{\delta X_1} = \mathbb{E}e^{(\delta+\omega)X_1}.$$

It is now elementary to check that $\mathbb{Q}$ also corresponds to a Lévy process, with triplet

$$\left(d + \sigma^2\omega + \int_{-1}^{1} x(e^{\omega x} - 1)\Pi(\mathrm{d}x), \sigma^2, e^{\omega x}\Pi(\mathrm{d}x) \right); \tag{10.1}$$

cf. [24, Example XII.6.2]. Observe the methodological similarity to the derivation of the Cramér–Lundberg asymptotics in Section 8.1.

Recall that the convexity of $\mathbb{E}e^{\delta X_1}$ implies that $\mathbb{E}_{\mathbb{Q}} X_1 = \mathbb{E}X_1 e^{\omega X_1} > 0$, so that the random variable $\sigma(u) := \inf\{t : X_t > u\}$ becomes non-defective under $\mathbb{Q}$. As we saw in identity (8.3),

$$\mathbb{P}(Q > u) = \mathbb{E}_{\mathbb{Q}} e^{-\omega X_{\sigma(u)}}.$$

In other words, we should simulate under $\mathbb{Q}$ until $\sigma(u)$, record the value L_i of $e^{-\omega X_{\sigma(u)}}$ in each run i, perform n runs, and estimate $\mathbb{P}(Q > u)$ by $n^{-1} \sum_{i=1}^{n} L_i$. It follows immediately from (8.3) that this estimator is unbiased; note that $\mathbb{E}_{\mathbb{Q}} L_i = \mathbb{P}(Q > u)$. In addition, due to the fact that each observation of $e^{-\omega X_{\sigma(u)}}$ is bounded by $e^{-\omega u}$, the estimator has excellent variance properties (in particular, it has bounded relative error; see [24, p. 159]).

Clearly, a prerequisite for applying this method is that one should be able to sample trajectories of Lévy processes; the state of the art on this issue is presented in [24, Chapter XII] and [63, Chapter VI], but see Section 10.1 as well. Notice that, in passing, we re-proved the fact that $\mathbb{P}(Q > u) \leq e^{-\omega u}$; see Cor. 8.1.

Example 10.1 Consider the case of X corresponding to $\mathbb{B}\text{m}(-1, 1)$, that is, a process with *negative drift*, characterized by a Laplace exponent of the form $\varphi(\alpha) = \alpha + \frac{1}{2}\alpha^2$. It is checked that $\omega = 2$. It entails that under $\mathbb{Q}$ we should sample the Brownian motion $\mathbb{B}\text{m}(1, 1)$, that is, a Brownian motion with a *positive* drift. ◇

Example 10.2 Another example concerns the case that the driving Lévy process X corresponds to $\mathbb{C}\text{P}(1, \lambda, b(\cdot))$ with the B sampled from an exponential distribution with mean μ^{-1}. Let $\varrho := \lambda/\mu < 1$. The decay rate $\omega > 0$ solves

$$\varphi(-\omega) = -\omega - \lambda + \lambda\,\frac{\mu}{\mu - \omega} = 0,$$

yielding $\omega = \mu - \lambda > 0$. The Laplace exponent of the process under $\mathbb{Q}$ is given by $\varphi(\alpha - \omega)$, that is,

$$\alpha - \mu + \mu\,\frac{\lambda}{\lambda + \alpha}.$$

In other words, we should let the jobs arrive according to a Poisson process with rate μ, with their sizes being sampled from an exponential distribution with mean λ^{-1}. Observe that under $\mathbb{Q}$ the drift is *positive*. ◇

For the case of heavy tails, we refer to [23, 27] and [24, Section VI.3]. In this context it is noted that the above ideas for $X \in \mathscr{L}$ do not carry over to the heavy-tailed case, basically because (most likely) overflow is not caused by several 'somewhat unlikely' events, but rather a single big jump.

10.3 Estimation of Busy-Period Asymptotics

We now aim at efficiently estimating the tail probability $p(t) = \mathbb{P}(\tau > t)$ for $X \in \mathscr{L}$. In this case the following alternative measure was proposed in [101]; for ease we concentrate on $X \in \mathscr{S}_+$, but the case $X \in \mathscr{S}_-$ can be dealt with similarly.

- In the interval $(0, t]$ let the Lévy process be twisted with $-\zeta = -\psi(\vartheta^\star) > 0$, as described above; $\vartheta^\star$ is as defined in Section 9.3. In this way we obtain that the Lévy process under this new measure has drift 0, making long busy periods more likely.
- In addition we twist the workload at time 0, denoted by Q_0; we do so by a factor $\kappa \geq 0$, for which we identify a suitable value later on. Here we recall that under $\mathbb{P}$ the workload is distributed as a random variable whose transform is given by Thm. 3.2. This effectively means that we sample Q_0 from a distribution with

Laplace–Stieltjes transform

$$\frac{\mathbb{E}e^{-(\alpha-\kappa)Q_0}}{\mathbb{E}e^{\kappa Q_0}} = \frac{\alpha-\kappa}{\varphi(\alpha-\kappa)}\frac{\varphi(-\kappa)}{-\kappa}.$$

From now on we denote this new measure, consisting of twisting Q_0 (with the yet unknown κ) as well as twisting $(X_s)_{s\in(0,t]}$ (with ζ, so that it has drift 0), by $\mathbb{Q}_\kappa$.

In each run we simulate the process under $\mathbb{Q}_\kappa$ till time t, so that we can check whether the event $\{\tau > t\}$ has occurred. Along these lines, we perform n independent runs. Then the estimator, based on these n runs, reads $n^{-1}\sum_{i=1}^n L_i 1_{\{\tau_i>t\}}$, where L_i is the likelihood ratio of run i. Let us write down this likelihood ratio more explicitly. First there is the contribution due to the twisted queue at time 0; using Thm. 3.2 we obtain

$$L_1 := e^{-\kappa Q_0}\cdot\mathbb{E}e^{\kappa Q_0} = e^{-\kappa Q_0}\cdot\frac{-\kappa\varphi'(0)}{\varphi(-\kappa)}. \tag{10.2}$$

Second there is the contribution due to the twisted Lévy process between 0 and t:

$$L_2 := e^{\psi(\vartheta^\star)X_t}\cdot\mathbb{E}e^{-\psi(\vartheta^\star)X_t} = e^{\psi(\vartheta^\star)X_t}\cdot e^{\vartheta^\star t}. \tag{10.3}$$

The 'total likelihood ratio' of a single run is thus $L := L_1 \times L_2$. As before, the resulting estimator is unbiased, as $\mathbb{E}_{\mathbb{Q}_\kappa} L1_{\{\tau>t\}}$ equals the probability of our interest, that is, $\mathbb{P}(\tau > t) = p(t)$.

As a consequence of the fact that $\mathbb{V}\mathrm{ar}_{\mathbb{Q}_\kappa}(L1_{\{\tau>t\}}) \geq 0$, we see that

$$\mathbb{E}_{\mathbb{Q}_\kappa} L^2 1_{\{\tau>t\}} \geq (\mathbb{E}_{\mathbb{Q}_\kappa} L1_{\{\tau>t\}})^2 = (p(t))^2.$$

In this sense, we could call our change of measure *logarithmically efficient* if

$$\lim_{t\to\infty}\frac{1}{t}\log\mathbb{E}_{\mathbb{Q}_\kappa} L^2 1_{\{\tau>t\}} \leq \lim_{t\to\infty}\frac{1}{t}\log(\mathbb{E}_{\mathbb{Q}_\kappa} L1_{\{\tau>t\}})^2 = 2\vartheta^\star;$$

here the equality is due to Eqn. (9.5) (or Eqn. (9.6) in the corresponding spectrally negative case). In this context, in which the probability of interest decays roughly exponentially, logarithmic efficiency essentially means that the number of replications needed to obtain an estimate with a certain fixed precision grows subexponentially in the 'rarity parameter' t; cf. Asmussen and Glynn [24, Chapter VI].

It is now easily seen that $\kappa = 0$ does *not* necessarily yield logarithmic efficiency. To this end, recall that a necessary condition for the event $\{\tau > t\}$ is $Q_0 + X_t > 0$,

$$\mathbb{E}_{\mathbb{Q}_\kappa} L^2 1_{\{\tau>t\}} \leq \left(-\frac{\kappa\varphi'(0)}{\varphi(-\kappa)}\right)^2 e^{2\vartheta^\star t}\mathbb{E}_{\mathbb{Q}_\kappa} e^{-2\kappa Q_0} e^{-2\psi(\vartheta^\star)Q_0}; \tag{10.4}$$

when picking $\kappa = 0$ we need to have

$$\mathbb{E}_{\mathbb{Q}_0} e^{-2\psi(\vartheta^\star)Q_0} < \infty$$

for logarithmic efficiency, and this is not a priori clear. But now we can see, realizing that $\varphi(\psi(\vartheta^\star))$ is finite (to see this, use that ζ is larger than the pole of $\varphi(\cdot)$), that picking $\kappa := -\psi(\vartheta^\star) = -\zeta$ *does* yield logarithmic efficiency! In other words, we have to exponentially twist Q_0 as well, and we have to do so with twist $\kappa = -\zeta > 0$.

The next question is, can we do better than twisting with $-\zeta$? Interestingly, using

$$\mathbb{E}_{\mathbb{Q}_\kappa} e^{-\alpha Q_0} = \frac{\alpha - \kappa}{\varphi(\alpha - \kappa)} \cdot \frac{\varphi(-\kappa)}{-\kappa},$$

the right-hand side of (10.4) can be rewritten as

$$(\varphi'(0))^2 \left(\frac{-\kappa}{\varphi(-\kappa)}\right)\left(\frac{2\zeta + \kappa}{\varphi(2\zeta + \kappa)}\right) e^{2\vartheta^\star t}. \tag{10.5}$$

Observe that it contains two factors in κ, the first of which increases in κ, the second decreases in κ: there is a trade-off. It is a straightforward exercise to show that the minimum is achieved for $\kappa = -\zeta$ (equate the derivative to 0, but it can also be seen using a symmetry argument). We conclude that, in the sense that it minimizes (10.5), the proposed change of measure is the best possible within the class of exponential twists of Q_0.

Example 10.3 For X corresponding to $\mathbb{B}\mathrm{m}(-1, 1)$, that is, with $\varphi(\alpha) = \alpha + \frac{1}{2}\alpha^2$ and $\zeta = -1$, we obtain that the optimal κ equals 1. It entails that we should sample Q_0 from an exponential distribution with mean 1; under the original measure it would be an exponential distribution with mean $\frac{1}{2}$. After time 0, we should sample the Brownian motion as $\mathbb{B}\mathrm{m}(0, 1)$, that is, a driftless Brownian motion. ◇

Example 10.4 We now consider the case that X represents $\mathbb{CP}(1, \lambda, b(\cdot))$ with the B sampled from an exponential distribution with mean μ^{-1}. Let $\varrho := \lambda/\mu < 1$. Recall that, with $E(\delta)$ denoting an exponentially distributed random variable with mean $1/\delta$,

$$Q_0 \stackrel{\mathrm{d}}{=} \begin{cases} 0 & \text{with probability } 1 - \varrho; \\ E(\mu - \lambda) & \text{with probability } \varrho. \end{cases}$$

It is now a matter of straightforward calculations to verify that the twisting with $-\zeta$ yields the distribution

$$\check{Q}_0 \stackrel{\mathrm{d}}{=} \begin{cases} 0 & \text{with probability } 1 - \sqrt{\varrho}; \\ E((\sqrt{\mu} - \sqrt{\lambda})^2) & \text{with probability } \sqrt{\varrho}. \end{cases}$$

After time 0 we should let the jobs arrive according to a Poisson process with rate $\sqrt{\lambda\mu}$, with their sizes being sampled from an exponential distribution with mean $1/\sqrt{\lambda\mu}$. ◇

10.4 Estimation of Workload Correlation Function

So far we have provided recipes to estimate two types of rare-event probabilities, that is, those related to a large workload, and those related to a long busy period. The workload correlation is also a rare-event-related quantity, but not a rare-event probability as such. As a result, inherently different techniques need to be developed, so as to efficiently estimate $r(t) = \mathbb{C}\mathrm{orr}(Q_0, Q_t)$. In this section we describe an approach that was developed in Glynn and Mandjes [101].

We again restrict ourselves to $X \in \mathscr{L} \cap \mathscr{S}_+$ (where it is noted that the corresponding spectrally negative case works similarly). Observe that it suffices to estimate $c(t) := \mathbb{C}\mathrm{ov}(Q_0, Q_t)$, as $v = \mathbb{V}\mathrm{ar}\, Q$ is known. The 'naïve estimator' of $c(t)$ is, in self-evident notation, and recalling that $\mathbb{E}Q$ is known,

$$c_n^{(\mathrm{NS})}(t) := \frac{1}{n}\sum_{i=1}^{n} Q_0^{(i)} Q_t^{(i)} - (\mathbb{E}Q)^2,$$

based on n independent runs. The variance of this naïve estimator reads $(n^{-1}) \cdot \mathbb{V}\mathrm{ar}(Q_0 Q_t)$. Now note that, as $t \to \infty$,

$$\mathbb{V}\mathrm{ar}(Q_0 Q_t) = \mathbb{E}(Q_0^2 Q_t^2) - (\mathbb{E}Q_0 Q_t)^2 \to (\mathbb{E}Q^2)^2 - (\mathbb{E}Q)^4,$$

which is positive due to the fact that $\mathbb{E}Q^2 > (\mathbb{E}Q)^2$. Suppose our goal is to simulate until our estimate has a certain given relative precision f (defined as the ratio between the width of the confidence interval and the estimate) and confidence α. The number of runs needed, say n, is roughly equal to the smallest n satisfying

$$t_\alpha \frac{\sqrt{\mathbb{V}\mathrm{ar}\, c_n^{(\mathrm{NS})}(t)}}{c(t)} < f,$$

with t_α as defined earlier. This yields that

$$n \geq \frac{t_\alpha^2}{f^2 (c(t))^2} \left((\mathbb{E}Q^2)^2 - (\mathbb{E}Q)^4 \right).$$

Now recall that in the situation at hand $c(t)$ decays roughly exponentially. We therefore obtain the following remarkable result for the naïve estimator: it says that the number of runs required blows up exponentially, but it is *quadratically* inversely proportional to $c(t)$, rather than just inversely proportional. This result underscores that efficient (simulation-based) computation of the workload correlation $c(t)$ poses fundamentally new questions (compared to the estimation of rare-event probabilities). This was perhaps not anticipated, given the fact that the decay of $r(t)$ (and hence of $c(t)$ as well) resembles that of the busy-period tail asymptotics $p(t)$, as was observed in Chapter 9.

To overcome this problem, we now consider a *coupling*-based algorithm, that reduces the number of runs needed from quadratically inversely proportional to $c(t)$, to just inversely proportional. We write

$$c(t) = \mathbb{E}(Q_0 \cdot (Q_t - Q_t^\star)),$$

where both Q and $Q^\star$ are stationary versions of the workload, and $Q_t^\star$ is *independent* of Q_0. We construct such a coupling as follows: generate Q_0 and $Q_0^\star$ independently, sampled from the stationary distribution of the workload. Now use exactly the same incoming Lévy process X_s over $(0,t]$ to drive both $(Q_s)_{s\in(0,t]}$ and $(Q_s^\star)_{s\in(0,t]}$ from their two independently generated initial conditions. This makes Q_t and Q_0 correlated, but leaves $Q_t^\star$ and Q_0 independent. The new estimator becomes, in self-evident notation,

$$c_n^{(\mathrm{CS})}(t) := \frac{1}{n}\sum_{i=1}^{n} Q_0^{(i)}\left(Q_t^{(i)} - Q_t^{\star\,(i)}\right),$$

based on n independent runs.

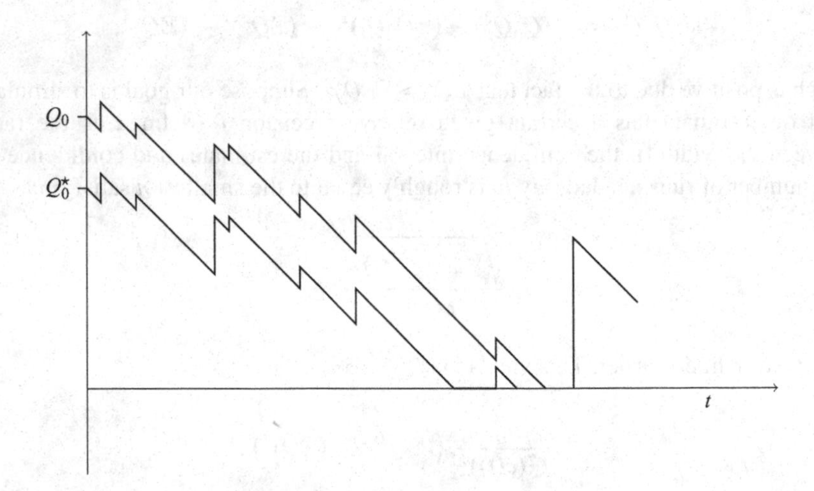

Fig. 10.1 Graphical illustration of the coupling technique, in a compound Poisson example. Top graph is Q_t, bottom graph is $Q_t^\star$. Observe that the distance between both graphs is non-increasing in t. The processes coincide at the end of the busy period of the workload process corresponding to the largest initial workload

A key observation is that $Q_t^{(i)} = Q_t^{\star\,(i)}$ if in both systems the busy period (that started at time 0) has ended; see also Fig. 10.1. In other words, we obtain a non-zero contribution only when at least one of the busy periods has not ended yet. Mainly due to this property, it is proved in [101] that the number of runs needed is roughly inversely proportional to $c(t)$. If this algorithm is augmented with importance sampling (very similarly to the way this was done in the algorithm to estimate $p(t)$ efficiently; see Section 10.3), one even obtains a logarithmically efficient algorithm [101, Section 4.3].

Exercises

Exercise 10.1 Verify Eqn. (10.1).

Exercise 10.2 Verify the formulas for the likelihood ratios (10.2) and (10.3).

Exercise 10.3 Prove that (10.5) is minimized for $\kappa = -\zeta$.

Exercise 10.4 Let X correspond to the gamma process $\mathbb{G}(\beta, \gamma)$, minus a deterministic drift of rate r.

(a) Show that the queue is stable if $\beta/\gamma < r$.
(b) Prove that the zero-drift change of measure, that can be used to efficiently estimate $\mathbb{P}(\tau > t)$ and $r(t)$, is such that X should be sampled as $\mathbb{G}(\beta, \beta/r)$ minus a deterministic drift of rate r.

Exercise 10.5 In the simulation algorithm to efficiently estimate $r(t)$, we construct a coupling as follows. First generate Q_0 and $Q_0^\star$ independently, sampled from the stationary distribution of the workload. Then use exactly the same incoming Lévy process X_s over $(0, t]$ to drive both $(Q_s)_{s\in(0,t]}$ and $(Q_s^\star)_{s\in(0,t]}$ from their two independently generated initial conditions. As a result, Q_t and Q_0 are correlated, but $Q_t^\star$ and Q_0 are independent.

(a) In the proof it is used that the distance between both processes, that is, $|\,Q_t - Q_t^\star\,|$, is non-increasing in t. Prove this property.
(b) Let τ (respectively, $\tau^\star$) denote the first epoch that $(Q_t)_t$ (respectively, $(Q_t^\star)_t$) hits 0. Show that for $t > \max\{\tau, \tau^\star\}$ the processes $(Q_t)_t$ and $(Q_t^\star)_t$ coincide.

Exercise 10.6 Consider a queueing process $(Q_t)_t$ driven by the Lévy process $(X_t)_t$. Assume that $Q_0 = 0$ a.s. Check that $\mathbb{P}(Q_s > x) \leq \mathbb{P}(Q_t > x)$ for each $x \geq 0$ if $s < t$, that is, Q_s, Q_t are *stochastically ordered*.

[illegible] such that Q_1 [illegible] = Q_2 [illegible] systems [illegible] and [illegible] time [illegible] edges [illegible] in Fig. 10.1. In other words, [illegible] ... time [illegible] property [illegible]. [illegible] proportional to [illegible] ... done in the algorithm [illegible] Section 10.[illegible] ... algorithm in [illegible] Section [illegible].

Exercises

Exercise 10.1 [illegible] Eqn. (10.1).

Exercise 10.2 [illegible] for [illegible] in (10.[illegible]) to (10.5).

Exercise 10.3 [illegible] (10.[illegible]) is minimized [illegible].

Exercise 10.4 [illegible] process [illegible]

(a) [illegible] the process is stable [illegible]

(b) [illegible] that [illegible] change [illegible]. This can be used to [illegible] estimate [illegible] and [illegible] is such that [illegible] ... and [illegible] unit of time.

Exercise 10.5 [illegible] the [illegible] algorithm for [illegible] as follows. [illegible] process Q_1 and Q_2 independently [illegible] from [illegible] ... Lévy [illegible] respective [illegible] $(Q_1$ [illegible] in (10.[illegible]) from [illegible] ... independently [illegible] condition [illegible] Q_1 and Q_2 [illegible] Q_1 and Q_2 are independent.

(a) In [illegible] that the distance between both processes [illegible] Q_1 [illegible] Q_2 [illegible]

(b) [illegible] respectively [illegible] first [illegible] that [illegible] Q_1 [illegible] Q_2 [illegible]

Exercise 10.6 Consider [illegible] process [illegible] [illegible] [illegible]

Chapter 11
Variants of the Standard Queue

So far we have considered the standard infinite-buffer queue with Lévy input. This chapter describes a number of variants of this standard model. The systems considered are (i) Lévy-driven finite-buffer queues, (ii) models in which the current workload level has impact on the input process ('feedback'), (iii) vacation and polling models, and (iv) queues with Markov additive input. In this chapter we typically sketch the state of the art in these areas, without giving all proofs in full detail.

11.1 Finite-Buffer Queues

In this section we consider a Lévy-driven queue in which the workload cannot exceed level $K > 0$; in the case that this upper boundary K would be exceeded by a positive jump, the part of the jump that fits into the buffer is accepted, and the rest is rejected (symmetrically to what happens to negative jumps at the lower boundary 0). The finite-buffer system that is thus defined is treated in detail in the monograph by Andersen et al. [8]; here we restrict ourselves to the main results.

Again we call the associated workload process $(Q_t)_t$. A corresponding Skorokhod problem can be formulated, in which Q_t is expressed in terms of the local time at 0 (as before), but now also the local time at K plays a role. Assuming for ease that $Q_0 = 0$, we have that $Q_t = X_t + L_t - \bar{L}_t$, with L_t (respectively, $\bar{L}_t$) the local time at 0 (respectively, at K); popularly speaking, L_t increases only when $Q_t = 0$, whereas $\bar{L}_t$ increases only when $Q_t = K$.

K. Dębicki, M. Mandjes, *Queues and Lévy Fluctuation Theory*, Universitext,
DOI 10.1007/978-3-319-20693-6_11

In Andersen and Mandjes [9] and Kruk et al. [142] it is shown how to solve $(Q_t)_t$ explicitly from the Skorokhod problem; the authors of [142] found the representation

$$Q_t = X_t - \sup_{s\in[0,t]}\left(\max\left\{\min\left\{X_s - K, \inf_{u\in[0,t]} X_u\right\}, \inf_{u\in[s,t]} X_u\right\}\right),$$

whereas the alternative solution in [9] is slightly simpler, and reads

$$Q_t = \sup_{s\in[0,t]} \max\left\{X_t - X_s, \inf_{u\in[s,t]}(K + X_t - X_u)\right\};$$

for ease we here consider the case that $Q_0 = 0$. It was proved for the infinite-buffer model that the mean $\mathbb{E}(Q_t \,|\, Q_0 = 0)$ is increasing and concave in t [119, 127], but, interestingly, this conclusion remains valid in the finite-buffer case as well [9].

The first part of the following result [153, 199] characterizes the steady-state workload Q in terms of a first-passage time; note that it is (obviously) now not necessary that $\mathbb{E}X_1 < 0$. The second part, which can be found in e.g. [43, Thm. 8, p. 194], assumes spectrally negative input, but realize that the spectrally positive case can be dealt with analogously (by considering the Lévy input $-X$). Recall the (implicit) definition of the scale function $W^{(0)}(\cdot)$ from Eqn. (4.7); write $\pi_K(u) := \mathbb{P}(Q \le u)$.

As we know the transform of $W^{(0)}(\cdot)$, the result below uniquely characterizes $\mathbb{P}(Q > u)$. For the case of Brownian input, it turns out that Q has a truncated exponential distribution, as is easily checked. We mention here that in [85] scale functions are also used to determine the busy-period distribution in a finite-buffer M/G/1 queue.

Proposition 11.1

(i) For $u \in [0, K]$,

$$1 - \pi_K(u) = \mathbb{P}(X_{\tau[u-K,u)} \ge u),$$

where $\tau[u, v) := \inf\{t \ge 0 : X_t \notin [u, v)\}$, for $u \le 0 \le v$.

(ii) Let $X \in \mathscr{S}_-$. Then, for $u \in [0, K]$,

$$1 - \pi_K(u) = \frac{W^{(0)}(K-u)}{W^{(0)}(K)}.$$

Proof We now prove both parts of Prop. 11.1. We start with part (i). The idea is that we find an alternative expression for the probability

$$\mathbb{P}_0(Q_T \le u) := \mathbb{P}(Q_T \le u \,|\, Q_0 = 0),$$

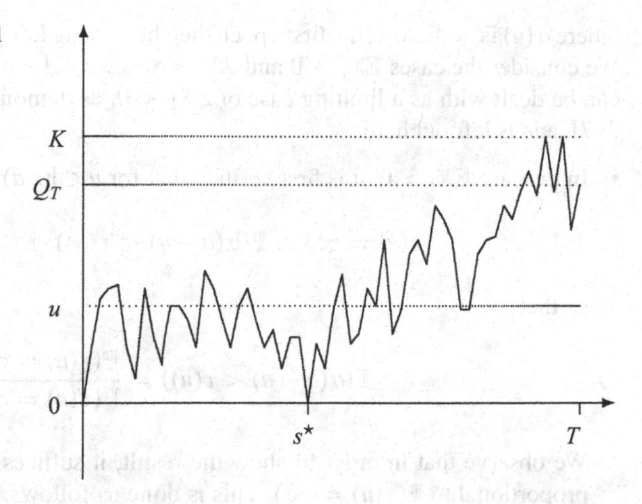

Fig. 11.1 Proof of Prop. 11.1

and then we let T grow to ∞, so as to obtain $\pi_K(u)$; the proof presented here follows essentially the same lines as Asmussen [19, Prop. XIV.3.7].

To this end, consider Fig. 11.1: we took $Q_0 = 0$, and we picked u such that $Q_T \geq u$. The following two claims hold.

(A) First, there is an s between 0 and T such that $X_T - X_s \geq u$. One such s for which this holds is $s^\star$, the last epoch before T that the system was empty (see Fig. 11.1). Indeed, it is verified that

$$u \leq Q_T = X_T - X_{s^\star} - \bar{L}_T + \bar{L}_{s^\star} \leq X_T - X_{s^\star},$$

using that $(\bar{L}_t)_t$ is an increasing process.

(B) Second, for all s between 0 and T, we have that $X_T - X_s \geq u - K$, because otherwise Q_T would be below u.

Now define $R_T(u) := u - X_T + X_{T-t}$ until this process hits $(-\infty, 0]$ (and then it is set equal to 0) or (K, ∞) (and then it is set equal to ∞). The observations (A) and (B) above now entail that the event $\{Q_T \geq u\}$ is equivalent to $\{R_T(u) = 0\}$. As a result,

$$\mathbb{P}_0(Q_T \leq u) = \mathbb{P}(\tau[u-K, u) \leq T, X_{\tau[u-K,u)} \geq u),$$

so that the stated result follows by sending T to ∞.

We continue with part (ii); for full details we refer to Kyprianou [146, Thm. 8.1]. We rewrite for $X \in \mathscr{S}_-$,

$$\mathbb{P}(X_{\tau[u-K,u)} \geq u) = \mathbb{P}(\sigma(u) < \tau(K-u)),$$

where $\sigma(u)$ is, as before, the first epoch that the driving Lévy process $(X_t)_t$ exceeds u. We consider the cases $\mathbb{E}X_1 > 0$ and $\mathbb{E}X_1 < 0$ separately; we mention that $\mathbb{E}X_1 = 0$ can be dealt with as a limiting case of $\mathbb{E}X_1 < 0$, as demonstrated in [146, pp. 216–217], and is left out here.

- In the case $\mathbb{E}X_1 > 0$, it is first verified that for $u \in [0, a)$,

$$\mathbb{P}(\tau(u) = \infty) = \mathbb{P}(\sigma(a-u) < \tau(u)) \cdot \mathbb{P}(\tau(a) = \infty),$$

so that

$$\mathbb{P}(\sigma(a-u) < \tau(u)) = \frac{\mathbb{P}(\tau(u) = \infty)}{\mathbb{P}(\tau(a) = \infty)}.$$

We observe that in order to show the result, it suffices to prove that $W^{(0)}(u)$ is proportional to $\mathbb{P}(\tau(u) = \infty)$. This is done as follows. Let X' equal $-X$, which is in $\mathscr{S}_+$ and has a negative drift. Then, due to integration by parts and Thm. 3.2 (i.e. the generalized Pollaczek–Khintchine formula),

$$\int_0^\infty e^{-\beta u}\mathbb{P}(\tau(u) = \infty)\mathrm{d}u = \int_0^\infty e^{-\beta u}\mathbb{P}\left(\sup_{t\geq 0} X'_t < u\right)\mathrm{d}u$$
$$= \frac{1}{\beta}\int_0^\infty e^{-\beta u}\mathbb{P}\left(\sup_{t\geq 0} X'_t \in \mathrm{d}u\right) = \frac{\Phi'(0)}{\Phi(\beta)}.$$

As the transform of $W^{(0)}(u)$ with respect to β is $1/\Phi(\beta)$, we proved the desired proportionality.

- Now consider the case $\mathbb{E}X_1 < 0$; the above approach does not work, as both $\mathbb{P}(\tau(u) = \infty)$ and $\mathbb{P}(\tau(a) = \infty)$ equal 0. Construct the measure $\mathbb{Q}$ as before, that is, with exponential twisting by ω, where ω solves $\Phi(\omega) = 0$.
Recall that $\mathbb{E}_{\mathbb{Q}}X_1 = \Phi'(\omega) > 0$. Because of what we found for the case $\mathbb{E}X_1 > 0$, it holds that

$$\mathbb{Q}(\sigma(a-u) < \tau(u)) = \frac{\mathbb{Q}(\tau(u) = \infty)}{\mathbb{Q}(\tau(a) = \infty)}.$$

Hence, using that $X_{\sigma(a-u)} = a - u$ for $X \in \mathscr{S}_-$,

$$\mathbb{P}(\sigma(a-u) < \tau(u)) = \mathbb{E}_{\mathbb{Q}}\left(e^{-\omega X_{\sigma(a-u)}} 1_{\{\sigma(a-u)<\tau(u)\}}\right)$$
$$= e^{-\omega(a-u)}\mathbb{Q}(\sigma(a-u) < \tau(u)) = \frac{e^{\omega u}\,\mathbb{Q}(\tau(u) = \infty)}{e^{\omega a}\,\mathbb{Q}(\tau(a) = \infty)}.$$

We observe that in this case it suffices to prove that $W^{(0)}(u)$ is proportional to $e^{\omega u}\mathbb{Q}(\tau(u) = \infty)$. Denote

$$\Phi_{\mathbb{Q}}(\beta) := \log \mathbb{E}_{\mathbb{Q}} e^{\beta X_1} = \Phi(\beta + \omega) - \Phi(\omega) = \Phi(\beta + \omega).$$

Analogously to the above argument, we find that

$$\int_0^\infty e^{-\beta u} e^{\omega u}\, \mathbb{Q}(\tau(u) = \infty)\mathrm{d}u = \frac{\Phi'_{\mathbb{Q}}(0)}{\Phi_{\mathbb{Q}}(\beta - \omega)} = \frac{\Phi'(\omega)}{\Phi(\beta)},$$

which is proportional to $1/\Phi(\beta)$, as desired. □

Remark 11.1 For the class of Lévy processes with phase-type jumps in both directions, that is, $X \in \mathscr{P}_+ \cup \mathscr{P}_-$, the quantity $\pi_K(u)$ can be determined using the theory of Section 3.4. ◇

Remark 11.2 It is noted that the analysis of dual-exit-related probabilities of the type $\mathbb{P}(\sigma(a) < \tau(b))$ (with a and b positive) plays a role in several other applications. One such application is *sequential analysis* [200]; the CUSUM method, which detects changes in the probability distribution underlying a stochastic process, requires the evaluation of the probability that an associated log likelihood process hits the upper (respectively, lower) barrier before hitting the lower (respectively, upper) barrier. ◇

Example 11.1 Let $(X_t)_t$ be a compound Poisson process with a *positive* drift of rate 1. The downward jumps arrive according to a Poisson process with rate λ and have an exponentially distributed size with mean μ^{-1}. Observe that this process is spectrally negative.

It is readily checked that

$$\frac{1}{\Phi(\beta)} = \frac{\mu + \beta}{\beta(\beta - \lambda + \mu)} = \frac{1}{\beta} + \frac{\lambda}{\beta(\beta - \lambda + \mu)}.$$

Realizing that $1/\Phi(\beta)$ is the Laplace transform of $W^{(0)}(x)$, this gives

$$W^{(0)}(x) = 1 + \frac{\lambda}{\mu - \lambda}\left(1 - e^{-(\mu-\lambda)x}\right),$$

so that, with $u \in [0, K)$, and assuming $\lambda \neq \mu$,

$$\mathbb{P}(Q > u) = \frac{\lambda e^{-(\mu-\lambda)(K-u)} - \mu}{\lambda e^{-(\mu-\lambda)K} - \mu}, \quad \mathbb{P}(Q = K) = \frac{\lambda - \mu}{\lambda e^{-(\mu-\lambda)K} - \mu}.$$

If $\mu > \lambda$, then $\mathbb{P}(Q = K) \to 1 - \lambda/\mu$ as $K \to \infty$, as also follows using trivial queueing-theoretic arguments. If $\mu < \lambda$, then $\mathbb{P}(Q = K) \sim \lambda e^{-(\lambda-\mu)K}(\lambda - \mu)$, as

$K \to \infty$. It is readily checked that in the case $\lambda = \mu$ we have $W^{(0)}(x) = 1 + \lambda x$, and therefore

$$\mathbb{P}(Q > u) = \frac{1 + \lambda(K - u)}{1 + \lambda K}$$

for $u \in [0, K)$, and $\mathbb{P}(Q = K) = 1/(1 + \lambda K) \sim 1/(\lambda K)$ as $K \to \infty$. ◇

In models with a finite buffer, there is the notion of a *loss rate* ℓ^K, which we define, in self-evident notation, by

$$\ell^K := \mathbb{E}_{\pi_K} \bar{L}_1.$$

In Asmussen and Pihlsgård [29] the following result was proved for general finite-buffer Lévy-driven queues.

Proposition 11.2 *If* $\int_1^\infty y\Pi(\mathrm{d}y) = \infty$, *then* $\ell^K = \infty$; *otherwise*

$$\ell^K = \frac{\mathbb{E}X_1}{K} \int_0^K x\,\pi_K(\mathrm{d}x) + \frac{\sigma^2}{2K} + \frac{1}{2K} \int_0^K \int_{-\infty}^\infty k(x, y)\Pi(\mathrm{d}y)\pi_K(\mathrm{d}x),$$

where $k(x, y) := -(x^2 + 2xy)$ *for* $y \le -x$, $k(x, y) := y^2$ *for* $-x < y < K - x$, *and* $k(x, y) := 2y(K - x) - (K - x)^2$ *for* $y \ge K - x$.

For $X \in \mathscr{L}$, [29] also studies the asymptotics of ℓ^K for K large. These are of the form $Ce^{-\omega K}$, for some rather complicated C, and ω solving $\mathbb{E}e^{\omega X_1} = 1$. Observe the similarity to the asymptotics of the tail distribution of the stationary workload in the model with infinite buffer.

11.2 Models with Feedback

In the queues we have studied so far, the input stream was not affected by the current level of the workload. In this section we *do* allow such dependencies, which we refer to as *feedback*.

We start by considering a queue whose input is $\mathbb{CP}(r(x), \lambda(x), b(\cdot))$ when the current workload level is $x \ge 0$; note that the distribution of the jobs B does *not* depend on x. Mimicking the procedure outlined in Section 3.1, a rate conservation argument shows that the density $f_Q(\cdot)$ of the stationary workload obeys the integral equation [36]

$$r(x)f_Q(x) = \int_{(0,x)} \lambda(y)f_Q(y)\mathbb{P}(B > x - y)\mathrm{d}y + \lambda(0)p_0\mathbb{P}(B > x),$$

with $p_0 := \mathbb{P}(Q = 0)$. In the special case that the jobs have an exponential distribution with mean $1/\mu$, multiplication by $e^{\mu x}$ yields the differential equation $g'(x) = g(x)\lambda(x)/r(x)$, with $g(x) := e^{\mu x} r(x) f_Q(x)$. For the case $p_0 > 0$ we obtain by an elementary separation of variables argument that

$$f_Q(x) = \frac{\lambda(0) p_0}{r(x)} \exp\left(\int_0^x \left(\frac{\lambda(y)}{r(y)} - \mu \right) \mathrm{d}y \right),$$

under appropriate integrability conditions; the case $p_0 = 0$ should be dealt with separately. Further details can be found in Bekker et al. [36].

In [38] attention is paid to a queue fed by a spectrally positive Lévy process, where feedback information about the workload level may lead to adaptation of the Lévy exponent. Among other models, the paper addresses the class of models in which the workload can only be observed at Poisson instants; at these Poisson instants, the Lévy exponent may be adapted based on the amount of work present at that time. In [37] a somewhat related model is studied: the focus is on a Lévy-driven queue, where the Lévy exponent of the input process alternates between two different forms (depending on the evolution of the workload process in the past). Classical related papers are [54, 110].

11.3 Vacation and Polling Models

In Boxma et al. [52] a Lévy-driven queue with *server vacations* is studied. It can be regarded as a stochastic storage process alternatingly experiencing active and passive (vacation) periods (see also [128]), and is described as follows.

During active periods, work is generated according to a Lévy process $X_D \in \mathscr{S}_+$ with negative drift, until the workload reaches 0 (i.e. the storage reservoir is empty). From then on, the storage level behaves according to a second Lévy process X_U, which is assumed to be non-decreasing. As during this period work accumulates in the queue, it may be interpreted as a vacation; it lasts $aI + bV$, where I is a function of the length of the preceding active period, and V is an independent vacation time, and a and b are given, non-negative scalars. The case in which the workload is still 0 after $aI + bV$ has to be treated separately: the vacation period is extended until work is generated by X_U. Subsequently a new active period starts; etc.

The steady-state workload in such a system can be found as follows. Consider the sequence of epochs right before an active period starts. The transform of the storage level at such an embedded epoch can be expressed in terms of the transform at the previous embedded epoch. As these transforms should be identical in equilibrium, we can thus obtain the transform of the stationary storage level at those embedded epochs [52, Section 3]. Relying on the Kella–Whitt martingale, they can be translated into the transform of the workload at an arbitrary epoch; see [52, Section 4]. Interestingly, these vacation models can be related to so-called *polling models*, in which a single server visits multiple queues according to some predefined discipline.

The topic of Lévy-driven polling systems is explored in full detail in Boxma et al. [51]. There the focus is on an N-queue polling model with switchover times. Each of the queues is fed by a non-decreasing Lévy process, which can be different during each of the consecutive periods within the server's cycle. The N-dimensional Lévy processes obtained in this fashion are described by their (joint) Laplace exponent, thus allowing for non-independent input streams. Again, as a first step the steady-state distribution of the workload is determined at embedded epochs (which are now polling and switching instants); importantly, the *joint* transform of all N workloads is found. As before, application of the Kella–Whitt martingale yields the steady-state distribution at an arbitrary epoch.

The analysis heavily relies on the link between the polling system and so-called *(multitype) Jiřina processes* (continuous-state discrete-time branching processes). The results are so general that they cover the most important polling disciplines, like exhaustive and gated.

11.4 Models with Markov-Additive Input

Markov-additive processes (MAPs) date back to Çinlar [61] and Neveu [169], and can be seen as the Markov-modulated version of Lévy processes; here we concentrate on MAPs in continuous time. We now give the definition of a MAP; for ease we restrict ourselves to the spectrally positive case (which we call $\mathscr{S}_+^{\rm MAP}$), but general MAPs can be introduced analogously.

A MAP is a bivariate Markovian process (X_t, J_t) that is defined as follows; see Fig. 11.2.

- Let $(J_t)_t$ be an irreducible continuous-time Markov chain with finite state space $E = \{1, \ldots, d\}$. Define by $(q_{ij})_{i,j=1}^d$ the $(d \times d)$ transition rate matrix of $(J_t)_t$ and by $\boldsymbol{\pi}$ the (unique) stationary distribution.
- For each state i that J_t can attain, let the process $(X_t^{(i)})_t$ be a Lévy process. As mentioned above, for the moment we assume these are in $\mathscr{S}_+$, and have Laplace exponents $\varphi_i(\alpha) := \log \mathbb{E} \exp(-\alpha X_1^{(i)})$, for $i = 1, \ldots, d$.
- Letting T_n and T_{n+1} be two successive transition epochs of J_t, and given that J_t jumps from state i to state j at $t = T_n$, we define the additive process X_t in the time interval $[T_n, T_{n+1})$ through

$$X_t := X_{T_n-} + U_{ij}^n + [X_t^{(j)} - X_{T_n}^{(j)}],$$

where the $(U_{ij}^n)_n$ constitute a sequence of i.i.d. random variables (each of which is distributed as a generic random variable U_{ij}) with Laplace–Stieltjes transform

$$b_{ij}(\alpha) = \mathbb{E} e^{-\alpha U_{ij}},$$

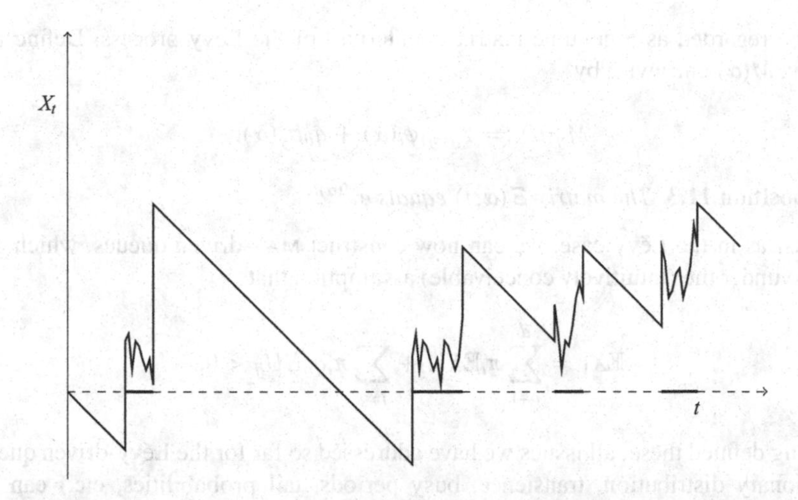

Fig. 11.2 Graphical illustration of the evolution of a MAP. In this case there are two states: one in which the process behaves as a Brownian motion (corresponding to the *solid parts* of the horizontal axis), and the other being a negative drift (corresponding to the *dashed parts* of the horizontal axis). In addition, there are upward jumps at the transition epochs of the process J_t

where $U_{ii} \equiv 0$, describing the jumps at transition epochs. To make the MAP spectrally positive, it is required that $U_{ij} \geq 0$ almost surely (for all $i, j \in \{1, \ldots, d\}$) and that the processes $X_t^{(i)}$ are allowed to have positive jumps only (for all $i \in \{1, \ldots, d\}$). As an aside we mention that the superposition of MAPs is again a MAP.

Observe that the modulating Markov chain does not jump in $[t, t+h)$ with probability $1 + q_{jj}h + o(h)$, given $J_t = j$ (recall that $q_{jj} < 0$), and jumps to k with probability $q_{jk}h + o(h)$, for $h \downarrow 0$. We therefore obtain, with the matrix $\Xi(\alpha, t)$ defined by

$$\Xi_{ij}(\alpha, t) := \mathbb{E}_i\left(e^{-\alpha X_t} 1_{\{J_t=j\}}\right) = \mathbb{E}\left(e^{-\alpha X_t} 1_{\{J_t=j\}} \mid J_0 = i\right),$$

the following equation (using that $b_{kk}(\alpha) = 1$ for all α):

$$\Xi_{ij}(\alpha, t+h) = (1 + q_{jj}h)\Xi_{ij}(\alpha, t)\mathbb{E}e^{-\alpha X_h^{(j)}} + \sum_{k \neq j} q_{kj}h \cdot \Xi_{ik}(\alpha, t)b_{kj}(\alpha) + o(h)$$

$$= (1 + \varphi_i(\alpha))\Xi_{ij}(\alpha, t) + h \sum_{k=1}^{d} \Xi_{ik}(\alpha, t)q_{kj}b_{kj}(\alpha) + o(h).$$

Subtracting $\Xi_{ij}(\alpha, t)$ from both sides and dividing by h, we obtain a system of linear differential equations. Its solution is given in the following proposition, which shows some sort of infinite divisibility, but *now at the matrix level*. In this sense, the MAP can be regarded as a genuine matrix counterpart of the Lévy process. Define the matrix $M(\alpha)$ entrywise by

$$M_{ij}(\alpha) := 1_{\{i=j\}}\varphi_i(\alpha) + q_{ij}b_{ij}(\alpha).$$

Proposition 11.3 *The matrix $\Xi(\alpha, t)$ equals $e^{M(\alpha)t}$.*

Just as in the Lévy case, we can now construct MAP-driven queues, which are stable under the (intuitively conceivable) assumption that

$$\mathbb{E}X_1 = \sum_{i=1}^{d} \pi_i \mathbb{E}X_1^{(i)} + \sum_{i \neq j} \pi_i q_{ij} \mathbb{E}U_{ij} < 0.$$

Having defined these, all issues we have addressed so far for the Lévy-driven queue (stationary distribution, transience, busy periods, tail probabilities, etc.) can be studied for the MAP-driven queue as well. We do not give an exhaustive overview of all results in this area here, as a vast body of literature has focused on this topic; we rather restrict ourselves to a relatively short account of the main findings concerning the stationary distribution.

In Asmussen and Kella [25] martingale methods have been developed in order to analyze, for $X \in \mathscr{S}_+^{\text{MAP}}$, the joint distribution of the steady-state workload Q and the steady state of the Markov chain J. Under the stability condition identified above, the transform of the Q reads

$$\mathbb{E}(e^{-\alpha Q}, J = j) = \left(\alpha \boldsymbol{\ell} (M(\alpha))^{-1}\right)_j, \tag{11.1}$$

where $\boldsymbol{\ell}$ is a row vector. It is interesting to compare the structure of this result with the Pollaczek–Khintchine formula of Thm. 3.2: observe that it is essentially its MAP counterpart.

Without formally proving (11.1), we now explain why a formula of this structure comes out. To this end, define $\kappa_j(\cdot)$ as the Laplace transform of the steady-state workload at epochs at which the modulating Markov chain enters state j. Suppose the modulating Markov chain just entered state i, and we consider what has happened in the exponentially distributed time (with mean $\hat{q}_j^{-1}$, with $\hat{q}_j := -q_{jj}$ if the modulating Markov chain came from state j) that it spent in the previous state. Relying on Thm. 4.1, $\kappa_i(\alpha)$ equals

$$\int_0^\infty \sum_{j \neq i} \frac{\hat{q}_j}{\hat{q}_j - \varphi_j(\alpha)} \left(e^{-\alpha x} - \frac{\alpha}{\psi_j(\hat{q}_j)} e^{-\psi_j(\hat{q}_j)x}\right) \left(\frac{q_{ji}}{\hat{q}_j}\right) b_{ji}(\alpha) f_j(\mathrm{d}x),$$

where $f_j(\cdot)$ is the density of the stationary workload at epochs that the modulating Markov chain enters state j. We thus obtain the identity

$$\kappa_i(\alpha) = \sum_{j \neq i} \left(\frac{q_{ji}}{\hat{q}_j}\right) b_{ji}(\alpha) \frac{\hat{q}_j}{\hat{q}_j - \varphi_j(\alpha)} \left(\kappa_j(\alpha) - \frac{\alpha}{\psi_j(\hat{q}_j)} \kappa_j(\psi_j(\hat{q}_j))\right).$$

Further manipulation of this equation leads to a matrix equation that has the structure of (11.1).

The authors of [25] do not succeed in uniquely characterizing the vector $\boldsymbol{\ell}$; it can be seen that $\sum_i \ell_i = \mathbb{E}X_1$ though. We also refer to [118] for related results. In D'Auria et al. [67] a method is developed that *does* determine $\boldsymbol{\ell}$. In this approach, an important role is played by the first passage time process $\tau(x) := \inf\{t \geq 0 : X(t) = -x\}$. It is readily seen that $J_{\tau(x)}$ is a time-homogeneous Markov process (as a function of x), say with generator Λ. The main finding of [67] is a way to identify this matrix, relying on the theory of Jordan chains. Then $\boldsymbol{\ell}$ can be expressed in terms of the invariant that is associated with Λ; in the proof of the key result a lemma on the number of zeros of the determinant of $M(\alpha)$ plays a crucial role [114]. A different approach is described in [83]. Reference [68] covers the special case that all the $X^{(i)}$ correspond to Brownian motions (with a specific focus on finite-buffer models).

Dieker and Mandjes [83] and Ivanovs et al. [114] also deal with the case of $X \in \mathscr{S}_-^{\rm MAP}$. Then Q has a phase-type distribution, whose parameters again follow directly with the techniques developed in [114]; this can be viewed as the MAP counterpart of the exponential distribution identified in Thm. 3.3. In that paper, the case of doubly reflected (i.e. finite-buffer-capacity) Markov-modulated Brownian motion is also dealt with. Other important papers are e.g. [53, 148, 186, 187], and various parts of the monograph [179].

The rest of this section addresses the case that, in self-evident notation, $X \in \mathscr{L}^{\rm MAP}$. This requires that all $X^{(i)}$ are in $\mathscr{L}$, and that all random variables U_{ij} are light tailed as well.

As we saw for the ordinary Lévy-driven queue, in order to obtain the workload asymptotics the alternative measure $\mathbb{Q}$ played a pivotal role. In that case the definition of $\mathbb{Q}$ rested on the (positive) solution ω of $\mathbb{E}e^{\omega X_1} = 1$. The first question is how such a measure $\mathbb{Q}$ can be constructed for $X \in \mathscr{L}^{\rm MAP}$. This can be done as follows.

- Let the eigenvalue/eigenvector pair $(e(\vartheta), \boldsymbol{h}(\vartheta))$ solve the following eigensystem:

$$M(-\vartheta)\,\boldsymbol{h}(\vartheta) = e(\vartheta)\,\boldsymbol{h}(\vartheta).$$

  Due to Perron–Frobenius theory [40, 195], it holds that $M(-\vartheta)$ has a *real* eigenvalue $e(\vartheta)$ with maximal real part, and the corresponding right eigenvector is componentwise positive; in the sequel, we refer to this (specific) eigenvalue/eigenvector pair as $(e(\vartheta), \boldsymbol{h}(\vartheta))$. With $\bar{\boldsymbol{h}}(\vartheta)$ denoting the left eigenvector of the above eigensystem, it is readily [19, Cor. XI.2.3] checked that $\Xi_{ij}(-\vartheta, t) \sim h_i(\vartheta)\bar{h}_j(\vartheta)e^{te(\vartheta)}$ for t large.
- Let $\omega > 0$ solve the equation $e(\omega) = 0$; this eigensystem-related equation is the MAP counterpart of $\mathbb{E}e^{\omega X_1} = 1$. Define the alternative measure $\mathbb{Q}$ as follows; it can be regarded as an exponentially twisted version of the original measure $\mathbb{P}$, with parameter ω.

- With the Lévy process $X^{(i)}$ under $\mathbb{P}$ being characterized by $\varphi_i(\alpha)$, under $\mathbb{Q}$ it corresponds to the Laplace exponent

$$\varphi_i^{\mathbb{Q}}(\alpha) = \varphi_i(\alpha - \omega) - \varphi_i(-\omega).$$

  More specifically, with $X^{(i)}$ under $\mathbb{P}$ being defined through (d_i, σ_i^2, Π_i), (10.1) tells us that under $\mathbb{Q}$ it corresponds to the triplet

$$\left(d_i + \sigma_i^2 \omega + \int_{-1}^{1} x(e^{\omega x} - 1)\Pi_i(\mathrm{d}x), \sigma_i^2, e^{\omega x}\Pi_i(\mathrm{d}x) \right).$$

- The distribution of the jumps U_{ij} is changed under $\mathbb{Q}$ such that the Laplace–Stieltjes transform of U_{ij} becomes

$$b_{ij}^{\mathbb{Q}}(\alpha) := \frac{b_{ij}(\alpha - \omega)}{b_{ij}(-\omega)}.$$

- Under $\mathbb{Q}$, the modulating Markov chain has transition rates

$$q_{ij}^{\mathbb{Q}} := q_{ij} \frac{h_j(\omega)}{h_i(\omega)} b_{ij}(-\omega)$$

  for $i \neq j$, and $q_{ii}^{\mathbb{Q}} := q_{ii} + \varphi_i(-\omega)$. It is readily verified that the row sums of the resulting generator matrix equal 0, as, using that $M(-\omega)\boldsymbol{h}(\omega) = 0$,

$$\begin{aligned}
\sum_{j \neq i} q_{ij}^{\mathbb{Q}} &= \sum_{j \neq i} q_{ij} \frac{h_j(\omega)}{h_i(\omega)} b_{ij}(-\omega) \\
&= \frac{1}{h_i(\omega)} \left(\varphi_i(-\omega) h_i(\omega) + \sum_{j=1}^{d} q_{ij}\, b_{ij}(-\omega)\, h_j(\omega) \right) - q_{ii} - \varphi_i(-\omega) \\
&= -q_{ii} - \varphi_i(-\omega) = -q_{ii}^{\mathbb{Q}}.
\end{aligned}$$

Observe that X under $\mathbb{Q}$ is again a MAP.

Now let $\omega > 0$ solve the equation $e(\omega) = 0$. We will show that this ω essentially determines the behavior of $\mathbb{P}(Q > u)$ for u large. To this end, it is first observed that

$$Q \stackrel{\mathrm{d}}{=} \sup_{t \leq 0}(-X_t) = -\inf_{t \leq 0} X_t,$$

where J_0 is distributed according to $\boldsymbol{\pi}$. This is only equal in distribution to $\sup_{t \geq 0} X_t$ if $(X_t)_t$ is reversible, which is the case if the modulating Markov chain is, and in addition $b_{ij}(\alpha) = b_{ji}(\alpha)$ for all pairs (i, j).

Based on the above, however, we *do* have, with $(\tilde{X}_t)_t$ denoting the time-reversed counterpart of $(X_t)_t$,

$$Q \stackrel{\mathrm{d}}{=} \sup_{t\geq 0} \tilde{X}_t.$$

As $(\tilde{X}_t)_t$ is again a MAP (with the transition rate matrix $(q_{ij})_{i,j=1}^d$ replaced by its time-reversed version, and U_{ij} by U_{ji}), from now on we focus on analyzing, for a given MAP $(X_t)_t$,

$$\mathbb{P}\left(\sup_{t\geq 0} X_t > u\right) = \sum_{i=1}^d \pi_i q_i(u), \quad \text{with} \quad q_i(u) := \mathbb{P}\left(\sup_{t\geq 0} X_t > u \,\middle|\, J_0 = i\right).$$

To this end, let us consider the probability $q_i(u)$ for a given $i \in \{1,\ldots,d\}$. It is first noted that under $\mathbb{Q}$ the MAP has a positive drift, so that $q_i(u)$ equals the value of the likelihood ratio at the epoch the level u is first crossed, which we denote as before by $\sigma(u)$. Now suppose we 'simulate' X under $\mathbb{Q}$ until $\sigma(u)$. Let the modulating Markov chain make N jumps before $\sigma(u)$, and let it visit states $i_0 = i, i_1, \ldots, i_N$, where the amount of time it stays in these states is $t_0, t_1, \ldots, t_N$, with $\sum_{n=0}^N t_n > \sigma(u)$. Also, let v_n be the jump size at the nth transition of the modulating Markov chain, and w_n be the increment of the Lévy process $X^{(i_n)}$ between t_n and t_{n+1}.

The likelihood ratio is the product of the following factors.

(i) First there is the contribution of the jumps of the modulating Markov chain:

$$\begin{aligned} L_1 &= \left(\frac{q_{i_0,i_1}\cdots q_{i_{N-1},i_N}}{q_{i_0}\cdots q_{i_{N-1}}}\right) \Big/ \left(\frac{q^{\mathbb{Q}}_{i_0,i_1}\cdots q^{\mathbb{Q}}_{i_{N-1},i_N}}{q^{\mathbb{Q}}_{i_0}\cdots q^{\mathbb{Q}}_{i_{N-1}}}\right) \\ &= \frac{h_{i_N}(\omega)}{h_{i_0}(\omega)} \prod_{n=0}^{N-1} \left(b_{i_n,i_{n+1}}(-\omega)\frac{q_{i_n}}{q^{\mathbb{Q}}_{i_n}}\right), \end{aligned}$$

where $q_i := -q_{ii}$.

(ii) Then there is the contribution of the exponentially distributed sojourn times in the states $i_0 = i, i_1, \ldots, i_N$. This yields, with $\bar{t}_N := \sigma_u - \sum_{n=0}^{N-1} t_n$,

$$L_2 = \left(\prod_{n=0}^{N-1} \frac{q^{\mathbb{Q}}_{i_n} e^{-q^{\mathbb{Q}}_{i_n} t_n}}{q_{i_n} e^{-q_{i_n} t_n}}\right) \frac{e^{-q^{\mathbb{Q}}_{i_N} \bar{t}_n}}{e^{-q_{i_N} \bar{t}_n}};$$

the last factor differs from the terms 0 up to $N-1$, as the only information used is that the level u is reached before the exponential clock (with rate $q^{\mathbb{Q}}_{i_N}$) expires.

(iii) The third contribution reflects the jumps at transition epochs of the modulating Markov chain:

$$L_3 = \prod_{n=1}^{N} \frac{e^{-\omega v_n}}{b_{i_{n-1},i_n}(-\omega)}.$$

(iv) Finally there is the contribution due to the increments of the Lévy processes: with $\bar{w}_N$ the increment of the Lévy process $X^{(i_N)}$ between $\sum_{n=0}^{N-1} t_n$ and the stopping time $\sigma(u)$,

$$L_4 = \left(\prod_{n=0}^{N-1} e^{-\omega w_n + \varphi_{i_n}(-\omega)t_n}\right) e^{-\omega \bar{w}_N + \varphi_{i_N}(-\omega)\bar{t}_N}.$$

It is elementary to verify that the product of these four factors equals

$$L = \frac{h_{i_N}(\omega)}{h_{i_0}(\omega)} \exp\left(-\omega\left(\sum_{n=1}^{N} v_n + \sum_{n=0}^{N-1} w_n + \bar{w}_N\right)\right).$$

Now realize that, due to the very definition of $\sigma(u)$,

$$\sum_{n=1}^{N} v_n + \sum_{n=0}^{N-1} w_n + \bar{w}_N > u.$$

We have now proved, with $H_i := \max_{j=1,\dots,d} h_j(\omega)/h_i(\omega) < \infty$, the following result, which can be viewed as the counterpart of Cor. 8.1.

Proposition 11.4 *Let $X \in \mathscr{L}^{\rm MAP}$. For any $u > 0$, and $i \in \{1, \dots, d\}$,*

$$q_i(u) \le H_i e^{-\omega u}.$$

It is even possible to derive the exact asymptotics of $q_i(u)$ using the standard change-of-measure relation $q_i(u) = \mathbb{E}_{\mathbb{Q}} L$:

$$q_i(u) = \frac{1}{h_i(\omega)} \mathbb{E}_{\mathbb{Q}}\left(h_{J_{\sigma(u)}}\, e^{-\omega X_{\sigma(u)}}\right);$$

this is the counterpart of identity (8.3). It can be verified [19, Thm. XIII.8.3] that under $\mathbb{Q}$ the overshoot $R_u := X_{\sigma(u)} - u$ has a proper limiting distribution as $u \to \infty$, say R, and so has $J_{\sigma(u)}$, say J. It leads to the following statement.

Proposition 11.5 *Let $X \in \mathscr{L}^{\rm MAP}$. For any $u > 0$,*

$$q_i(u) \sim \frac{1}{h_i(\omega)}\, e^{-\omega u} \mathbb{E}_{\mathbb{Q}}\left(h_J\, e^{-\omega R}\right).$$

As mentioned earlier, the literature on light-tailed MAPs is vast. Several special cases have been dealt with in great detail. The most prominent among these is the case of the *Markov fluid* input model. In this model the Lévy processes $X^{(1)}, \dots, X^{(d)}$ correspond to deterministic drifts (with rates, say, $r_1, \dots, r_d$), and the jumps U_{ij} are absent; this model was predominantly motivated by applications in communication networks. Under the obvious stability constraint $\sum_{i=1}^{d} \pi_i r_i < 0$, techniques have been developed to compute the stationary workload $\mathbb{P}(Q > u)$; the computational effort needed amounts to solving a d-dimensional eigensystem, and in addition a system of linear equations should be solved to identify a set of unknown coefficients.

In the first contributions (see Anick et al. [10] and Kosten [140]), specific attention was paid to the situation that the Markov fluid source corresponds to the superposition of multiple (stochastically identical) two-state Markov fluids; the idea is that this models the situation of multiple users feeding traffic into a network element. In later contributions (see Kesidis et al. [132] and Mitra [165]), a substantial amount of attention has been paid to the asymptotics of $\mathbb{P}(Q > u)$ for u large, which are of the form $Ce^{-\omega u}$, as we saw above. Reference [159] is an early paper on identifying the exponentially twisted version of a MAP. In e.g. [87, 99, 131, 132, 214] the relation

$$e(\vartheta) = \lim_{t\to\infty} \frac{1}{t} \log \mathbb{E} e^{\vartheta X_t}$$

is explored in great detail, giving rise to the concept of *effective bandwidth*.

Exercises

Exercise 11.1 Let X correspond to $\mathbb{B}\mathrm{m}(d, \sigma^2)$.

(a) Prove that, for $d \neq 0$,

$$W^{(0)}(x) = \frac{1}{d}\left(1 - e^{-2(d/\sigma^2)x}\right).$$

(b) Consider the case of a finite-buffer queue with buffer size $K > 0$. Determine, for u between 0 and K, $\mathbb{P}(Q < u)$ for $d \neq 0$.

(c) Prove that for $d = 0$ it holds that $\mathbb{P}(Q < u) = u/K$ for $u \in [0, K]$. Give an intuitive explanation of this fact.

Exercise 11.2 Let $(X_t)_t$ be a compound Poisson process with a *positive* drift of rate 1. The downward jumps arrive according to a Poisson process with rate λ and have an exponentially distributed size with mean μ^{-1}. Prove that, under $\lambda \neq \mu$,

$$W^{(0)}(x) = 1 + \frac{\lambda}{\mu - \lambda}\left(1 - e^{-(\mu-\lambda)x}\right).$$

Exercise 11.3 In this exercise we analyze $(\underline{Q} \mid \underline{Q} > 0)$ for $X \in \mathscr{S}_-$, where it is assumed that the initial workload Q_0 equals $x > 0$; see Section 6.4. As it turns out, the steady-state workload in the finite-buffer queue plays an important role here. Let T be an exponential random variable with mean $1/q$, independent of the driving Lévy process.

(a) Use the identity (6.8) to show that

$$\mathbb{E}_x\left(e^{-\beta Q_T} 1_{\{\underline{Q}_T > 0\}}\right) = q\left(\frac{W^{(q)}(x)}{\beta + \Psi(q)} - e^{-\beta x}\int_0^x e^{\beta y} W^{(q)}(y)\mathrm{d}y\right).$$

(b) Use Prop. 11.1 to show that

$$\lim_{t\to\infty} \mathbb{E}_x\left(e^{-\beta \underline{Q}_t} \mid \underline{Q}_t > 0\right) = \frac{\zeta(\beta)}{\zeta(0)},$$

where

$$\zeta(\beta) := \frac{1}{\beta + \Psi(0)} - \int_0^x e^{-\beta y}(1 - \pi_x(y))\mathrm{d}y.$$

Exercise 11.4 Prove that the superposition of MAPs is again a MAP.

Exercise 11.5 Let X correspond to a d-dimensional Markov fluid with $\varphi_i(\alpha) = -r_i\alpha$. Define

$$F_i(t,x) = \mathbb{P}(Q_t < x, J_t = i).$$

(a) Prove that

$$F_i(t+\Delta t, x) = F_i(t, x - r_i\,\Delta t)\left(1 - \sum_{j\neq i} q_{ij}\Delta t\right) + \sum_{j\neq i} F_j(t, x - r_{ji}\Delta t)\, q_{ji}\Delta t + o(\Delta t),$$

for $\Delta t \downarrow 0$, with r_{ji} the input rate in the interval $[t, t+\Delta t)$

(*Note*: The parameter r_{ji} will turn out to be irrelevant in (b) and (c)).

(b) Show that this entails that

$$\frac{\partial}{\partial t}F_i(t,x) + r_i\frac{\partial}{\partial x}F_i(t,x) = \sum_{j=1}^d q_{ji}F_j(t,x).$$

(c) Now we consider the stationary workload. Let $F_i(x) := \mathbb{P}(Q < x, J = i)$. Argue that under the stationarity condition $\sum_{i=1}^d \pi_i r_i < 0$ the distribution functions

$F_i(x)$ satisfy the following system of linear differential equations:

$$r_i F_i'(x) = \sum_{j=1}^{d} q_{ji} F_j(x),$$

or, in matrix notation, with $R := \text{diag}\{r_1, \ldots, r_d\}$,

$$\boldsymbol{F}'(x) = R^{-1} Q' \boldsymbol{F}(x)$$

(with Q' denoting the transpose of Q), assuming that none of the r_i's equal 0.

(d) Show that

$$\boldsymbol{F}(x) = \sum_{j=1}^{d} c_j e^{\xi_j x} \boldsymbol{v}^{(j)},$$

where the $(\xi_j, \boldsymbol{v}^{(j)})$ are the eigenvalue–eigenvector pairs of the matrix $M := R^{-1}Q'$ and c_j are constants.

(e) Argue why $\text{Re}(\xi_j) > 0$ implies that $c_j = 0$.

(f) Argue why for all i such that $r_i > 0$ we have that $F_i(0) = 0$.

(*Note*: Let the number of states with $r_i < 0$ be N_-, so that the number of states with $r_i > 0$ is $d - N_-$. It turns out that the number of j such that $\text{Re}(\xi_j) > 0$ equals N_-, so that there are as many constraints as coefficients c_j; see e.g. [67, 114, 203].)

Exercise 11.6 Let X correspond to a two-state Markov fluid, that is,

$$Q = \begin{pmatrix} -\lambda & \lambda \\ \mu & -\mu \end{pmatrix}, \quad \begin{pmatrix} \varphi_1(\alpha) \\ \varphi_2(\alpha) \end{pmatrix} = \begin{pmatrix} -r_1\alpha \\ -r_2\alpha \end{pmatrix}.$$

(a) Find the stability condition.

(b) Determine, under the stability condition, $\mathbb{P}(Q > u)$ for $u > 0$, using the preceding exercise. And what is $\mathbb{P}(Q = 0)$?

Chapter 12
Lévy-Driven Tandem Queues

In this chapter we analyze a system consisting of two *concatenated* Lévy-driven queues, a so-called *Lévy-driven tandem queue*. This model, being a natural extension of the one-node queueing system, can be regarded as a building block for more complex network architectures that will be concentrated on in Chapter 13.

We first informally describe what we mean by a Lévy-driven tandem queue. We consider a two-node system, in which the output of the first (upstream) queue is fed into the second (downstream) queue. Let the service rate at the upstream node be r_1, and at the downstream node be r_2; both rates are positive and constant. In order to avoid the downstream node becoming degenerate, it is assumed throughout that $r_2 < r_1$. We suppose that a Lévy process J_t feeds into the first queue, with $\mathbb{E}J_1 < r_2$ to ensure stability. We assume that no additional work enters the second queue. The tandem system is depicted in Fig. 12.1.

In a queue with compound Poisson input for instance, there is a logical and intuitive concept of output: traffic enters the second queue at a rate r_1 during busy periods of the first queue, and at a rate 0 during idle periods. Notice, however, that such a notion cannot always be naturally defined; for instance, this is true of the Brownian case as a consequence of the fact that the notion of busy periods and idle periods is problematic.

This conceptual issue problem can be remedied as follows. Since both the transient and the stationary scenarios follow the same reasoning, we now focus on the argument for the system in steady state (the transient case being treated in Section 12.3). Let $Q^{(1)}$ and $Q^{(2)}$ be the stationary workloads at the first and second nodes respectively, and let Q denote the *total* stationary workload present in stations 1 and 2 together. The stationary workload of the upstream queue can be defined in

K. Dębicki, M. Mandjes, *Queues and Lévy Fluctuation Theory*, Universitext,
DOI 10.1007/978-3-319-20693-6_12

Fig. 12.1 Tandem network

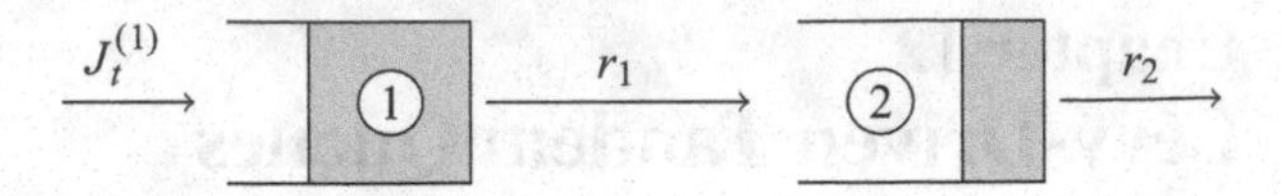

the usual way: with $X_t^{(1)} := J_t - r_1 t$, we have due to (2.5) that $Q^{(1)}$ is distributed as $\sup_{t\geq 0} X_t^{(1)}$. A crucial observation is that, in addition, the total queue behaves as a single queue fed by J_t, *but emptied at rate* r_2 [31, 94, 190]:

$$Q^{(1)} + Q^{(2)} \stackrel{\mathrm{d}}{=} \sup_{t\geq 0} X_t^{(2)},$$

where $X_t^{(2)} := J_t - r_2 t$. Then we can reconstruct $Q^{(2)}$ as the difference between the total workload and the workload in the upstream queue, so as to obtain the distributional equality

$$(Q^{(1)}, Q^{(2)}) \stackrel{\mathrm{d}}{=} \left(\sup_{t\geq 0} X_t^{(1)}, \sup_{t\geq 0} X_t^{(2)} - \sup_{t\geq 0} X_t^{(1)} \right) = (\bar{X}^{(1)}, \bar{X}^{(2)} - \bar{X}^{(1)}); \quad (12.1)$$

here we use the short notation $\bar{X}^{(i)} = \sup_{t\geq 0} X_t^{(i)}$. In line with the notation introduced in Chapter 2, we define for $J \in \mathscr{S}_+$, the Laplace exponents

$$\phi_i(\alpha) := \log \mathbb{E} e^{-\alpha X_1^{(i)}},$$

for $i = 1, 2$, and $\psi_i(\cdot) = \phi_i^{-1}(\cdot)$. Analogously, for $J \in \mathscr{S}_-$, we let

$$\Phi_i(\beta) := \log \mathbb{E} e^{\beta X_1^{(i)}},$$

for $i = 1, 2$ be the cumulants, and $\Psi_i(q) := \sup\{\beta \geq 0 : \Phi_i(\beta) = q\}$ be their right inverses.

We start this chapter by providing in Section 12.1 a useful representation for the stationary workload distribution of the downstream queue; this result enables us to find closed-form expressions for the corresponding Laplace transform in terms of the model primitives, as shown in Section 12.2. Then, in Section 12.3 we consider the transient case. Again distinguishing between light-tailed and heavy-tailed scenarios, the workload asymptotics are presented in Section 12.4. The chapter is concluded in Section 12.5 with results on the joint distribution of the stationary workloads in the upstream and downstream queue.

12.1 Representation for Stationary Downstream Workload

In this section we focus on distributional properties of the stationary downstream queue; in particular, we derive a reduction property describing the distribution of $Q^{(2)}$. To make the notation more compact, in the sequel we let, for $S \subset \mathbb{R}$,

$$\bar{X}_S^{(i)} := \sup_{t \in S} X_t^{(i)}.$$

Based on (12.1), we have that

$$\mathbb{P}(Q^{(2)} > u) = \mathbb{P}\left(\bar{X}^{(2)} - \bar{X}^{(1)} > u\right) = \mathbb{P}\left(\bar{X}^{(2)}_{[0,\infty)} - \bar{X}^{(1)}_{[0,\infty)} > u\right). \tag{12.2}$$

We note that despite this explicit formula, its direct applicability is limited, since $(X_t^{(1)})_{t\geq 0}$ and $(X_t^{(2)})_{t\geq 0}$ are highly dependent; e.g. note that $X_t^{(1)} - X_t^{(2)} = (r_2 - r_1)t$. However, under the assumption that J is a Lévy process, a compact alternative representation can be deduced, as we show in this section.

In the first place, it can be shown that we can 'shrink' the intervals over which both suprema in (12.2) are taken: rather than a difference of two suprema over $[0, \infty)$, we thus obtain the difference of two suprema over disjoint, adjacent intervals. This is done as follows.

For given $u > 0$, we define $t_u := u/(r_1 - r_2)$, to be interpreted as the minimal time needed for the second queue to exceed level u, starting empty. Let $t_i^\star := \arg\sup_{t\in[0,\infty)} X_t^{(i)}$, for $i = 1, 2$.

- We first show that $\bar{X}^{(2)}_{[0,\infty)} - \bar{X}^{(1)}_{[0,\infty)} > u$ implies $t_2^\star \geq t_u$. To show this, let us suppose that $t_2^\star < t_u$. But then we obtain a contradiction:

$$\begin{aligned}\bar{X}^{(2)}_{[0,\infty)} - \bar{X}^{(1)}_{[0,\infty)} &= \sup_{t\in[0,t_u)} (J_t - r_2 t) - \sup_{s\geq 0}(J_s - r_1 s) \\ &\leq \sup_{t\in[0,t_u)} ((J_t - r_2 t) - (J_t - r_1 t)) = u.\end{aligned}$$

  As a consequence, we replace $\bar{X}^{(2)}_{[0,\infty)}$ by $\bar{X}^{(2)}_{[t_u,\infty)}$.
- Notice that

$$\{t_1^\star > t_u\} \subseteq \left\{\bar{X}^{(2)}_{[t_u,\infty)} - \bar{X}^{(1)}_{[0,\infty)} > u\right\} \subseteq \left\{\bar{X}^{(2)}_{[t_u,\infty)} - \bar{X}^{(1)}_{[0,t_u]} > u\right\}, \tag{12.3}$$

  where the second inclusion is trivial, and the first inclusion an immediate consequence of the fact that, given that $t_1^\star > t_u$,

$$\bar{X}^{(2)}_{[t_u,\infty)} - \bar{X}^{(1)}_{[0,\infty)} \geq (J_{t_1^\star} - r_2 t_1^\star) - (J_{t_1^\star} - r_1 t_1^\star) > u.$$

- The first inclusion in (12.3) implies that

$$\mathbb{P}\left(\bar{X}^{(2)}_{[t_u,\infty)} - \bar{X}^{(1)}_{[0,\infty)} > u;\ t_1^\star > t_u\right) = \mathbb{P}\left(t_1^\star > t_u\right).$$

Due to the second inclusion in (12.3),

$$\mathbb{P}\left(\bar{X}^{(2)}_{[t_u,\infty)} - \bar{X}^{(1)}_{[0,\infty)} > u;\ t_1^\star > t_u\right) \le \mathbb{P}\left(\bar{X}^{(2)}_{[t_u,\infty)} - \bar{X}^{(1)}_{[0,t_u]} > u;\ t_1^\star > t_u\right) \le \mathbb{P}\left(t_1^\star > t_u\right).$$

Combining the previous two displays yields

$$\mathbb{P}\left(\bar{X}^{(2)}_{[t_u,\infty)} - \bar{X}^{(1)}_{[0,\infty)} > u;\ t_1^\star > t_u\right) = \mathbb{P}\left(\bar{X}^{(2)}_{[t_u,\infty)} - \bar{X}^{(1)}_{[0,t_u]} > u;\ t_1^\star > t_u\right). \tag{12.4}$$

- From (12.4) and the trivial relation

$$\mathbb{P}\left(\bar{X}^{(2)}_{[t_u,\infty)} - \bar{X}^{(1)}_{[0,\infty)} > u;\ t_1^\star \le t_u\right) = \mathbb{P}\left(\bar{X}^{(2)}_{[t_u,\infty)} - \bar{X}^{(1)}_{[0,t_u]} > u;\ t_1^\star \le t_u\right)$$

we find

$$\begin{aligned}
\mathbb{P}\left(Q^{(2)} > u\right) &= \mathbb{P}\left(\bar{X}^{(2)}_{[t_u,\infty)} - \bar{X}^{(1)}_{[0,\infty)} > u\right)\\
&= \mathbb{P}\left(\bar{X}^{(2)}_{[t_u,\infty)} - \bar{X}^{(1)}_{[0,\infty)} > u;\ t_1^\star \le t_u\right) + \mathbb{P}\left(\bar{X}^{(2)}_{[t_u,\infty)} - \bar{X}^{(1)}_{[0,\infty)} > u;\ t_1^\star > t_u\right)\\
&= \mathbb{P}\left(\bar{X}^{(2)}_{[t_u,\infty)} - \bar{X}^{(1)}_{[0,t_u]} > u;\ t_1^\star \le t_u\right) + \mathbb{P}\left(\bar{X}^{(2)}_{[t_u,\infty)} - \bar{X}^{(1)}_{[0,t_u]} > u;\ t_1^\star > t_u\right)\\
&= \mathbb{P}\left(\bar{X}^{(2)}_{[t_u,\infty)} - \bar{X}^{(1)}_{[0,t_u]} > u\right).
\end{aligned} \tag{12.5}$$

Realize that the reduction property (12.5) holds irrespective of the underlying process J being Lévy or not. Assuming J is Lévy, the probability $\mathbb{P}(Q^{(2)} > u)$ can be simplified even further. Recalling that $X^{(1)}_{t_u} - X^{(2)}_{t_u} = -u$, we have

$$\bar{X}^{(2)}_{[t_u,\infty)} - \bar{X}^{(1)}_{[0,t_u]} = \left(\bar{X}^{(2)}_{[t_u,\infty)} - X^{(2)}_{t_u}\right) - \left(\bar{X}^{(1)}_{[0,t_u]} - X^{(1)}_{t_u}\right) + u.$$

In view of the stationarity and independence of the increments of J,

$$\bar{X}^{(2)}_{[t_u,\infty)} - X^{(2)}_{t_u} \overset{\mathrm{d}}{=} \sup_{t\in[0,\infty)} X^{(2)}_t$$

is independent of

$$\bar{X}^{(1)}_{[0,t_u]} - X^{(1)}_{t_u} \overset{\mathrm{d}}{=} \sup_{t\in[0,t_u]} -X^{(1)}_t.$$

This leads to the following representation, which is originally due to Dębicki et al. [76].

Theorem 12.1 *For each* $u > 0$, *and* $(\check{X}_t^{(1)})_{t\geq 0}$, $(\check{X}_t^{(2)})_{t\geq 0}$ *denoting independent copies of* $(X_t^{(1)})_{t\geq 0}$, $(X_t^{(2)})_{t\geq 0}$ *respectively,*

$$\mathbb{P}(Q^{(2)} > u) = \mathbb{P}\left(\sup_{t\in[0,\infty)} \check{X}_t^{(2)} > \sup_{t\in[0,t_u]} -\check{X}_t^{(1)} \right).$$

12.2 Steady-State Workload of the Downstream Queue

As we demonstrate now, direct application of Thm. 12.1 to the class of spectrally one-sided input processes yields an expression for the Laplace–Stieltjes transform $\mathbb{E}e^{-\alpha Q^{(2)}}$. In the case that J is spectrally positive, we in addition obtain a representation in the spirit of (3.2). We also briefly comment on the case that the input process is not necessarily spectrally one sided.

We first derive a result that holds for any Lévy process J, that is, at the moment it is not yet required that X be spectrally one sided. Define $\tau^{(1)}(x) := \inf\{t \geq 0 : X_t^{(1)} \leq -x\}$. Then, for each $x \geq 0$, using the notation of Thm. 12.1,

$$\mathbb{P}\left(\sup_{t\in[0,t_u]} -X_t^{(1)} < x \right) = \mathbb{P}\left(\tau^{(1)}(x) > t_u\right).$$

Obviously, $\sup_{t\in[0,\infty)} X_t^{(2)} \stackrel{\mathrm{d}}{=} Q$, as we saw above; recall that Q denotes the total workload. Application of the above to Thm. 12.1 leads, after a few elementary steps, to

$$\begin{aligned}\int_0^\infty e^{-\alpha u}\mathbb{P}(Q^{(2)} > u)\mathrm{d}u &= \int_0^\infty e^{-\alpha u} \int_0^\infty \mathbb{P}\left(\tau^{(1)}(x) > t_u\right)\mathbb{P}(Q \in \mathrm{d}x)\mathrm{d}u \\ &= (r_1 - r_2) \int_0^\infty \int_0^\infty e^{-\alpha(r_1-r_2)v}\mathbb{P}\left(\tau^{(1)}(x) > v\right)\mathrm{d}v\, \mathbb{P}(Q \in \mathrm{d}x) \\ &= \frac{1}{\alpha}\left(1 - \int_0^\infty \int_0^\infty e^{-\alpha(r_1-r_2)v}\mathbb{P}\left(\tau^{(1)}(x) \in \mathrm{d}v\right)\mathbb{P}(Q \in \mathrm{d}x)\right) \\ &= \frac{1}{\alpha}\left(1 - \mathbb{E}e^{-\alpha(r_1-r_2)\tau^{(1)}(Q)}\right).\end{aligned}$$

As a consequence we obtain the following representation for the steady-state workload in the downstream queue.

Theorem 12.2 *For $\alpha \geq 0$,*

$$\mathbb{E}e^{-\alpha Q^{(2)}} = \mathbb{E}e^{-\alpha(r_1-r_2)\tau^{(1)}(Q)}.$$

It turns out that for one-sided Lévy inputs one can express $\mathbb{E}e^{-\alpha Q^{(2)}}$ more explicitly. To this end, in the rest of this section we tacitly assume that

$$\tau^{(1)}(x) \stackrel{\mathrm{d}}{=} \inf\{t \geq 0 : X_t^{(1)} < -x\}; \tag{12.6}$$

cf. $\tau(x)$ as introduced in Chapter 2. Assumption (12.6) is satisfied for all $J \in \mathscr{S}_+$, since 0 is *regular* for $(0,\infty)$ for the class $\mathscr{S}_+$; this is easily checked from the definition of regularity [146, Def. 6.4]. The case $J \in \mathscr{S}_-$ is more subtle, although (12.6) still holds for a wide class of spectrally negative Lévy processes, including α-stable Lévy motions. We refer e.g. to [60] or [161] for explicit criteria under which (12.6) holds.

Spectrally positive case—Recall that in the spectrally positive case we have that, by combining (12.6) with Lemma 6.2, $\mathbb{E}e^{-\vartheta\tau^{(1)}(x)} = e^{-x\psi_1(\vartheta)}$. As a consequence we obtain

$$\mathbb{E}e^{-\alpha Q^{(2)}} = \mathbb{E}e^{-\psi_1(\alpha(r_1-r_2))Q}, \tag{12.7}$$

which, in view of Thm. 3.2, gives the following result.

Theorem 12.3 *Let $J \in \mathscr{S}_+$. For $\alpha \geq 0$,*

$$\mathbb{E}e^{-\alpha Q^{(2)}} = \frac{-\mathbb{E}X_1^{(2)}}{r_1 - r_2}\frac{\psi_1(\alpha(r_1-r_2))}{\alpha - \psi_1(\alpha(r_1-r_2))}.$$

Now define $\bar{\tau}^{(1)}(x) := (r_1-r_2)\tau^{(1)}(x)$. It follows from Lemma 6.2 that the process $(\bar{\tau}^{(1)}(x))_{x\geq 0}$ is an *increasing* Lévy process with $\mathbb{E}e^{-\vartheta\bar{\tau}^{(1)}(x)} = e^{-x\xi(\vartheta)}$, where $\xi(\vartheta) := \psi_1((r_1-r_2)\vartheta)$. Thm. 12.3 can be written in the form [76]

$$\mathbb{E}e^{-\alpha Q^{(2)}} = (1-\varrho)\sum_{i=1}^{\infty}\varrho^{i-1}\left(\ell_H(\alpha)\right)^i,$$

where $H(\cdot)$ is a distribution function such that $H(x) = 0$ for $x < 0$ and

$$\ell_H(\alpha) := \int_0^\infty e^{-\alpha v}\mathrm{d}H(v) = \frac{\xi(\alpha)}{\varrho\alpha}, \quad \text{with} \quad \varrho := \lim_{\alpha\downarrow 0}\frac{\xi(\alpha)}{\alpha} = \frac{r_1-r_2}{-\mathbb{E}X_1^{(1)}};$$

cf. [209, Eqn. (23)]. As a result, we get the following counterpart of (3.2) for the downstream queue.

Proposition 12.1 *Let $J \in \mathscr{S}_+$. For $u \geq 0$,*

$$\mathbb{P}(Q^{(2)} \leq u) = (1-\varrho)\sum_{i=1}^{\infty} \varrho^{i-1} H^{\star i}(u).$$

Remark 12.1 The distribution $H(\cdot)$ has a natural representation in the language of the Lévy measure associated with $(\bar{\tau}^{(1)}(x))_{x\geq 0}$. As it is an increasing process, there is no Brownian term. In other words, we can let $(d, 0, \Pi)$ be the characteristic triplet corresponding to this Lévy process, so that

$$\xi(\alpha) = \alpha d + \alpha \int_0^{\infty} e^{-\alpha x} \bar{\Pi}(x)\mathrm{d}x,$$

where $\bar{\Pi}(x) := \Pi((x,\infty))$ is the tail of the Lévy measure and

$$H(t) = \frac{d}{\varrho} + \frac{1}{\varrho}\int_0^t \bar{\Pi}(x)\mathrm{d}x$$

for all $t \geq 0$. In addition, $\varrho = d + \int_0^{\infty} \bar{\Pi}(x)\mathrm{d}x$. ◇

Following [76], Thm. 12.1 enables us to find the exact distribution function of the downstream workload for several specific input processes.

Example 12.1 Suppose J corresponds to $\mathbb{B}\mathrm{m}(0,1)$. Then the density function of $\sup_{t\in[0,t_u]} -X_t^{(1)}$ equals

$$\frac{\mathrm{d}}{\mathrm{d}x}\mathbb{P}\left(\sup_{t\in[0,t_u]} -X_t^{(1)} \leq x\right) = \sqrt{\frac{2}{\pi t_u}}\exp\left(-\frac{(x-r_1 t_u)^2}{2t_u}\right) - 2r_1 e^{2r_1 x}\left(1 - \Phi_{\mathrm{N}}\left(\frac{x + r_1 t_u}{\sqrt{t_u}}\right)\right);$$

see e.g. [35]. Combining this with Example 3.1 and Thm. 12.1 yields, after standard calculus, for $u \geq 0$,

$$\mathbb{P}(Q^{(2)} > u) = \frac{r_1 - 2r_2}{r_1 - r_2} e^{-2r_2 u}\Phi_{\mathrm{N}}\left(\frac{r_1 - 2r_2}{\sqrt{r_1 - r_2}}\sqrt{u}\right) + \frac{r_1}{r_1 - r_2}\left(1 - \Phi_{\mathrm{N}}\left(\frac{r_1}{\sqrt{r_1 - r_2}}\sqrt{u}\right)\right), \tag{12.8}$$

with, $\Phi_{\mathrm{N}}(\cdot)$, as before, the distribution function of a standard normal random variable. ◇

A similar argument also works for the case of $J \in \mathbb{C}\mathrm{P}(0,\lambda,b(\cdot))$ (so that $X^{(i)} \in \mathbb{C}\mathrm{P}(r_i,\lambda,b(\cdot))$ for $i = 1,2$). However, in this case, $\mathbb{P}(Q^{(2)} > u)$ is expressed in terms of a convolution involving Q and $\sup_{t\in[0,t_u]} -X_t^{(1)}$, and for the corresponding distribution functions only series representations are available; see [76].

Spectrally negative case—Using the same line of reasoning for $X \in \mathscr{S}_-$ as for $X \in \mathscr{S}_+$, and using Thm. 3.3, we obtain

$$\mathbb{E}e^{-\beta Q^{(2)}} = \beta_0 \int_0^\infty \mathbb{E}e^{-\beta(r_1-r_2)\tau^{(1)}(x)} e^{-\beta_0 x} \mathrm{d}x,$$

where $\beta_0 = \Psi_2(0) > 0$ solves $\Phi_2(\beta_0) = 0$, recalling that $\Psi_i(\cdot)$ denotes the right inverse of $\Phi_i(\cdot)$. Now invoking Lemma 6.3, with $\gamma := \beta(r_1 - r_2)$ we obtain

$$\mathbb{E}e^{-\beta Q^{(2)}} = \beta_0 \cdot \frac{1}{\beta_0}\left(1 - \frac{\gamma}{\Psi_1(\gamma)} \cdot \frac{\Psi_1(\gamma) - \beta_0}{\gamma - \Phi_1(\beta_0)}\right);$$

recall that we assume (12.6). We thus find

$$\mathbb{E}e^{-\beta Q^{(2)}} = \frac{\beta_0 \beta(r_1 - r_2) - \Psi_1(\beta(r_1 - r_2))\Phi_1(\beta_0)}{\Psi_1(\beta(r_1 - r_2))(\beta(r_1 - r_2) - \Phi_1(\beta_0))}.$$

Realizing that $\Phi_1(\beta_0) = \Phi_2(\beta_0) - (r_1 - r_2)\beta_0 = -(r_1 - r_2)\beta_0$, this eventually leads to the following result.

Theorem 12.4 *Let $J \in \mathscr{S}_-$. For $\beta \geq 0$,*

$$\mathbb{E}e^{-\beta Q^{(2)}} = \frac{\Psi_1(\beta(r_1 - r_2)) + \beta}{\Psi_1(\beta(r_1 - r_2))} \frac{\beta_0}{\beta_0 + \beta}.$$

General case—The results that can be obtained using the representation found in Thm. 12.1 for the spectrally two-sided case are somewhat implicit; later, in Section 12.5 we develop an alternative technique that leads to an explicit formula for the transform of the steady-state workload in the downstream queue in terms of the Wiener–Hopf factors.

Here we use the representation of Thm. 12.1, and propose the following approach. Let the random variable V be distributed as the all-time maximum $\bar{X}^{(2)} = \sup_{t\in[0,\infty)} X_t^{(2)}$, and let

$$W(u) \stackrel{\mathrm{d}}{=} \sup_{t\in[0,\gamma u]} -X_t^{(1)},$$

independently of V, with $\gamma := (r_1 - r_2)^{-1}$. Then, appealing to Thm. 12.1, we have that $\mathbb{P}(Q^{(2)} > u) = \mathbb{P}(V - W(u) > 0)$. Observe now that this means that if we are able to compute

$$F(\alpha, u) := \mathbb{E}e^{\alpha i(V - W(u))} = \mathbb{E}e^{\alpha i V}\, \mathbb{E}e^{-\alpha i W(u)},$$

then we can obtain $\mathbb{P}(Q^{(2)} > u)$ by performing numerical Fourier inversion. Unfortunately, $F(\alpha, u)$ cannot be determined in closed form, but as we point out below, we *can* compute the double transform

$$\hat{F}(\alpha, \beta) := \int_0^\infty e^{-\beta u} \mathbb{E} e^{\alpha i V} \, \mathbb{E} e^{-\alpha i W(u)} \mathrm{d}u = \mathbb{E} e^{\alpha i V} \int_0^\infty e^{-\beta u} \mathbb{E} e^{-\alpha i W(u)} \mathrm{d}u,$$

from which (obviously) $\mathbb{P}(Q^{(2)} > u)$ can be determined by performing a *double* numerical inversion.

We now evaluate $\hat{F}(\alpha, \beta)$. Using Thm. 3.4, we can express $\mathbb{E} e^{\alpha i V} = \mathbb{E} e^{\alpha i \bar{X}^{(2)}}$ in terms of the Wiener–Hopf functions. In addition,

$$\begin{aligned}
\int_0^\infty e^{-\beta u} \mathbb{E} e^{-\alpha i W(u)} \mathrm{d}u &= \frac{1}{\beta} \int_0^\infty \frac{\beta}{\gamma} e^{-(\beta/\gamma)v} \, \mathbb{E} \exp\left(\alpha i \inf_{t \in [0,v]} X_t^{(1)} \right) \mathrm{d}v \\
&= \frac{1}{\beta} \int_0^\infty \frac{\beta}{\gamma} e^{-(\beta/\gamma)v} \, \mathbb{E} \exp\left(\alpha i \left(X_v^{(1)} - \sup_{t \in [0,v]} X_t^{(1)} \right) \right) \mathrm{d}v \\
&= \frac{1}{\beta} \mathbb{E} \exp\left(\alpha i \left(X_T^{(1)} - \sup_{t \in [0,T]} X_t^{(1)} \right) \right),
\end{aligned}$$

with T exponentially distributed with mean γ/β (independent of the driving Lévy process). Again using Thm. 3.4, this quantity can be expressed in terms of Wiener–Hopf functions.

12.3 Transient Downstream Workload

In this section we study the transient workload of the downstream queue in our tandem system. The goal is to analyze $Q_t^{(2)}$, assuming that at time 0 the system starts off empty (i.e. $Q_0^{(1)} = Q_0^{(2)} = 0$). Again we primarily focus on spectrally one-sided cases.

Let $Q_t^{(1)}$ and $Q_t^{(2)}$ denote, respectively, the up- and downstream workloads of the tandem network at time $t > 0$, provided that $(Q_0^{(1)}, Q_0^{(2)}) = (0, 0)$. By

$$Q_t := Q_t^{(1)} + Q_t^{(2)}$$

we denote the total workload in the system at time $t > 0$. Then, repeating the argument that was used for the stationary case (cf. Section 12.1), and using that

$$\left(Q_t^{(1)}, Q_t \right) \stackrel{\mathrm{d}}{=} \left(\sup_{s \in [0,t]} X_s^{(1)}, \sup_{s \in [0,t]} X_s^{(2)} \right),$$

we have the following counterpart of Thm. 12.1; see also Dębicki et al. [75].

Theorem 12.5 *For each $u > 0$, $t > t_u := u/(r_1 - r_2)$, and $(\check{X}_t^{(1)})_{t\geq 0}$, $(\check{X}_t^{(2)})_{t\geq 0}$ being independent copies of $(X_t^{(1)})_{t\geq 0}$, $(X_t^{(2)})_{t\geq 0}$ respectively,*

$$\mathbb{P}\left(Q_t^{(2)} > u\right) = \mathbb{P}\left(\sup_{s\in[0,t-t_u]} \check{X}_s^{(2)} > \sup_{s\in[0,t_u]} -\check{X}_s^{(1)}\right).$$

The above representation allows us to characterize the transient downstream workload in terms of the double transform $\mathbb{E}e^{-\alpha Q_T^{(2)}}$, where T is exponentially distributed with mean $1/\vartheta > 0$ and $\alpha > 0$. This approach is consistent with the analysis of the transient workload of the single queue, as given in Chapter 4.

Indeed, following the line of reasoning of [75], it is convenient to first write

$$\mathbb{E}e^{-\alpha Q_T^{(2)}} = 1 - \alpha\vartheta \int_0^\infty \int_0^\infty e^{-\alpha u - \vartheta t}\mathbb{P}\left(Q_t^{(2)} > u\right) \mathrm{d}u\,\mathrm{d}t.$$

Now observe that by Thm. 12.5,

$$\begin{aligned}
&\int_0^\infty \int_0^\infty e^{-\alpha u-\vartheta t}\mathbb{P}\left(Q_t^{(2)} > u\right) \mathrm{d}t\,\mathrm{d}u = \\
&= \int_0^\infty \int_{t_u}^\infty e^{-\alpha u-\vartheta t}\mathbb{P}\left(\sup_{s\in[0,t-t_u]} \check{X}_s^{(2)} > \sup_{s\in[0,t_u]} -\check{X}_s^{(1)}\right) \mathrm{d}t\,\mathrm{d}u \\
&= \int_0^\infty \int_0^\infty e^{-\vartheta w} e^{-u(\alpha+\vartheta/(r_1-r_2))} \\
&\qquad \int_0^\infty \mathbb{P}\left(\tau^{(1)}(z) > t_u\right)\mathbb{P}(Q_w \in \mathrm{d}z)\mathrm{d}w\,\mathrm{d}u, \qquad (12.9)
\end{aligned}$$

where we applied a change of variables (i.e. $w := t - t_u$). Again changing variables (now $v := u/(r_1 - r_2)$), the expression (12.9) turns out to equal

$$\begin{aligned}
&\frac{r_1 - r_2}{\vartheta + \alpha(r_1 - r_2)} \\
&\quad\times \int_0^\infty e^{-\vartheta w}\left(1 - \int_0^\infty \int_0^\infty e^{-v(\vartheta+\alpha(r_1-r_2))}\mathbb{P}\left(\tau^{(1)}(z) \in \mathrm{d}v\right)\mathbb{P}(Q_w \in \mathrm{d}z)\right)\mathrm{d}w \\
&= \frac{r_1 - r_2}{\vartheta + \alpha(r_1 - r_2)}\frac{1}{\vartheta}\left(1 - \mathbb{E}e^{-(\vartheta+\alpha(r_1-r_2))\tau^{(1)}(Q_T)}\right).
\end{aligned}$$

This leads to the following counterpart of (12.7), which was originally derived in [75].

Theorem 12.6 *Let J be a general Lévy process, and let T be exponentially distributed with mean $1/\vartheta$, independently of J. For $\alpha \geq 0$,*

$$\mathbb{E}e^{-\alpha Q_T^{(2)}} = \frac{\vartheta}{\vartheta + \alpha(r_1 - r_2)} + \frac{\alpha(r_1 - r_2)}{\vartheta + \alpha(r_1 - r_2)} \mathbb{E}e^{-(\vartheta + \alpha(r_1 - r_2))\tau^{(1)}(Q_T)}.$$

Similarly to the stationary case considered in Section 12.2, we can now derive more explicit formulas for $\mathbb{E}e^{-\alpha Q_T^{(2)}}$ in the case that the Lévy input process J is spectrally one sided. For this, in what follows, we again assume (12.6).

Spectrally positive case—Assume that $J \in \mathscr{S}_+$. Following Thm. 12.6 it suffices to focus on $\mathbb{E}e^{-(\vartheta + \alpha(r_1 - r_2))\tau^{(1)}(Q_T)}$, which by Lemma 6.2, equals

$$\mathbb{E}e^{-\psi_1(\vartheta + \alpha(r_1 - r_2))Q_T}.$$

Now, following Thm. 4.1 (recall that we assume $x = 0$), the above equals

$$\frac{\psi_2(\vartheta) - \psi_1(\vartheta + \alpha(r_1 - r_2))}{(r_1 - r_2)(\psi_1(\vartheta + \alpha(r_1 - r_2)) - \alpha)} \frac{\vartheta}{\psi_2(\vartheta)},$$

which straightforwardly leads to the following result [75].

Theorem 12.7 *Let $J \in \mathscr{S}_+$, and let T be exponentially distributed with mean $1/\vartheta$, independently of J. For $\alpha \geq 0$,*

$$\mathbb{E}e^{-\alpha Q_T^{(2)}} = \frac{\vartheta}{\vartheta + \alpha(r_1 - r_2)} \frac{\psi_1(\vartheta + \alpha(r_1 - r_2))}{\psi_2(\vartheta)} \frac{\psi_2(\vartheta) - \alpha}{\psi_1(\vartheta + \alpha(r_1 - r_2)) - \alpha}.$$

Spectrally negative case—Now consider $J \in \mathscr{S}_-$; let T be exponentially distributed with mean $1/q$, independently of J. Observe that Q_T is exponentially distributed with mean $1/\Psi_2(q)$, as follows from the theory of Chapter 4. Hence, relying on Thm. 12.6,

$$\mathbb{E}e^{-\beta Q_T^{(2)}} = \frac{q}{q + \beta(r_1 - r_2)} + \frac{\beta(r_1 - r_2)}{q + \beta(r_1 - r_2)} \int_0^\infty \Psi_2(q) e^{-\Psi_2(q)s} \mathbb{E}e^{-(q + \beta(r_1 - r_2))\tau^{(1)}(s)} \mathrm{d}s. \qquad (12.10)$$

On the other hand, Lemma 6.1 implies that, for $\beta > 0$, $q \geq 0$,

$$\int_0^\infty e^{-\beta x} \mathbb{E}e^{-q\tau^{(1)}(x)} \mathrm{d}x = \frac{1}{\beta}\left(1 - \frac{q}{\Psi_1(q)} \frac{\Psi_1(q) - \beta}{q - \Phi_1(\beta)}\right).$$

Inserting the above into (12.10), we obtain the following result [75].

Theorem 12.8 *Let $J \in \mathscr{S}_-$, and let T be exponentially distributed with mean $1/q$, independently of J. For $\beta \geq 0$,*

$$\mathbb{E}e^{-\beta Q_T^{(2)}} = 1 - \frac{\beta(r_1 - r_2)}{\Psi_1(q + \beta(r_1 - r_2))} \frac{\Psi_1(q + \beta(r_1 - r_2)) - \Psi_2(q)}{q + \beta(r_1 - r_2) - \Phi_1(\Psi_2(q))}.$$

12.4 Stationary Downstream Workload Asymptotics

The objective of this section is to characterize $\mathbb{P}(Q^{(2)} > u)$ as $u \to \infty$. We analyze two regimes: light- and heavy-tailed input (leaving out the intermediate regime).

Light-tailed regime—To get a feel for the general form of the asymptotics, we start by focusing on the special case of J corresponding to $\mathbb{B}\mathrm{m}(0, 1)$. The more general case of $J \in \mathscr{L}$ is considered later.
Suppose $J \in \mathbb{B}\mathrm{m}(0, 1)$ and $r_1 > r_2$. Then, after some lengthy but standard calculus, formula (12.8) leads to the following asymptotics, as $u \to \infty$:

(i) if $r_1 > 2r_2$, then

$$\mathbb{P}\left(Q^{(2)} > u\right) e^{2r_2 u} \to \frac{r_1 - 2r_2}{r_1 - r_2};$$

(ii) if $r_1 = 2r_2$, then

$$\mathbb{P}\left(Q^{(2)} > u\right) \sqrt{u} e^{2r_2 u} \to \frac{1}{\sqrt{2\pi r_2}};$$

(iii) if $r_1 < 2r_2$, then

$$\mathbb{P}\left(Q^{(2)} > u\right) \left(\frac{u}{r_1 - r_2}\right)^{3/2} \exp\left(\frac{r_1^2}{2(r_1 - r_2)} u\right) \to \frac{1}{\sqrt{2\pi}} \frac{4r_2}{r_1^2(r_1 - 2r_2)^2}.$$

One now distinguishes between two situations: with $r_1^\star := 2r_2$, there is qualitatively different behavior for $r_1 \geq r_1^\star$ and $r_1 < r_1^\star$. In the former case, that is, $r_1 \geq r_1^\star$, the most likely overflow scenario of the downstream queue is that, upon overflow, the upstream queue remains essentially empty while the downstream queue fills (roughly at a rate r_2 during $1/(2r_2)$ units of time; that is, the input rate of the tandem queue is roughly $2r_2$). Thus the asymptotics in cases (i)–(ii) have essentially the same shape as those of $\mathbb{P}(Q > u) = e^{-2r_2 u}$. In the latter case, that is, $r_1 < r_1^\star$, the most likely scenario is that J feeds into the first queue at a rate of about r_1 during t_u units of time.

The observed dichotomy extends to the more general class of light-tailed inputs. Assuming $J \in \mathscr{S}_+ \cap \mathscr{L}$, following the setup given in Lieshout and Mandjes [152], the asymptotics of $Q^{(2)}$ can be analyzed by applying the Heaviside technique of Recipe 8.1 to the Laplace–Stieltjes transform $\mathbb{E}e^{-\alpha Q^{(2)}}$. Let $\bar{t}$ be the (non-zero) root of $\varphi_1(\alpha) = (r_1 - r_2)\alpha$,

$$t_b := \frac{\inf_\alpha \varphi_1(\alpha)}{r_1 - r_2},$$

and $\bar{\alpha} := \arg\inf \varphi_1(\alpha)$. Then 'Heaviside' gives that

(i) if $\varphi_1'(\bar{t}) > 0$, then, as $u \to \infty$,

$$\mathbb{P}(Q^{(2)} > u)e^{-\bar{t}u} \to \frac{-\mathbb{E}X_1^{(2)}\varphi_1'(\bar{t})}{(r_1 - r_2)(r_1 - r_2 - \varphi_1'(\bar{t}))};$$

(ii) if $\varphi_1'(\bar{t}) = 0$, then, as $u \to \infty$,

$$\mathbb{P}(Q^{(2)} > u)\sqrt{u}e^{-\bar{t}u} \to \frac{1}{\sqrt{2\pi}} \frac{-\mathbb{E}X_1^{(2)}}{r_1 - r_2} \sqrt{\frac{\varphi_1''(\bar{t})}{r_1 - r_2}};$$

(iii) if $\varphi_1'(\bar{t}) < 0$, then, as $u \to \infty$,

$$\mathbb{P}(Q^{(2)} > u)u^{3/2}e^{-t_b u} \to \frac{1}{\sqrt{2\pi}} \frac{-\mathbb{E}X_1^{(2)}}{(t_b - \bar{\alpha})^2} \sqrt{\frac{1}{(r_1 - r_2)\varphi_1''(\bar{\alpha})}}.$$

Heavy-tailed regime—We now study the asymptotics of the workload of the downstream queue in the case $J \in \mathscr{S}_+ \cap \mathscr{R}$. Before we state the main result, we first relate these asymptotics to those of $\mathbb{P}(Q > u)$. In view of the results of Chapter 8, let us assume for the moment that

$$\mathbb{P}(Q > u) = u^{1-\nu}L(u)(1 + o(1))$$

as $u \to \infty$, where $L(\cdot)$ is slowly varying at ∞, with $\nu \in (1, 2)$. Then, by (12.7), Thm. 3.2, and 'Tauber',

$$\begin{aligned}
\mathbb{E}e^{-\alpha Q^{(2)}} - 1 &= \mathbb{E}e^{-\psi_1(\alpha(r_1 - r_2))Q} - 1 \\
&= \Gamma(2 - \nu)(\psi_1(\alpha(r_1 - r_2)))^{\nu-1} L\left(\frac{1}{\psi_1(\alpha(r_1 - r_2))}\right)(1 + o(1)) \\
&= \Gamma(2 - \nu)\varrho^{\nu-1}\alpha^{\nu-1} L\left(\frac{1}{\alpha}\right)(1 + o(1)),
\end{aligned}$$

as $\alpha \to 0$, since $\lim_{\alpha\to 0} \psi_1(\alpha(r_1 - r_2))/\alpha = \varrho$ with $\varrho = (r_1 - r_2)/(-\mathbb{E}X_1^{(1)})$. Thus, again using 'Tauber', we obtain the asymptotics, as $u \to \infty$,

$$\mathbb{P}(Q^{(2)} > u) = \varrho^{\nu-1}u^{1-\nu}L(u)(1 + o(1)).$$

The following theorem generalizes the above findings to the case of $J \in \mathscr{S}_+ \cap \mathscr{R}$ (so we do not necessarily have that $\nu \in (1,2)$, but rather that $\nu > 1$), with $\varphi_1(\alpha) \in \mathscr{R}_\nu(n,\eta)$; see [152, Thm. 4.7].

Theorem 12.9 *Let $J \in \mathscr{S}_+ \cap \mathscr{R}$, with $\varphi_1(\alpha) \in \mathscr{R}_\nu(n,\eta)$. Then, as $u \to \infty$,*

$$\mathbb{P}(Q^{(2)} > u) \sim \left(\frac{-\mathbb{E}X_1^{(1)}}{r_1 - r_2}\right)^{1-\nu} \mathbb{P}(Q > u)$$

$$\sim \frac{(-1)^{n+1}}{\Gamma(2-\nu)} \frac{\eta}{-\mathbb{E}X_1^{(2)}} \left(\frac{-\mathbb{E}X_1^{(1)}}{r_1 - r_2}\right)^{1-\nu} u^{1-\nu} L(u).$$

Example 12.2 Consider the case of $J \in \mathbb{S}(\alpha, 1, 0)$, with $\alpha \in (1,2)$. Then Prop. 8.2, in combination with Thm. 12.9, straightforwardly provides us with the asymptotics

$$\mathbb{P}(Q^{(2)} > u) \sim \frac{1}{\Gamma(2-\alpha)\cos(\pi(\alpha-2)/2)} \frac{1}{r_2} \left(\frac{r_1}{r_1 - r_2}\right)^{1-\alpha} u^{1-\alpha},$$

as $u \to \infty$. ◇

Example 12.3 Suppose $J \in \mathbb{CP}(0, \lambda, b(\cdot))$ and $\mathbb{P}(B > x) = x^{-\delta} L(x)$ with $\delta > 1$. The combination of Example 8.1 with Thm. 12.9 immediately implies

$$\mathbb{P}(Q^{(2)} > u) \sim \left(\frac{r_1 - \lambda\,\mathbb{E}B}{r_1 - r_2}\right)^{1-\delta} \mathbb{P}(Q > u)$$

$$\sim \frac{\lambda}{r_2 - \lambda\,\mathbb{E}B} \left(\frac{r_1 - \lambda\,\mathbb{E}B}{r_1 - r_2}\right)^{1-\delta} \frac{1}{\delta - 1} u^{1-\delta} L(u),$$

as $u \to \infty$. ◇

12.5 Bivariate Distribution

So far we have studied the distribution of the downstream workload. In this section we set ourselves a more ambitious goal, that is, the joint distribution of $Q^{(1)}$ and $Q^{(2)}$. It turns out that in order to derive the associated bivariate Laplace–Stieltjes transform, the notion of *splitting times* is particularly useful.

Recall that $(Q^{(1)}, Q^{(2)}) \stackrel{\mathrm{d}}{=} (\bar{X}^{(1)}, \bar{X}^{(2)} - \bar{X}^{(1)})$, so that

$$\mathbb{E}e^{-\alpha Q^{(1)} - \tilde{\alpha} Q^{(2)}} = \mathbb{E}e^{-(\alpha - \tilde{\alpha})\bar{X}^{(1)} - \tilde{\alpha}\bar{X}^{(2)}}. \tag{12.11}$$

As a consequence, in order to study the joint distribution of both stationary workloads, it suffices to characterize the joint distribution of $(\bar{X}^{(1)}, \bar{X}^{(2)})$. Also, we let $G^{(i)} := \arg\sup_{t\geq 0} X_t^{(i)}$ be the (first) epoch that $(X_t^{(i)})_{t\geq 0}$ attains its maximum, for $i = 1, 2$. Before proceeding, we note that it is convenient to distinguish between two scenarios:

- 0 is *irregular* for $(\sup_{s\in[0,t]} X_s^{(1)} - X_t^{(1)})_t$, which means that

$$R^{(1)} := \inf\left\{ t > 0 : \sup_{s\in[0,t]} X_s^{(1)} = X_t^{(1)} \right\} > 0$$

  almost surely, or $X^{(1)}$ is a compound Poisson process;
- 0 is *regular* for $(\sup_{s\in[0,t]} X_s^{(1)} - X_t^{(1)})_t$, that is, $R^{(1)} = 0$ almost surely.

Since both the cases follow the same idea and lead to the same formula, we analyze only the first scenario, referring to Dębicki et al. [71] for details on the regular case.

Trivially,

$$\mathbb{E}e^{-\alpha\bar{X}^{(1)}-\bar{\alpha}\bar{X}^{(2)}} = \mathbb{E}e^{-\alpha X_{G^{(1)}}^{(1)}-\bar{\alpha}X_{G^{(1)}}^{(2)}} e^{-\bar{\alpha}\left(\bar{X}^{(2)}-X_{G^{(1)}}^{(2)}\right)}.$$

First consider the first exponential term in the expression in the right-hand side of the previous display. Using that $X_t^{(2)} - X_t^{(1)} = (r_1 - r_2)t > 0$, we have

$$-\alpha X_{G^{(1)}}^{(1)} - \bar{\alpha}X_{G^{(1)}}^{(2)} = -(\alpha+\bar{\alpha})\bar{X}^{(1)} - \bar{\alpha}(r_1 - r_2)G^{(1)}.$$

In addition, with respect to the second exponential term, we remark that

$$\bar{X}^{(2)} - X_{G^{(1)}}^{(2)} = \sup_{t\geq G^{(1)}} X_t^{(2)} - X_{G^{(1)}}^{(2)},$$

where it is used that almost surely $G^{(2)} \geq G^{(1)}$ (why?).

Then it is a crucial step to note that $-\alpha X_{G^{(1)}}^{(1)} - \bar{\alpha}X_{G^{(1)}}^{(2)}$ and $\bar{X}^{(2)} - X_{G^{(1)}}^{(2)}$ are independent; this property is justified in e.g. Bertoin [43, Lemma VI.6], or in [71, Lemma 2.1]. These considerations straightforwardly yield

$$\mathbb{E}e^{-\alpha\bar{X}^{(1)}-\bar{\alpha}\bar{X}^{(2)}} = \mathbb{E}e^{-(\alpha+\bar{\alpha})\bar{X}^{(1)}-\bar{\alpha}(r_1-r_2)G^{(1)}}\mathbb{E}e^{-\bar{\alpha}(\bar{X}^{(2)}-X_{G^{(1)}}^{(2)})}.$$

The factor $\mathbb{E}e^{-\bar{\alpha}(\bar{X}^{(2)}-X_{G^{(1)}}^{(2)})}$ can be computed upon choosing $\alpha = 0$ in the above equality. Hence we find

$$\mathbb{E}e^{-\alpha\bar{X}^{(1)}-\bar{\alpha}\bar{X}^{(2)}} = \mathbb{E}e^{-(\alpha+\bar{\alpha})\bar{X}^{(1)}-\bar{\alpha}(r_1-r_2)G^{(1)}} \frac{\mathbb{E}e^{-\bar{\alpha}\bar{X}^{(2)}}}{\mathbb{E}e^{-\bar{\alpha}\bar{X}^{(1)}-\bar{\alpha}(r_1-r_2)G^{(1)}}}.$$

This expression can be further evaluated by using Thm. 3.4. To this end, let $k^{(i)}(\vartheta,\alpha)$ be defined as $k(\vartheta,\alpha)$ in (3.10), but with X_t replaced by $X_t^{(i)}$, for $i = 1,2$. We obtain

$$\mathbb{E}e^{-\alpha\bar{X}^{(1)}-\bar{\alpha}\bar{X}^{(2)}} = \frac{k^{(1)}(\bar{\alpha}(r_1-r_2),\alpha+\bar{\alpha})}{k^{(1)}(\bar{\alpha}(r_1-r_2),\bar{\alpha})}\frac{k^{(2)}(0,\bar{\alpha})}{k^{(2)}(0,0)}.$$

By replacing α by $\alpha-\bar{\alpha}$, we thus find the following result, relying on the relation between the all-time supremum and stationary workloads (12.11).

Theorem 12.10 *Let J be a general Lévy process. For $\alpha,\bar{\alpha}\geq 0$,*

$$\mathbb{E}e^{-\alpha Q^{(1)}-\bar{\alpha}Q^{(2)}} = \frac{k^{(1)}(\bar{\alpha}(r_1-r_2),\alpha)\,k^{(2)}(0,\bar{\alpha})}{k^{(1)}(\bar{\alpha}(r_1-r_2),\bar{\alpha})\,k^{(2)}(0,0)}.$$

This theorem yields the following transform for the stationary workload in the downstream queue:

$$\mathbb{E}e^{-\alpha Q^{(2)}} = \frac{k^{(1)}(\alpha(r_1-r_2),0)\,k^{(2)}(0,\alpha)}{k^{(1)}(\alpha(r_1-r_2),\alpha)\,k^{(2)}(0,0)},$$

or, using Thm. 3.4 in combination with the fact that $\bar{X}^{(2)}$ and $Q = Q^{(1)}+Q^{(2)}$ have the same distribution,

$$\mathbb{E}e^{-\alpha Q} = \frac{k^{(1)}(\alpha(r_1-r_2),\alpha)}{k^{(1)}(\alpha(r_1-r_2),0)}\,\mathbb{E}e^{-\alpha Q^{(2)}}.$$

Now realize, again by Thm. 3.4, that

$$\frac{k^{(1)}(\alpha(r_1-r_2),\alpha)}{k^{(1)}(\alpha(r_1-r_2),0)} = \mathbb{E}e^{-\alpha\bar{X}^{(1)}_{T_\alpha}},$$

with T_α having an exponential distribution with mean $(\alpha(r_1-r_2))^{-1}$, independently of the driving Lévy process J. We find the appealing identity, valid for general Lévy input X,

$$\mathbb{E}e^{-\alpha Q} = \mathbb{E}e^{-\alpha(Q^{(1)}+Q^{(2)})} = \mathbb{E}e^{-\alpha\bar{X}^{(1)}_{T_\alpha}}\,\mathbb{E}e^{-\alpha Q^{(2)}}.$$

Observe the connection with the representation of the stationary downstream workload identified in Thm. 12.1.

Spectrally positive case—Combining the above with the known fact that for $X^{(i)}\in\mathscr{S}_+$,

$$\mathbb{E}e^{-\alpha G^{(i)}-\bar{\alpha}\bar{X}^{(i)}} = -\mathbb{E}X_1^{(i)}\frac{\psi_i(\alpha)-\bar{\alpha}}{\alpha-\varphi_i(\bar{\alpha})}, \tag{12.12}$$

for $\alpha, \bar{\alpha} \geq 0$, $(\alpha, \bar{\alpha}) \neq (0,0)$, $\bar{\alpha} \neq \psi_i(\alpha)$, $i = 1,2$ (see e.g. [43, Thm. VII.4]) directly leads to the following result; see [71]. Plugging in $\alpha = 0$, Thm. 12.3 is recovered.

Theorem 12.11 *Let $J \in \mathcal{S}_+$. For $\alpha, \bar{\alpha} \geq 0$,*

$$\mathbb{E}e^{-\alpha Q^{(1)} - \bar{\alpha} Q^{(2)}} = \frac{-\mathbb{E}X_1^{(2)}\bar{\alpha}}{\bar{\alpha} - \psi_1(\bar{\alpha}(r_1 - r_2))} \frac{\psi_1(\bar{\alpha}(r_1 - r_2)) - \alpha}{(r_1 - r_2)\bar{\alpha} - \varphi_1(\alpha)}.$$

Spectrally negative case—The corresponding spectrally negative case can be dealt with as well; cf. Section 12.2. To this end, it is recalled that [43, Thm. VII.4] for $X \in \mathcal{S}_-$,

$$\mathbb{E}e^{-\beta G^{(i)} - \bar{\beta} \bar{X}^{(i)}} = \frac{\Psi_i(0)}{\Psi_i(\beta) + \bar{\beta}}.$$

We find the following result, in line with Thm. 12.4.

Theorem 12.12 *Let $J \in \mathcal{S}_-$. For $\beta, \bar{\beta} \geq 0$,*

$$\mathbb{E}e^{-\beta Q^{(1)} - \bar{\beta} Q^{(2)}} = \frac{\Psi_1(\bar{\beta}(r_1 - r_2)) + \bar{\beta}}{\Psi_1(\bar{\beta}(r_1 - r_2)) + \beta} \frac{\beta_0}{\beta_0 + \bar{\beta}}.$$

In Chapter 13 we show that the idea leading to Thm. 12.10 can be generalized to considerably more complex network structures, including n-node tandem networks and networks with a *tree-type* structure.

Having the formula for the bivariate transform of the workload, one may try to use it to explicitly obtain the joint distribution of the workloads in steady state. Due to the complexity of this task, it was solved in only a few special cases; see Lieshout and Mandjes [151] and Mandjes [156]. In the following proposition we consider the Brownian tandem case; see [151].

Proposition 12.2 *Let $J \in \mathbb{B}\mathrm{m}(0, 1)$. For $u, v \geq 0$,*

$$\mathbb{P}\left(Q^{(1)} > u, Q^{(2)} > v\right) =$$

$$= \frac{r_2}{r_1 - r_2}\left(1 - \Phi_{\mathrm{N}}\left(\frac{u + vr_1/(r_1 - r_2)}{\sqrt{v/(r_1 - r_2)}}\right)\right)$$

$$+ \left(1 - \Phi_{\mathrm{N}}\left(\frac{-u + vr_1/(r_1 - r_2)}{\sqrt{v/(r_1 - r_2)}}\right)\right) e^{-2r_1 u}$$

$$+ \frac{r_1 - 2r_2}{r_1 - r_2}\Phi_{\mathrm{N}}\left(\frac{-u + v(r_1 - 2r_2)/(r_1 - r_2)}{\sqrt{v/(r_1 - r_2)}}\right) e^{-2((r_1 - r_2)u + r_2 v)}.$$

We refer to [151, 152] for a full analysis of the joint buffer overflow probabilities of the type $\mathbb{P}(Q^{(1)} > Au, Q^{(2)} > (1-A)u)$ as $u \to \infty$, for a given $A \in (0,1)$, but we present here the intuition for the case $J \in \mathscr{L}$. To this end, define

$$F := \left\{(\alpha, \bar{\alpha}) \in \mathbb{R}^2 : \mathbb{E}e^{-\alpha Q^{(1)} - \bar{\alpha} Q^{(2)}} < \infty\right\},$$

and $\bar{F} := F \cap \mathbb{R}^2_-$. Then it is easy to see, as follows, that F and $\bar{F}$ are convex. Take $(\alpha_1, \bar{\alpha}_1)$ and $(\alpha_2, \bar{\alpha}_2)$ in F. Take a $\lambda \in (0,1)$. Then

$$\begin{aligned}&\mathbb{E}\exp\left(-(\lambda\alpha_1 + (1-\lambda)\alpha_2)Q^{(1)} - (\lambda\bar{\alpha}_1 + (1-\lambda)\bar{\alpha}_2)Q^{(2)}\right)\\ &\quad \leq \lambda\mathbb{E}e^{-\alpha_1 Q^{(1)} - \bar{\alpha}_1 Q^{(2)}} + (1-\lambda)\mathbb{E}e^{-\alpha_2 Q^{(1)} - \bar{\alpha}_2 Q^{(2)}} < \infty,\end{aligned}$$

due to straightforward convexity arguments. The convexity of $\bar{F}$ follows immediately.

Given that the set $\bar{F}$ is convex, due to the Chernoff bound, we have, for any $(\alpha, \bar{\alpha}) \in \bar{F}$,

$$\mathbb{P}\left(Q^{(1)} > Au, Q^{(2)} > (1-A)u\right) \leq \mathbb{E}e^{-\alpha Q^{(1)} - \bar{\alpha} Q^{(2)}} e^{A\alpha u + (1-A)\bar{\alpha} u}.$$

Now taking logs of both sides and dividing by u, we obtain

$$\limsup_{u\to\infty} \frac{1}{u} \log \mathbb{P}\left(Q^{(1)} > Au, Q^{(2)} > (1-A)u\right) \leq \inf_{(\alpha,\bar{\alpha})\in\bar{F}} (A\alpha u + (1-A)\bar{\alpha} u).$$

We conclude that finding an upper bound on the decay rate reduces to a *convex programming problem*, which has attractive numerical properties. Interestingly, in [152] it is shown that the upper bound on the decay rate identified above is actually tight, relying on sample-path large deviations [69].

This concludes the chapter on tandem queues. Several other issues concerning tandem Lévy systems, including steady-state characteristics and correlation analysis, can be found in [120, 129].

Exercises

Exercise 12.1 Let J correspond to $\mathbb{B}\text{m}(0,1)$.

(a) Check Eqn. (12.8).
(b) Verify the asymptotics of $\mathbb{P}(Q^{(2)} > u)$ as $u \to \infty$.
(c) Find $\mathbb{E}\, Q^{(2)}$ and $\mathbb{V}\text{ar}\, Q^{(2)}$.

Exercise 12.2 Compute the Laplace–Stieltjes transform of $Q^{(2)}$ for the case that J corresponds to $\mathbb{B}\mathrm{m}(0,1)$, both by using the result for spectrally positive J and spectrally negative J.

Exercise 12.3 Use 'Heaviside' to identify the asymptotics of $\mathbb{P}(Q^{(2)} > u)$ for $X \in \mathscr{S}_{-}$.

Exercise 12.4 Use 'Tauber' to verify Thm. 12.9.

Exercise 12.5 Argue that almost surely $G^{(2)} \geq G^{(1)}$.

Exercise 12.6 Let $J \in \mathscr{S}_{+}$, and let T be exponentially distributed with mean $1/\vartheta$, independently of J. Find $\mathbb{E}\, e^{-\alpha Q_T^{(2)}}$, as $\vartheta \downarrow 0$.

Exercise 12.7 Consider $J \in \mathbb{S}(\alpha, 1, 0)$, with $\alpha \in (1,2)$. Find the exact asymptotics of $\mathbb{P}(Q^{(2)} > u)$, as $u \to \infty$.

Exercise 12.8 Consider $\mathbb{P}(Q^{(1)} > Au, Q^{(2)} > (1-A)u)$ for $A \in (0,1)$ and u large. Determine the set $\bar{F}$ for J corresponding to $\mathbb{B}\mathrm{m}(0,1)$.

Chapter 13
Lévy-Driven Queueing Networks

In this chapter we consider a general class of Lévy-driven queueing networks, which can be regarded as the natural extension of the tandem networks studied in the previous chapter. First, in Section 13.1, we formally introduce these networks through a Skorokhod formulation. Restricting ourselves to the class of tree networks, the solution of the corresponding Skorokhod problem can be found explicitly (see Section 13.2), leading to the representation (in terms of transforms) for the joint distribution of the stationary workloads, ages of busy periods, and ages of idle periods (Section 13.3). Finally, we apply these findings to obtain insightful expressions for the joint Laplace–Stieltjes transform of the workloads and ages of busy periods for (i) multinode tandem networks (see Section 13.4), (ii) individual nodes within a tree network (see Section 13.5), and (iii) priority queueing models (see Section 13.6).

13.1 Definition, Multidimensional Skorokhod Problem

In this section we introduce the framework studied in this chapter. We consider a network of n infinite-buffer queues, where 'queues' are to be interpreted as reservoirs that can temporarily store workload, as previously in this monograph. Queue i is externally fed by the process $J_t^{(i)}$, where it is assumed that

$$\boldsymbol{J} := (\boldsymbol{J}_t)_t = \left((J_t^{(1)}, \ldots, J_t^{(n)})'\right)_t$$

is an n-dimensional Lévy process, with $\boldsymbol{J}_0 = 0$ and $\mathbb{E}|\boldsymbol{J}_1| < \infty$. We denote by $\boldsymbol{r}$ the vector $(r_1, \ldots, r_n)'$, where $r_i > 0$ is the output rate of queue i. The interaction between the queues is described by the so-called *routing matrix* $P = \left(p_{ij}\right)_{i,j=1,\ldots,n}$; here $p_{ij} \in [0, 1]$ is the fraction of output of station i that is immediately transferred to

K. Dębicki, M. Mandjes, *Queues and Lévy Fluctuation Theory*, Universitext,
DOI 10.1007/978-3-319-20693-6_13

station j, where a fraction $1-\sum_{j\neq i} p_{ij}$ leaves the system (after having been served by queue i). We assume that $p_{ii} = 0$ (no 'self-loops') and, for evident reasons, $\sum_{j=1}^{n} p_{ij} \leq 1$, for all $i = 1, \ldots, n$. We represent the resulting network by the triplet $(\boldsymbol{J}, \boldsymbol{r}, P)$.

Having defined the queueing network's input and routing, the next step is to point out how to construct the associated n-dimensional workload process. We do so in a way that is analogous to the procedure developed in Chapter 2 for the single node, that is, setting up a Skorokhod formulation (which is done in this section), and explicitly solving this (done in the next section).

Following Harrison and Reiman [109] (see also Robert [185]), the workload process corresponding to the 'driving triplet' $(\boldsymbol{J}, \boldsymbol{r}, P)$, in the sequel denoted by

$$\boldsymbol{Q} := (\boldsymbol{Q}_t)_{t\geq 0} = \left(Q_t^{(1)}, \ldots, Q_t^{(n)}\right)'_{t\geq 0},$$

is the solution of the following *multidimensional Skorokhod problem*:

(A$^+$) $\boldsymbol{Q}_t$ is given by $\boldsymbol{Q}_0 = \boldsymbol{x}$ and, for $t \geq 0$,

$$\boldsymbol{Q}_t = \boldsymbol{x} + \boldsymbol{J}_t - (I - P')\boldsymbol{r}t + (I - P')\boldsymbol{L}_t,$$

is non-negative for all $t \geq 0$;

(B$^+$) $\boldsymbol{L}_0 = 0$ and $\boldsymbol{L}_t$ is non-decreasing, and

$$\sum_{i=1}^{n} \int_0^T Q_t^{(i)} \, \mathrm{d}L_t^{(i)} = 0, \quad \text{for all } T > 0,$$

where $\boldsymbol{x} \geq \boldsymbol{0}$, I is the $n \times n$ identity matrix, and

$$\boldsymbol{L} = (\boldsymbol{L}_t)_{t\geq 0} = \left(L_t^{(1)}, \ldots, L_t^{(n)}\right)'_{t\geq 0}$$

is the so-called *reflecting process* or *regulator*. The reflecting process $L_t^{(i)}$ has the informal interpretation of cumulative amount (in the time interval $[0,t]$) of unused capacity at node i, for $i = 1, \ldots, n$.

It is known that a pair $(\boldsymbol{Q}, \boldsymbol{L})$ satisfying (A$^+$) and (B$^+$) exists and that it is unique; for a detailed treatment, we refer e.g. to [185].

Example 13.1 Consider, as introduced in Section 2.4, a single-node Lévy queue driven by $(X_t)_{t\geq 0}$. This system is described by the triplet $(J^{(1)}, r_1, P)$, with $P = (p_{11}) = (0)$; here $X_t = J_t^{(1)} - r_1 t$. Then Q_t solves the corresponding Skorokhod problem. It is easily seen that the conditions associated with the single-dimensional Skorokhod problem (A)–(B), as introduced in Section 2.4, are in agreement with conditions (A$^+$)–(B$^+$) above. ◇

Example 13.2 Consider the two-node Lévy tandem network analyzed in Chapter 12. This network is described by the triplet $(\boldsymbol{J}, \boldsymbol{r}, P)$, with

$$\boldsymbol{J}_t = \begin{pmatrix} J_t \\ 0 \end{pmatrix}, \quad \boldsymbol{r} = \begin{pmatrix} r_1 \\ r_2 \end{pmatrix}, \quad \text{and } P = \begin{pmatrix} 0 & 1 \\ 0 & 0 \end{pmatrix}.$$

It requires a simple verification to conclude that the bivariate workload process $\boldsymbol{Q}_t = (Q_t^{(1)}, Q_t^{(2)})'$ solves the associated Skorokhod problem. ◇

13.2 Lévy-Driven Tree Networks

Above we observed that, both for single-node queues and tandem networks, it is possible to solve the corresponding Skorokhod problem (A$^+$)–(B$^+$), leading to a representation for the workload process in terms of the driving triplet $(\boldsymbol{J}, \boldsymbol{r}, P)$. For general networks, however, it is not clear how to explicitly express the pair $(\boldsymbol{Q}, \boldsymbol{L})$ in terms of the driving triplet. An important large class for which this *is* possible is the class of so-called *tree-type* networks, to be considered in this section.

In the rest of this chapter we suppose that $(\boldsymbol{J}, \boldsymbol{r}, P)$ obeys the following properties (see [71]):

(T$_1$) P is strictly *upper triangular* (i.e. $p_{ij} = 0$ if $j \leq i$) and the jth column of P contains exactly one strictly positive element for $j = 2, \ldots, n$;
(T$_2$) the processes $(J_t^{(j)})_{t \geq 0}$ are *non-decreasing*, for $j = 2, \ldots, n$;
(T$_3$) if $p_{ij} > 0$, then $p_{ij} > r_j / r_i$.

The resulting network can be represented by a graph, with a directed vertex from node i to node j if $p_{ij} > 0$; due to (T$_1$) such a graph corresponds to a 'tree network' in which every node is fed by at most one predecessor node. Importantly, we do *not* impose the requirement that $J^{(1)}$ be non-decreasing. Observe that both the single-node and tandem Lévy-driven queues, as described in earlier sections, satisfy the three properties (T$_1$)–(T$_3$). An example of a tree network is given in Fig. 13.1.

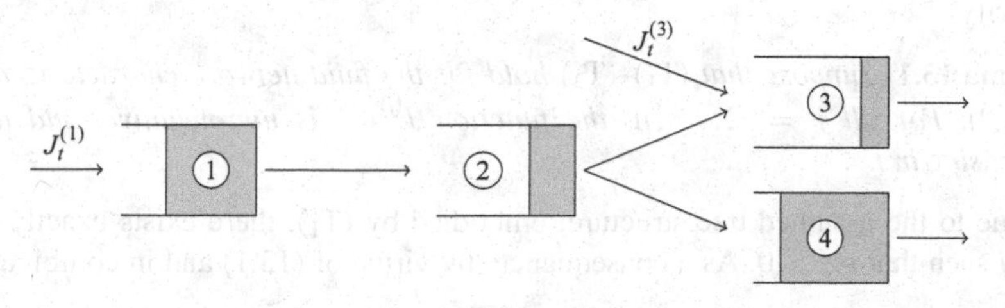

Fig. 13.1 Example of a tree network; $n = 4$. Queues 1 and 3 have external input, output of queue 1 is routed to queue 2, output of queue 2 is split and routed to queues 3 and 4

In order to find the solution of the Skorokhod problem (A^+)–(B^+) under (T_1)–(T_3), and hence to identify the explicit representation for the workload process, we analyze $\boldsymbol{Q}_t$ coordinate by coordinate.

Recalling that $\boldsymbol{Q}_0 = \boldsymbol{x}$, for each $i = 1, \dots, n$, and defining $\check{P} := P'$ throughout the remainder of this section, by (A^+)–(B^+) we have

$$Q_t^{(i)} = (\boldsymbol{x} + \boldsymbol{J}_t - (I - \check{P})\boldsymbol{r}t + (I - \check{P})\boldsymbol{L}_t)_i = x_i + J_t^{(i)} - ((I - \check{P})\boldsymbol{r}t - \check{P}\boldsymbol{L}_t)_i + L_t^{(i)}$$

and $\int_0^T Q_t^{(i)} \, \mathrm{d}L_t^{(i)} = 0$, for all $T > 0$. The above reduces, for each $i = 1, \dots, n$, to a one-dimensional Skorokhod problem (as introduced in Chapter 2), with $L_t^{\star(i)} := L_t^{(i)} + x_i$, and the driving process $J_t^{(i)} - ((I - \check{P})\boldsymbol{r})_i t - (\check{P}\boldsymbol{L}_t)_i$. As a consequence, the following set of fixed-point equations holds: with

$$\boldsymbol{J}_t(\boldsymbol{x}) := \boldsymbol{J}_t - (I - \check{P})\boldsymbol{r}t + \boldsymbol{x}$$

and $J_t^{(j)}(\boldsymbol{x})$ its jth component, we have

$$L_t^{(i)} = \max\left\{0, \sup_{s \in [0,t]} \left((\check{P}\boldsymbol{L}_s)_i - J_s^{(i)}(\boldsymbol{x}) \right) \right\}, \tag{13.1}$$

for $i = 1, \dots, n$; see e.g. Prop. 2.3.

We find the solution of (13.1) following the approach given in Dębicki et al. [71]. Under (T_1)–(T_2), iterating Eqn. (13.1) yields

$$L_t^{(i)} = \max\left\{0, \sup_{s \in [0,t]} \left(-\sum_{k=0}^{i-1} \left(\check{P}^k \boldsymbol{J}_s(\boldsymbol{x}) \right)_i \right) \right\}; \tag{13.2}$$

the summation is up to $k = i - 1$ as a consequence of the fact that P is strictly upper triangular. Indeed, Eqns. (13.1) and (13.2) are the same for $i = 1$. We now present a proof by induction: assuming that (13.2) holds true for $i = 1, \dots, j-1$ with $j = 2, \dots, n$, we shall now verify that (13.2) holds for $i = j$ as well. Before doing that, we first state a lemma; we do not include a proof, as the claim is straightforward to verify.

Lemma 13.1 *Suppose that* (T_1)–(T_3) *hold for the fluid network characterized by* $(\boldsymbol{J}, \boldsymbol{r}, P)$. *For all* $j = 2, \dots, n$, *the function* $J_t^{(j)}(\boldsymbol{x})$ *is non-negative, and non-decreasing in* t.

Due to the assumed tree structure, embodied by (T_1), there exists exactly one $\ell < j$ such that $p_{\ell j} > 0$. As a consequence, by virtue of (13.1) and in combination with the induction hypothesis,

$$L_t^{(j)} = \max\left\{0, \sup_{s\in[0,t]} \left(p_{\ell j} L_s^{(\ell)} - J_s^{(j)}(x)\right)\right\}$$

$$= \max\left\{0, \sup_{s\in[0,t]} \left(p_{\ell j} \max\left\{0, \sup_{v\in[0,s]} \left(-\sum_{k=0}^{\ell-1}\left(\check{P}^k J_v(x)\right)_\ell\right)\right\} - J_s^{(j)}(x)\right)\right\}. \tag{13.3}$$

Now consider a general function $F(\cdot)$ and a non-increasing function $G(\cdot)$; then it is obvious that

$$\max\left\{0, \sup_{s\in[0,t]} \max\left\{0, \sup_{v\in[0,s]} (F(v)+G(s))\right\}\right\} = \max\left\{0, \sup_{s\in[0,t]} \sup_{v\in[0,s]} (F(v)+G(s))\right\}$$

$$= \max\left\{0, \sup_{v\in[0,t]} \sup_{s\in[v,t]} (F(v)+G(s))\right\} = \max\left\{0, \sup_{v\in[0,t]} (F(v)+G(v))\right\}.$$

As a consequence, expression (13.3) equals (use Lemma 13.1!)

$$\max\left\{0, \sup_{s\in[0,t]} \left(p_{\ell j}\left(-\sum_{k=0}^{\ell-1}\left(\check{P}^k J_s(x)\right)_\ell\right) - J_s^{(j)}(x)\right)\right\}$$

$$= \max\left\{0, \sup_{s\in[0,t]} \left(-\sum_{k=1}^{\ell}\left(\check{P}^k J_s(x)\right)_\ell - J_s^{(j)}(x)\right)\right\}$$

$$= \max\left\{0, \sup_{s\in[0,t]} \left(-\sum_{k=0}^{\ell}\left(\check{P}^k J_s(x)\right)_j\right)\right\}.$$

Since the jth column of $\check{P}^k$ consists of 0s for $k = \ell+1,\ldots,j-1$, it follows that the above expression equals (13.2) (for $i = j$).

We now combine (13.2) with the fact that, due to the (strict) upper triangular structure of P assumed in (T_1),

$$(I-\check{P})^{-1} = \sum_{k=0}^{\infty} \check{P}^k = I + \check{P} + \check{P}^2 + \cdots + \check{P}^{n-1};$$

in addition, again because of (T_1), the jth row of $(I - \check{P})^{-1}$ equals the jth row of $I+\check{P}+\check{P}^2+\cdots+\check{P}^{j-1}$. This immediately leads to the following result [71, Thm. 5.1]:

$$\begin{aligned} \boldsymbol{Q}_t &= \boldsymbol{J}_t(\boldsymbol{x}) + \max\left\{\boldsymbol{0}, -\inf_{s\in[0,t]} \boldsymbol{J}_s(\boldsymbol{x})\right\} \\ &= \boldsymbol{J}_t - (I - \check{P})\boldsymbol{r}t + \boldsymbol{x} + \max\left\{\boldsymbol{0}, -\inf_{s\in[0,t]}\left(\boldsymbol{J}_s - (I - \check{P})\boldsymbol{r}s + \boldsymbol{x}\right)\right\} \\ &= \boldsymbol{J}_t - (I - \check{P})\boldsymbol{r}t + \max\left\{\boldsymbol{x}, -\inf_{s\in[0,t]}\left(\boldsymbol{J}_s - (I - \check{P})\boldsymbol{r}s\right)\right\}. \end{aligned}$$

Our findings are summarized in the following theorem.

Theorem 13.1 *Suppose that* (T_1)–(T_3) *hold for the fluid network characterized by* $(\boldsymbol{J}, \boldsymbol{r}, P)$. *Then*

$$\boldsymbol{L}_t = \max\left\{\boldsymbol{0}, \sup_{s\in[0,t]}\left(-(I - \check{P})^{-1}(\boldsymbol{x} + \boldsymbol{J}_s) + \boldsymbol{r}s\right)\right\},$$

and

$$\boldsymbol{Q}_t = \boldsymbol{J}_t - (I - \check{P})\boldsymbol{r}t + \max\left\{\boldsymbol{x}, -\inf_{s\in[0,t]}\left(\boldsymbol{J}_s - (I - \check{P})\boldsymbol{r}s\right)\right\},$$

where the suprema should be interpreted componentwise.

It is readily verified that Thm. 13.1 is a true generalization of the single-node and tandem queues that we discussed before.

13.3 Representation for the Stationary Workload

In Section 13.2 we found a representation for the transient workload for the class of fluid networks that satisfy the three assumptions (T_1)–(T_3). In order to get a representation for the stationary workload, we additionally assume that

(T_4) $\quad (I - P')^{-1}\,\mathbb{E}\boldsymbol{J}_1 < \boldsymbol{r}.$

Condition (T_4), where the inequality is understood to hold componentwise, ensures stability of the network, being (obviously) necessary for the existence of the stationary workload distribution.

As it turns out, under (T_1)–(T_4) we can find the joint distribution of the steady-state workload $\boldsymbol{Q}$, the *age of the busy period* $\boldsymbol{B}$, and the *age of the idle period* $\boldsymbol{I}$. More precisely, let $\boldsymbol{B}_t$ denote the n-dimensional vector of the ages of the busy periods at time t, in all n queues of the network: $\boldsymbol{B}_t = (B_t^{(1)}, \ldots, B_t^{(n)})'$ with

$$B_t^{(i)} := t - \sup\left\{s \leq t : Q_s^{(i)} = 0\right\}. \tag{13.4}$$

Analogously, we define the idle period process $\boldsymbol{I}_t = (I_t^{(1)}, \ldots, I_t^{(n)})'$ by

$$I_t^{(i)} := t - \sup\left\{s \leq t : Q_s^{(i)} > 0\right\}. \tag{13.5}$$

To make our notation compact, we denote (with a mild abuse of notation) the system of equations (13.4) by

$$\boldsymbol{B}_t := t\boldsymbol{1} - \sup\left\{s \leq t : \boldsymbol{Q}_s = 0\right\}$$

and (13.5) by

$$\boldsymbol{I}_t := t\boldsymbol{1} - \sup\left\{s \leq t : \boldsymbol{Q}_s > 0\right\}$$

(where we note that there is 'abuse of notation', as one should realize that for any component the supremum in the right-hand side may be attained for another value of s).

Our objective in this section is to characterize the joint stationary distribution $(\boldsymbol{Q}, \boldsymbol{B}, \boldsymbol{I})'$, corresponding to $(\boldsymbol{Q}_t, \boldsymbol{B}_t, \boldsymbol{I}_t)'$ as $t \to \infty$; we do so by following the setup given in [71]. The idea is to consider a transformed version of the workload process, that is,

$$\tilde{\boldsymbol{Q}}_t := (I - \check{P})^{-1}\boldsymbol{Q}_t.$$

Similarly to $B_t^{(i)}$ and $I_t^{(i)}$ we define $\tilde{B}_t^{(i)}$ and $\tilde{I}_t^{(i)}$, with $Q_s^{(i)}$ replaced by $\tilde{Q}_s^{(i)}$ in each case.

In the sequel, we let $\tilde{\boldsymbol{Q}}_t^{(i)}$ represent the total workload of all the buffers on the path from the root of the tree (labeled 1) up to (and including) station i. Likewise, $\tilde{B}_t^{(i)}$ and $\tilde{I}_t^{(i)}$ describe the ages of the busy and idle periods of the aggregate buffer on the path between stations 1 and i. To this end we introduce the process

$$\boldsymbol{X}_t := (I - \check{P})^{-1}\boldsymbol{J}_t - \boldsymbol{r}t.$$

As in the tandem case, let $\bar{\boldsymbol{X}} = (\bar{X}^{(1)}, \ldots, \bar{X}^{(n)})'$ be defined by $\bar{X}^{(i)} := \sup_{t \geq 0} X_t^{(i)}$.

A key observation is that $Q_t^{(i)} = 0$ if and only if $\tilde{Q}_t^{(i)} = 0$ (use that $(I - \check{P})^{-1}$ is non-negative and lower triangular), which combined with (A$^+$), leads to

($\tilde{\text{A}}^+$) $\tilde{\boldsymbol{Q}}_0 = \tilde{\boldsymbol{x}}$, and

$$\tilde{\boldsymbol{Q}}_t = \tilde{\boldsymbol{x}} + \boldsymbol{X}_t + \boldsymbol{L}_t$$

is non-negative for all $t \geq 0$, with $\tilde{\boldsymbol{x}} := (I - \check{P})^{-1}\boldsymbol{x}$;

($\tilde{\mathrm{B}}^+$) $\boldsymbol{L}_0 = 0$ and $\boldsymbol{L}_t$ is non-decreasing, and

$$\sum_{i=1}^{n} \int_0^T \tilde{Q}_t^{(i)}\, \mathrm{d}L_t^{(i)} = 0, \quad \text{for all } T > 0.$$

It is noted that the requirements ($\tilde{\mathrm{A}}^+$)–($\tilde{\mathrm{B}}^+$), analyzed coordinatewise, correspond to a set of n (one-dimensional) Skorokhod problems. As a consequence, by virtue of Thm. 13.1, we have, for any $t \geq 0$,

$$\tilde{\boldsymbol{Q}}_t = \boldsymbol{X}_t + \max\left\{\tilde{\boldsymbol{x}}, -\inf_{s\in[0,t]} \boldsymbol{X}_s\right\} = \max\left\{(\tilde{\boldsymbol{x}} + \boldsymbol{X}_t),\ \sup_{s\in[0,t]} (\boldsymbol{X}_t - \boldsymbol{X}_s)\right\}.$$

Moreover, again using that $Q_t^{(i)} = 0$ if and only if $\tilde{Q}_t^{(i)} = 0$ for $i = 1, \ldots, n$, we have $\boldsymbol{B}_t = \tilde{\boldsymbol{B}}_t$ and $\boldsymbol{I}_t = \tilde{\boldsymbol{I}}_t$. Hence for the age of the busy period we find

$$\begin{aligned}\boldsymbol{B}_t &= t\mathbf{1} - \sup\left\{s \leq t : \tilde{\boldsymbol{x}} + \boldsymbol{X}_s = \min\left\{0,\ \inf_{u\in[0,s]} (\tilde{\boldsymbol{x}} + \boldsymbol{X}_u)\right\}\right\}\\ &= t\mathbf{1} - \sup\left\{s \leq t : \tilde{\boldsymbol{x}} + \boldsymbol{X}_s = \min\left\{0,\ \inf_{u\in[0,t]} (\tilde{\boldsymbol{x}} + \boldsymbol{X}_u)\right\}\right\}\end{aligned}$$

(where a sample-path argument is used to validate the last equality), whereas for the age of the idle period,

$$\boldsymbol{I}_t = t\mathbf{1} - \sup\left\{s \leq t : \tilde{\boldsymbol{x}} + \boldsymbol{X}_s \neq \min\left\{0,\ \inf_{u\in[0,t]} (\tilde{\boldsymbol{x}} + \boldsymbol{X}_u)\right\}\right\}.$$

Due to the stationarity of the increments of the n-dimensional Lévy process $\boldsymbol{X}$, we thus find the following distributional equality:

$$\begin{pmatrix}\tilde{\boldsymbol{Q}}_t\\ \tilde{\boldsymbol{B}}_t\\ \tilde{\boldsymbol{I}}_t\end{pmatrix} \stackrel{\mathrm{d}}{=} \begin{pmatrix}\max\left\{(\tilde{\boldsymbol{x}} - \boldsymbol{X}_{-t}),\ \sup_{s\in[-t,0]} (-\boldsymbol{X}_s)\right\}\\ -\sup\left\{s \in [-t,0] : -\boldsymbol{X}_s = \max\left\{\tilde{\boldsymbol{x}} - \boldsymbol{X}_{-t},\ \sup_{u\in[-t,0]} (-\boldsymbol{X}_u)\right\}\right\}\\ -\sup\left\{s \in [-t,0] : -\boldsymbol{X}_s \neq \max\left\{\tilde{\boldsymbol{x}} - \boldsymbol{X}_{-t},\ \sup_{u\in[-t,s]} (-\boldsymbol{X}_u)\right\}\right\}\end{pmatrix}.$$

Since $\tilde{\boldsymbol{x}} - \boldsymbol{X}_{-t} \to -\infty$, as $t \to \infty$ (due to requirement (T_4)), we arrive at the following result [71], where $\boldsymbol{G} := (G^{(1)}, \ldots, G^{(n)})'$, with

$$G^{(i)} := \arg\sup_{t\geq 0} X_t^{(i)} = \inf\left\{s \geq 0 : X_s^{(i)} = \sup_{t\geq 0} X_t^{(i)}\right\}$$

(cf. Section 12.5) and $\boldsymbol{H} := (H^{(1)}, \ldots, H^{(n)})'$, with

$$H^{(i)} := \inf\left\{s \geq 0 : X_s^{(i)} \neq \sup_{t\geq s} X_t^{(i)}\right\}.$$

Theorem 13.2 *Suppose that* (T_1)–(T_4) *hold for the fluid network characterized by* $(\boldsymbol{J}, \boldsymbol{r}, P)$. *Then for any initial condition* $\boldsymbol{Q}_0 = \boldsymbol{x}$, *the triplet of vectors* $(\boldsymbol{Q}_t, \boldsymbol{B}_t, \boldsymbol{I}_t)'$ *converges in distribution to* $((I - \check{P})\bar{\boldsymbol{X}}, \boldsymbol{G}, \boldsymbol{H})'$, *as* $t \to \infty$.

As we demonstrate in detail in the next sections, for a number of specific network structures Thm. 13.2 allows the derivation of an explicit formula for the Laplace–Stieltjes transform

$$\mathbb{E}e^{-\langle\boldsymbol{\alpha},\boldsymbol{Q}\rangle-\langle\boldsymbol{\vartheta},\boldsymbol{B}\rangle}$$

of the stationary workloads and ages of busy periods, jointly for all queues in the network; $\boldsymbol{\alpha}, \boldsymbol{\vartheta} \in \mathbb{R}_+^n$ and $\langle\cdot,\cdot\rangle$ denotes the inner product. The network architectures that are covered include multinode tandem systems (Section 13.4) and single nodes in tree networks (Section 13.5). Interestingly, the techniques that we have developed also facilitate the analysis of the *priority queue*, that is, a single queue fed by two traffic streams, of which one stream has service priority over the other stream (Section 13.6).

13.4 Multinode Tandem Networks

In this section, we revisit the tandem network that was analyzed in Chapter 12, but now in a substantially more general setting. We now allow any number $n \in \mathbb{N}$ of nodes (whereas in the previous chapter we restricted ourselves to the two-node case); in addition, we may have independent external inputs to nodes $2, \ldots, n$, and at each station some output may leave the system. Put differently, the network is characterized by $(\boldsymbol{J}, \boldsymbol{r}, P)$, where the routing matrix P is such that $p_{i,i+1} \in (0, 1]$ for $i = 1, \ldots, n-1$, and $p_{ij} = 0$ otherwise.

To this end, we assume that, as before, (T_1)–(T_4) apply, but in addition also

(T_5) $\boldsymbol{J}$ has mutually independent components.

The aim of this section is to find an explicit formula for $\mathbb{E}e^{-\langle\boldsymbol{\alpha},\boldsymbol{Q}\rangle-\langle\boldsymbol{\vartheta},\boldsymbol{B}\rangle}$, with $\boldsymbol{\alpha}, \boldsymbol{\vartheta} \in \mathbb{R}_+^n$. We essentially follow the line of reasoning of [71].

As a first step we note that, as a direct consequence of Thm. 13.2, we have the identity

$$\begin{aligned}\mathbb{E}e^{-\langle\boldsymbol{\alpha},\boldsymbol{Q}\rangle-\langle\boldsymbol{\vartheta},\boldsymbol{B}\rangle} &= \mathbb{E}e^{-\langle(I-\check{P})\boldsymbol{\alpha},(I-\check{P})^{-1}\boldsymbol{Q}\rangle-\langle\boldsymbol{\vartheta},\boldsymbol{G}\rangle} \\ &= \mathbb{E}e^{-\langle\tilde{\boldsymbol{\alpha}},\bar{\boldsymbol{X}}\rangle-\langle\boldsymbol{\vartheta},\boldsymbol{G}\rangle},\end{aligned}$$

with $\tilde{\boldsymbol{\alpha}} = (\tilde{\alpha}_1, \ldots, \tilde{\alpha}_n)' = (I - \check{P})\boldsymbol{\alpha}$; here $\tilde{\boldsymbol{\alpha}} \in \mathbb{R}^n_+$. In this tandem setting with external inputs (at any node), we have that

$$X_t^{(1)} = J_t^{(1)} - r_1 t,$$
$$X_t^{(i+1)} = p_{i,i+1} X_t^{(i)} + J_t^{(i+1)} + (p_{i,i+1} r_i - r_{i+1})t, \tag{13.6}$$

for $i = 1, \ldots, n-1$. In the sequel we use the, by now common, notation $\bar{X}_t^{(i)} := \sup_{s \in [0,t]} X_s^{(i)}$ and $\bar{X}^{(i)} := \sup_{s \geq 0} X_s^{(i)}$.

It is convenient to distinguish between (i) the case that 0 is *irregular* for the process $(\bar{X}_t^{(i)} - X_t^{(i)})_t$, meaning that

$$R^{(i)} := \inf\left\{t > 0 : \bar{X}_t^{(i)} = X_t^{(i)}\right\} > 0$$

almost surely (cf. Section 12.5), or $X^{(i)}$ is a compound Poisson process, for all $i = 1, \ldots, n-1$, and (ii) the case that among the nodes 1 to $n-1$ there is a node i for which 0 is *regular* for the process $(\bar{X}_t^{(i)} - X_t^{(i)})_t$, meaning that $R^{(i)} = 0$ almost surely, but $X^{(i)}$ is not compound Poisson. We start our exposition with the former case.

Case (i): Any node i from 1 *to* $n-1$ *is irregular, or* $X^{(i)}$ *is compound Poisson*—Suppose first that for all $i = 1, \ldots, n-1$, we have that $R^{(i)} > 0$ almost surely or $X^{(i)}$ is a compound Poisson process. Then $\bar{X}^{(i)} = X^{(i)}_{G^{(i)}}$. Moreover, since $J_t^{(i+1)} + (p_{i,i+1} r_i - r_{i+1})t$ is increasing in t for all $i = 1, \ldots, n-1$, we have that $G^{(1)} \leq G^{(2)} \leq \cdots \leq G^{(n)}$ almost surely (use the same argument as in Section 12.5). As a result, for $\ell > i$,

$$\bar{X}^{(\ell)} - X^{(\ell)}_{G^{(i)}} = \left(\sup_{s \geq G^{(i)}} X_s^{(\ell)}\right) - X^{(\ell)}_{G^{(i)}}.$$

A key observation is now that

$$\sum_{\ell=i}^{n} \tilde{\alpha}_\ell \bar{X}^{(\ell)} = \sum_{\ell=i}^{n} \tilde{\alpha}_\ell \left(\sup_{s \geq G^{(i)}} X_s^{(\ell)}\right) \quad \text{and} \quad \sum_{\ell=i+1}^{n} \tilde{\alpha}_\ell (\bar{X}^{(\ell)} - X^{(\ell)}_{G^{(i)}})$$

are independent random variables, as is justified by [71, Lemma 2.1]. This implies that, for each $i = 1, \ldots, n-1$,

$$\begin{aligned}
&\mathbb{E}\left(e^{-\sum_{\ell=i}^n \vartheta_\ell G^{(\ell)} - \sum_{\ell=i}^n \tilde{\alpha}_\ell \bar{X}^{(\ell)}}\right) \\
&= \mathbb{E}\left(\left(e^{-(\sum_{\ell=i}^n \vartheta_\ell) G^{(i)} - \sum_{\ell=i}^n \tilde{\alpha}_\ell X^{(\ell)}_{G^{(i)}}}\right)\left(e^{-\sum_{\ell=i+1}^n \vartheta_\ell (G^{(\ell)} - G^{(i)}) - \sum_{\ell=i+1}^n \tilde{\alpha}_\ell (\bar{X}^{(\ell)} - X^{(\ell)}_{G^{(i)}})}\right)\right) \\
&= \mathbb{E}\left(e^{-(\sum_{\ell=i}^n \vartheta_\ell) G^{(i)} - \sum_{\ell=i}^n \tilde{\alpha}_\ell X^{(\ell)}_{G^{(i)}}}\right) \mathbb{E}\left(e^{-\sum_{\ell=i+1}^n \vartheta_\ell (G^{(\ell)} - G^{(i)}) - \sum_{\ell=i+1}^n \tilde{\alpha}_\ell (\bar{X}^{(\ell)} - X^{(\ell)}_{G^{(i)}})}\right).
\end{aligned}$$

We now point out how to determine the second factor in the last expression in the previous display. Notice that this factor does not depend on ϑ_i and $\tilde{\alpha}_i$. As a consequence, upon choosing in the above equality $\vartheta_i = \tilde{\alpha}_i = 0$, we readily conclude that

$$\mathbb{E}\left(e^{-\sum_{\ell=i+1}^n \vartheta_\ell (G^{(\ell)}-G^{(i)})-\sum_{\ell=i+1}^n \tilde{\alpha}_\ell (\bar{X}^{(\ell)}-X^{(\ell)}_{G^{(i)}})}\right) = \frac{\mathbb{E}\left(e^{-\sum_{\ell=i+1}^n \vartheta_\ell G^{(\ell)}-\sum_{\ell=i+1}^n \tilde{\alpha}_\ell \bar{X}^{(\ell)}}\right)}{\mathbb{E}\left(e^{-(\sum_{\ell=i+1}^n \vartheta_\ell)G^{(i)}-\sum_{\ell=i+1}^n \tilde{\alpha}_\ell X^{(\ell)}_{G^{(i)}}}\right)}.$$

Combining the above findings, we thus obtain the following recurrence equation:

$$\mathbb{E}\left(e^{-\sum_{\ell=i}^n \vartheta_\ell G^{(\ell)}-\sum_{\ell=i}^n \tilde{\alpha}_\ell \bar{X}^{(\ell)}}\right)$$
$$= \frac{\mathbb{E}\left(e^{-(\sum_{\ell=i}^n \vartheta_\ell)G^{(i)}-\sum_{\ell=i}^n \tilde{\alpha}_\ell X^{(\ell)}_{G^{(i)}}}\right)}{\mathbb{E}\left(e^{-(\sum_{\ell=i+1}^n \vartheta_\ell)G^{(i)}-\sum_{\ell=i+1}^n \tilde{\alpha}_\ell X^{(\ell)}_{G^{(i)}}}\right)} \mathbb{E}\left(e^{-\sum_{\ell=i+1}^n \vartheta_\ell G^{(\ell)}-\sum_{\ell=i+1}^n \tilde{\alpha}_\ell \bar{X}^{(\ell)}}\right).$$

Iterating this recursion repeatedly immediately leads to the following *quasi-product form*:

$$\mathbb{E}e^{-\langle \boldsymbol{\alpha},\boldsymbol{Q}\rangle-\langle \boldsymbol{\vartheta},\boldsymbol{B}\rangle} = \mathbb{E}\left(e^{-\sum_{\ell=i}^n \vartheta_\ell G^{(\ell)}-\sum_{\ell=i}^n \tilde{\alpha}_\ell \bar{X}^{(\ell)}}\right)$$
$$= \prod_{i=1}^{n-1} \frac{\mathbb{E}\left(e^{-(\sum_{\ell=i}^n \vartheta_\ell)G^{(i)}-\sum_{\ell=i}^n \tilde{\alpha}_\ell X^{(\ell)}_{G^{(i)}}}\right)}{\mathbb{E}\left(e^{-(\sum_{\ell=i+1}^n \vartheta_\ell)G^{(i)}-\sum_{\ell=i+1}^n \tilde{\alpha}_\ell X^{(\ell)}_{G^{(i)}}}\right)} \mathbb{E}\left(e^{-\vartheta_n G^{(n)}-\tilde{\alpha}_n \bar{X}^{(n)}}\right). \tag{13.7}$$

It is noted that we have *not* used (T_5) yet. Application of the condition (T_5) allows us to get a more explicit form for $\mathbb{E}e^{-\langle \boldsymbol{\alpha},\boldsymbol{Q}\rangle-\langle \boldsymbol{\vartheta},\boldsymbol{B}\rangle}$. Relation (13.6) immediately implies that, for $\ell \geq i$ and $i = 1, \ldots, n-1$,

$$X_t^{(\ell)} = \left(\prod_{j=i+1}^{\ell} p_{j-1,j}\right) X_t^{(i)} + \sum_{j=i+1}^{\ell} \left(\prod_{k=j+1}^{\ell} p_{k-1,k}\right) \left(J_t^{(j)} + \tilde{r}_j t\right), \tag{13.8}$$

with $\tilde{r}_j := p_{j-1,j} r_{j-1} - r_j$, and following the convention that $\prod_{k=j+1}^{j} p_{k-1,k} := 1$. Now consider the ith numerator in (13.7). We have

$$\sum_{\ell=i}^n \vartheta_\ell t + \sum_{\ell=i}^n \tilde{\alpha}_\ell X_t^{(\ell)}$$
$$= \sum_{\ell=i}^n \vartheta_\ell t + \sum_{\ell=i}^n \tilde{\alpha}_\ell \left(\prod_{j=i+1}^{\ell} p_{j-1,j}\right) X_t^{(i)} + \sum_{\ell=i}^n \sum_{j=i+1}^{\ell} \tilde{\alpha}_\ell \left(\prod_{k=j+1}^{\ell} p_{k-1,k}\right) \left(J_t^{(j)} + \tilde{r}_j t\right).$$

Now changing the order of summation in the double sum, and swapping the roles of j and ℓ, this quantity turns out to equal

$$\sum_{\ell=i}^{n} \vartheta_\ell t + \sum_{\ell=i}^{n} \tilde{\alpha}_\ell \left(\prod_{j=i+1}^{\ell} p_{j-1,j} \right) X_t^{(i)} + \sum_{\ell=i+1}^{n} \sum_{j=\ell}^{n} \tilde{\alpha}_j \left(\prod_{k=\ell+1}^{j} p_{k-1,k} \right) \left(J_t^{(\ell)} + \tilde{r}_\ell t \right).$$

It is a matter of straightforward algebra to check that $\sum_{\ell=i}^{n} \tilde{\alpha}_\ell \prod_{j=i+1}^{\ell} p_{j-1,j} = \alpha_i$ (using the structure of the routing matrix P), which implies that the expression in the previous display equals

$$\sum_{\ell=i}^{n} \vartheta_\ell t + \alpha_i X_t^{(i)} + \sum_{\ell=i+1}^{n} \alpha_\ell \left(J_t^{(\ell)} + \tilde{r}_\ell t \right).$$

As a consequence, we have that

$$\begin{aligned} &\mathbb{E}\left(e^{-(\sum_{\ell=i}^{n} \vartheta_\ell) G^{(i)} - \sum_{\ell=i}^{n} \tilde{\alpha}_\ell X^{(\ell)}_{G^{(i)}}} \right) \\ &\quad = \mathbb{E}\left(e^{-(\sum_{\ell=i}^{n} \vartheta_\ell + \sum_{\ell=i+1}^{n} \alpha_\ell \tilde{r}_\ell) G^{(i)} - \alpha_i X^{(i)}_{G^{(i)}}} \mathbb{E}\left(e^{-(\sum_{\ell=i+1}^{n} \alpha_\ell J^{(\ell)}_{G^{(i)}})} \,\middle|\, G^{(i)} \right) \right) \\ &\quad = \mathbb{E}\left(e^{-(\sum_{\ell=i}^{n} \vartheta_\ell + \sum_{\ell=i+1}^{n} \alpha_\ell \tilde{r}_\ell + \sum_{\ell=i+1}^{n} \theta_\ell^J(\alpha_\ell)) G^{(i)} - \alpha_i \bar{X}^{(i)}} \right), \end{aligned} \tag{13.9}$$

where $\theta_\ell^J(\alpha) := -\log \mathbb{E} e^{-\alpha J_1^{(\ell)}}$, for $\ell = 2, \ldots, n$.

In a similar way, for the ith denominator in (13.7) we obtain

$$\begin{aligned} &\mathbb{E}\left(e^{-(\sum_{\ell=i+1}^{n} \vartheta_\ell) G^{(i)} - \sum_{\ell=i+1}^{n} \tilde{\alpha}_\ell X^{(\ell)}_{G^{(i)}}} \right) \\ &\quad = \mathbb{E}\left(e^{-(\sum_{\ell=i+1}^{n} \vartheta_\ell + \sum_{\ell=i+1}^{n} \alpha_\ell \tilde{r}_\ell + \sum_{\ell=i+1}^{n} \theta_\ell^J(\alpha_\ell)) G^{(i)} - p_{i,i+1} \alpha_{i+1} \bar{X}^{(i)}} \right), \end{aligned} \tag{13.10}$$

for $i = 1, \ldots, n-1$, where we used that $\sum_{\ell=i+1}^{n} \tilde{\alpha}_\ell \prod_{j=i+1}^{\ell} p_{j-1,j} = p_{i,i+1} \alpha_{i+1}$.

Combining (13.10) and (13.9) with (13.7) leads to the following explicit expression for the joint transform $\mathbb{E} e^{-\langle \alpha, Q \rangle - \langle \vartheta, B \rangle}$ under study:

$$\begin{aligned} &\left(\prod_{i=1}^{n-1} \frac{\mathbb{E}\left(e^{-(\sum_{\ell=i}^{n} \vartheta_\ell + \sum_{\ell=i+1}^{n} \alpha_\ell \tilde{r}_\ell + \sum_{\ell=i+1}^{n} \theta_\ell^J(\alpha_\ell)) G^{(i)} - \alpha_i \bar{X}^{(i)}} \right)}{\mathbb{E}\left(e^{-(\sum_{\ell=i+1}^{n} \vartheta_\ell + \sum_{\ell=i+1}^{n} \alpha_\ell \tilde{r}_\ell + \sum_{\ell=i+1}^{n} \theta_\ell^J(\alpha_\ell)) G^{(i)} - p_{i,i+1} \alpha_{i+1} \bar{X}^{(i)}} \right)} \right) \\ &\qquad \times \left(\mathbb{E} e^{-\tilde{\alpha}_n \bar{X}^{(n)} - \vartheta_n G^{(n)}} \right). \end{aligned} \tag{13.11}$$

Case (ii): Among nodes 1 *to* $n-1$ *there is a regular node i, but* $X^{(i)}$ *is not compound Poisson*—In the above analysis we assumed that for all $i = 1,\dots,n-1$, $R^{(i)} > 0$ almost surely, or $X^{(i)}$ is a compound Poisson process. The remaining scenario, when there exists $i = 1,\dots,n-1$ such that $R^{(i)} = 0$ almost surely but $(X_t^{(i)})_t$ is not a compound Poisson process, follows the same line of reasoning as given above, with the tiny subtlety that in (13.7) one has to change $X^{(\ell)}_{G^{(i)}}$ into $X^{(\ell)}_{G^{(i)}-}$ for all $i = 1,\dots,n-1$ for which $R^{(i)} = 0$ almost surely. However, the resulting formula for $\mathbb{E}e^{-\langle\boldsymbol{\alpha},\boldsymbol{Q}\rangle-\langle\boldsymbol{\vartheta},\boldsymbol{B}\rangle}$ is exactly the same as the one found in (13.11), since $X^{(j)}_{G^{(j)}-}$ can be replaced by $\bar{X}^{(j)}$, as argued in the proof of [43, Thm. VI.5(i)]. As a consequence, (13.11) holds for *all* Lévy tandem networks that satisfy (T_1)–(T_5).

Similarly to the procedure followed in Section 12.5, let $k^{(i)}(\vartheta,\alpha)$ be defined as $k(\vartheta,\alpha)$ in (3.10), with X_t replaced by $X_t^{(i)}$, for $i = 1,\dots,n$. Relying on Thm 3.4, we obtain the following result.

Theorem 13.3 *Suppose that* (T_1)–(T_5) *hold for the fluid tandem network characterized by* $(\boldsymbol{J},\boldsymbol{r},P)$. *For* $\boldsymbol{\alpha},\boldsymbol{\vartheta} \in \mathbb{R}^n_+$,

$$\mathbb{E}e^{-\langle\boldsymbol{\alpha},\boldsymbol{Q}\rangle-\langle\boldsymbol{\vartheta},\boldsymbol{B}\rangle} = \left(\prod_{i=1}^{n-1} \frac{k^{(i)}\left(\sum_{\ell=i}^{n}\vartheta_\ell + \sum_{\ell=i+1}^{n}\alpha_\ell\tilde{r}_\ell + \sum_{\ell=i+1}^{n}\theta_\ell^J(\alpha_\ell),\alpha_i\right)}{k^{(i)}\left(\sum_{\ell=i+1}^{n}\vartheta_\ell + \sum_{\ell=i+1}^{n}\alpha_\ell\tilde{r}_\ell + \sum_{\ell=i+1}^{n}\theta_\ell^J(\alpha_\ell),p_{i,i+1}\alpha_{i+1}\right)}\right)\frac{k^{(n)}(\vartheta_n,\alpha_n)}{k^{(n)}(0,0)}.$$

Example 13.3 We consider the case of spectrally positive inputs. Assume that (T_1)–(T_5) hold and $J^{(1)} \in \mathscr{S}_+$ in an n-node tandem system $(\boldsymbol{J},\boldsymbol{r},P)$. Then, (T_5) combined with (13.8) implies that $X^{(i)} \in \mathscr{S}_+$ for all $i = 1,\dots,n$. Thus, applying (12.12) to Thm. 13.3, we obtain for $\boldsymbol{\alpha},\boldsymbol{\vartheta} \in \mathbb{R}^n_+$,

$$\begin{aligned}\mathbb{E}e^{-\langle\boldsymbol{\alpha},\boldsymbol{Q}\rangle-\langle\boldsymbol{\vartheta},\boldsymbol{B}\rangle} &= -\mathbb{E}X_1^{(n)}\frac{\psi_n(\vartheta_n)-\alpha_n}{\vartheta_n-\varphi_n(\alpha_n)}\\ &\times\prod_{i=1}^{n-1}\frac{\psi_i\left(\sum_{\ell=i+1}^{n}\theta_\ell^J(\alpha_\ell)+\sum_{\ell=i+1}^{n}\tilde{r}_\ell\alpha_\ell+\sum_{\ell=i}^{n}\vartheta_\ell\right)-\alpha_i}{\psi_i\left(\sum_{\ell=i+1}^{n}\theta_\ell^J(\alpha_\ell)+\sum_{\ell=i+1}^{n}\tilde{r}_\ell\alpha_\ell+\sum_{\ell=i+1}^{n}\vartheta_\ell\right)-p_{i,i+1}\alpha_{i+1}}\\ &\times\prod_{i=1}^{n-1}\frac{\sum_{\ell=i+1}^{n}\theta_\ell^J(\alpha_\ell)+\sum_{\ell=i+1}^{n}\tilde{r}_\ell\alpha_\ell+\sum_{\ell=i+1}^{n}\vartheta_\ell-\varphi_i(p_{i,i+1}\alpha_{i+1})}{\sum_{\ell=i+1}^{n}\theta_\ell^J(\alpha_\ell)+\sum_{\ell=i+1}^{n}\tilde{r}_\ell\alpha_\ell+\sum_{\ell=i}^{n}\vartheta_\ell-\varphi_i(\alpha_i)},\end{aligned}$$

where we have defined $\varphi_i(\alpha) := \log \mathbb{E}e^{-\alpha X_1^{(i)}}$, and $\psi_i(\cdot) = \varphi_i^{-1}(\cdot)$; see also [71, Thm. 6.1]. In addition, we note that if $n = 2$ and

$$J_t = \begin{pmatrix} J_t^{(1)} \\ 0 \end{pmatrix}, \quad P = \begin{pmatrix} 0 & 1 \\ 0 & 0 \end{pmatrix},$$

and choosing $\vartheta_1 = \vartheta_2 = 0$, then we recover Thm. 12.11. ◇

13.5 Tree Networks: Stationary Distribution at a Specific Node

In this section we demonstrate how the findings of Section 13.4 allow us to derive a closed-form expression for the joint Laplace–Stieltjes transform of $(Q^{(i)}, B^{(i)})'$ at any specific node $i = 1, \ldots, n$ of a tree fluid network $(\boldsymbol{J}, \boldsymbol{r}, P)$, as introduced in Section 13.2, which we assume to satisfy (T_1)–(T_5).

To this end, we first note that the dynamics of the workload at a given node i of this network (where $i = 2, \ldots, n$) can be described as the workload in a corresponding tandem system which consists of nodes that connect the root of $(\boldsymbol{J}, \boldsymbol{r}, P)$ (i.e. node 1) with node i. Let ℓ be the only (by (T_1)) node for which $p_{\ell i} > 0$. Then a straightforward application of Thm. 13.3, leads to the following formula.

Theorem 13.4 *Suppose that* (T_1)–(T_5) *hold for the fluid network characterized by* $(\boldsymbol{J}, \boldsymbol{r}, P)$. *For* $\alpha, \vartheta \in \mathbb{R}_+$,

$$\mathbb{E}e^{-\alpha Q^{(i)} - \vartheta B^{(i)}} = \frac{k^{(\ell)}\left(\vartheta + \alpha(p_{\ell i} r_\ell - r_i) + \theta_i^J(\alpha), 0\right)}{k^{(\ell)}\left(\vartheta + \alpha(p_{\ell i} r_\ell - r_i) + \theta_i^J(\alpha), p_{\ell i}\alpha\right)} \frac{k^{(i)}(\vartheta, \alpha)}{k^{(i)}(0,0)}.$$

Example 13.4 Consider the spectrally positive case, that is, we study a tree network $(\boldsymbol{J}, \boldsymbol{r}, P)$ such that $J^{(1)} \in \mathscr{S}_+$. Using that $X^{(i)} \in \mathscr{S}_+$ for all $i = 1, \ldots, n$, combination of (12.12), Thms. 13.3 and 13.4, straightforwardly implies that

$$\mathbb{E}e^{-\alpha Q^{(i)} - \vartheta B^{(i)}} = -\mathbb{E}X_1^{(i)} \frac{\psi_i(\vartheta) - \alpha}{\vartheta - \varphi_i(\alpha)}$$

$$\times \frac{\psi_\ell\left(\vartheta + \alpha(p_{\ell i} r_\ell - r_i) + \theta_i^J(\alpha)\right)}{\vartheta + \alpha(p_{\ell i} r_\ell - r_i) + \theta_i^J(\alpha)} \frac{\vartheta + \alpha(p_{\ell i} r_\ell - r_i) + \theta_i^J(\alpha) - \varphi_\ell(p_{\ell i}\alpha)}{\psi_\ell\left(\vartheta + \alpha(p_{\ell i} r_\ell - r_i) + \theta_i^J(\alpha)\right) - p_{\ell i}\alpha}$$

for $\alpha, \vartheta \in \mathbb{R}_+$. ◇

Besides the Laplace–Stieltjes transforms of the age $\boldsymbol{B}$ of the busy periods, Thm. 13.3 also enables the Laplace–Stieltjes transforms of the length of the steady-state running busy periods to be found. We refer to [71, Cor. 6.1] for details.

13.6 Priority Fluid Queues

Consider a single station that is fed by n external Lévy inputs

$$\boldsymbol{J} := (\boldsymbol{J}_t)_t = \left(J_t^{(1)}, \ldots, J_t^{(n)}\right)_t',$$

each equipped with its own buffer. The system is emptied at a constant rate $r > 0$. The queue discipline allows, for each $i = 1, \ldots, n$, the ith buffer to be continuously drained only if buffers 1 to $i-1$ do not require the full capacity r. We call such a system a *priority fluid queue*.

We assume that the individual components of $(\boldsymbol{J}_t)_t$, that is, $(J_t^{(1)})_t, \ldots, (J_t^{(n)})_t$, are mutually independent. In addition, we require that their Lévy measures are concentrated on $(0, \infty)$ and $J_0^{(i)} = 0$ for any $i = 1, \ldots, n$. Also, the processes $(J_t^{(i)})_t$ are assumed to be non-decreasing for $i = 2, \ldots, n$. To guarantee the stability of the system, we impose the condition $\sum_{i=1}^n \mathbb{E}J_1^{(i)} < r$. The aim of this section is to find the Laplace transform of $\boldsymbol{Q} = (Q^{(1)}, \ldots, Q^{(n)})'$, where $Q^{(i)}$ is the stationary workload corresponding to buffer i, for $i = 1, \ldots, n$.

The key observation is that one can establish an association between the considered priority system and a specifically chosen tandem fluid network, in the sense that their respective workload processes have the same dynamics. Indeed, consider a tandem fluid network $(\boldsymbol{J}, \boldsymbol{r}, P)$ with $\boldsymbol{r} = (r, \ldots, r)'$ and P such that $p_{i,i+1} = 1$ for $i = 1, \ldots, n-1$ and $p_{i,j} = 0$ otherwise. Then the solution of the Skorokhod problem for $(\boldsymbol{J}, \boldsymbol{r}, P)$ evolves in the same manner as the buffer content process of the priority system; see also [88]. Moreover, $(\boldsymbol{J}, \boldsymbol{r}, P)$ satisfies (T_1)–(T_5), with the exception that in (T_3), $p_{i,i+1} = r_i/r_{i+1} = 1$. However, as remarked in [71, Section 6.1], Thm. 13.3 still holds. The following theorem follows; see[71].

Theorem 13.5 *Suppose that* (T_1)–(T_5) *hold for the priority fluid network characterized by* $(\boldsymbol{J}, \boldsymbol{r}, P)$. *For* $\boldsymbol{\alpha} \in \mathbb{R}_+^n$,

$$\mathbb{E}e^{-\langle \boldsymbol{\alpha}, \boldsymbol{Q} \rangle} = \left(\prod_{i=1}^{n-1} \frac{k^{(i)}\left(\sum_{\ell=i+1}^n \theta_\ell^J(\alpha_\ell), \alpha_i\right)}{k^{(i)}\left(\sum_{\ell=i+1}^n \theta_\ell^J(\alpha_\ell), \alpha_{i+1}\right)} \right) \frac{k^{(n)}(0, \alpha_n)}{k^{(n)}(0, 0)}.$$

Other interesting problems related to Lévy-driven networks, such as stability and the applicability of (quasi-)product form solutions, are analyzed in a series of papers by Kella and Whitt [120–123, 130].

Exercises

Exercise 13.1 Consider a two-node Lévy network described by the triplet $(\boldsymbol{J}, \boldsymbol{r}, P)$ with

$$\boldsymbol{J}_t = \begin{pmatrix} J_t^{(1)} \\ at \end{pmatrix}, \quad \boldsymbol{r} = \begin{pmatrix} r_1 \\ r_2 \end{pmatrix}, \quad \text{and} \quad P = \begin{pmatrix} 0 & 1 \\ 0 & 0 \end{pmatrix},$$

and $r_1 > r_2 > a > 0$. Suppose that $(Q_0^{(1)}, Q_0^{(2)})' = (0,0)'$. State and solve the corresponding Skorokhod problem.

Exercise 13.2 Assume that (T_1)–(T_5) hold and assume $J^{(1)} \in \mathscr{S}_+$, in an n-node tandem system $(\boldsymbol{J}, \boldsymbol{r}, P)$ with routing matrix P such that $p_{i,i+1} = 1$ for $i = 1, \ldots, n-1$; see Example 13.3. Calculate $\mathbb{E}e^{-\vartheta B^{(k)}}$, for $k \in \{1, \ldots, n\}$, and compare this with the corresponding result for a single-node queue with input $\sum_{i=1}^{k} J^{(k)}$ and output rate r_k.

Exercise 13.3 Suppose that (T_1)–(T_4) hold for the tree fluid network $(\boldsymbol{J}, \boldsymbol{r}, P)$. Let μ be the distribution of $(I-\check{P})\bar{X}$. Prove that μ is the only stationary distribution that satisfies the corresponding Skorokhod problem.

Hint: Suppose (on the contrary) that there exists another stationary distribution $\hat{\mu} \neq \mu$ and $(\hat{Q}_t)_t$ is the corresponding stationary workload process. Observe that for any Borel $B \subset \mathbb{R}_+^n$,

$$\mathbb{P}(\hat{Q}_0 \in B) = \lim_{t\to\infty} \mathbb{P}(\hat{Q}_t \in B) = \lim_{t\to\infty} \int_0^\infty \mathbb{P}(\hat{Q}_t \in B \mid \hat{Q}_0 = x)\mathbb{P}(\hat{Q}_t \in \mathrm{d}x).$$

Now use Thm. 13.2, to show that the above equals $\mathbb{P}((I-\check{P})\bar{X} \in B)$; cf. [71, Cor. 5.1].

Exercise 13.4 Consider a two-queue priority system; a service rate r is shared between the two queues. Traffic stream $J^{(1)}$ feeding into buffer 1 has strict service priority over traffic stream $J^{(2)}$ (assumed to be non-decreasing) feeding into buffer 2.

(a) Show that this network is described by the triplet $(\boldsymbol{J}, \boldsymbol{r}, P)$, with

$$\boldsymbol{J}_t = \begin{pmatrix} J_t^{(1)} \\ J_t^{(2)} \end{pmatrix}, \quad \boldsymbol{r} = \begin{pmatrix} r \\ r \end{pmatrix}, \quad \text{and} \quad P = \begin{pmatrix} 0 & 1 \\ 0 & 0 \end{pmatrix}.$$

(b) Compute the Laplace transform of the joint workload distribution (in stationarity).

Exercise 13.5 Compute the counterpart of Example 13.3 for the case $J^{(1)} \in \mathscr{S}_-$.

Chapter 14
Applications in Communication Networks

Statistical analyses show that, in certain circumstances, traffic aggregates in modern communication networks can be accurately described by specific classes of Lévy processes. In particular, there is widespread consensus that network traffic exhibits properties like *self-similarity* (at least up to a certain timescale threshold) and heavy-tailed traffic bursts. In addition, the Pareto-type marginal distribution of α-stable Lévy motions matches quite well the distribution empirically observed. As a result, a network element in a communication network can be modeled reasonably accurately as a queue driven by an α-stable Lévy motion.

In traffic theory, one distinguishes between traffic models at the user level (in which the behavior of individual users is modeled), and traffic models associated with large numbers of users (in which the individual users are abstract). Besides aggregation over large groups of users (also sometimes referred to as *vertical aggregation*) there can be aggregation over long timescales (referred to as *horizontal aggregation*).

Two commonly used models at the user level are (i) the superposition of *on–off-type sources* (see e.g. Anick et al. [10] or Heath et al. [111]) and (ii) the *infinite-source Poisson model* (see e.g. Resnick and van den Berg [183], Mikosch et al. [163] and references therein). In model (i), it is assumed that each user generates a traffic pattern that alternates between transmitting data at a constant rate (if it is in the on-state) and remaining silent (if it is in the off-state). In model (ii), transmissions by users start at times governed by a homogeneous Poisson process, where the transmission durations form a sequence of i.i.d. random variables (independent of the Poisson process), and during its transmission each user generates traffic at a constant rate (where it is remarked that in some models a variable rate is also considered).

In this chapter we provide a formal justification that both traffic models introduced above, under appropriately chosen scaling of the number of customers and time (i.e. vertical and horizontal aggregation), can be approximated by a self-similar Lévy process. Importantly, this convergence carries over to the workload

K. Dębicki, M. Mandjes, *Queues and Lévy Fluctuation Theory*, Universitext,
DOI 10.1007/978-3-319-20693-6_14

process, which confirms the applicability of Lévy-driven queues in the performance evaluation of communication networks.

It turns out that, by and large, results for convergence of model (ii) (i.e. the infinite-source Poisson model) are in line with their analogues for model (i) (i.e. the on–off model); see e.g. [163, 217]. This motivates, in the rest of this chapter, our focus on just one of the two variants: we choose to present the analysis for the case that traffic is modeled by a superposition of on–off sources.

14.1 Construction of Stationary On–Off Source

We start by considering a single on–off source. The objective of this section is to introduce the notation needed in the chapter, and to explicitly indicate how a stationary version of such an on–off source can be constructed.

In the on–off model it is assumed that the rate at which traffic is generated alternates between an on-mode and an off-mode. During the on-times traffic is fed into the queue at a constant peak rate; without loss of generality we can normalize this rate to 1. During the off-periods the input rate is 0. The durations of the activity periods $\{T_{\text{on},i}, i \geq 0\}$ are i.i.d. random variables, distributed as a non-negative random variable T_{on} attaining values in $\mathbb{R}_+$. The silence periods, $\{T_{\text{off},i}, i \geq 0\}$ are also i.i.d., distributed as a random variable T_{off} with values in $\mathbb{R}_+$. Both sequences are mutually independent and the generic random variables $T_{\text{on}}, T_{\text{off}}$ have finite densities. It is throughout assumed that $\mu_{\text{on}} := \mathbb{E}T_{\text{on}} < \infty$ and $\mu_{\text{off}} := \mathbb{E}T_{\text{off}} < \infty$.

For future use we first provide a construction of the *stationary* on–off process; cf. Heath et al. [111], or [224]. To this end, we start by defining the long-run fraction of time the source is on:

$$\mu := \frac{\mu_{\text{on}}}{\mu_{\text{on}} + \mu_{\text{off}}} \in (0, 1).$$

Additionally, let I be an independent random variable such that $\mathbb{P}(I = 1) = 1 - \mathbb{P}(I = 0) = \mu$. We introduce the *delayed renewal sequence*

$$\{T_i, i \geq 0\} := \left\{ T_0, T_0 + \sum_{k=1}^{i} (T_{\text{on},k} + T_{\text{off},k}), i \geq 1 \right\},$$

where

$$T_0 := I(T_{\text{on},0}^{\text{res}} + T_{\text{off},0}) + (1 - I)T_{\text{off},0}^{\text{res}}; \tag{14.1}$$

here the random variables $T_{\text{on},0}^{\text{res}}$ and $T_{\text{off},0}^{\text{res}}$ follow the usual residual lifetime distribution. It is seen that, informally, the T_i's represent the epochs of the starts of the individual on–off cycles. Then the *stationary* on–off process is defined as

$$\eta_t := I \cdot 1_{\{t<T_{\text{on},0}^{\text{res}}\}} + \sum_{i=0}^{\infty} 1_{\{t\in[T_i;\,T_i+T_{\text{on},i+1})\}};$$

$\eta_t = 1$ if the source is on at time t, and 0 otherwise. It can be verified that, owing to its very construction, $\mathbb{E}\eta_t \equiv \mu$, for all $t \geq 0$.

Let $K_t := \min\{k \geq 0 : t \leq T_k\}$ be the counter of the number of renewal epochs until time t (i.e. starts of on–off cycles). It is clear that the accumulated input by time t can be represented by

$$J_t := \int_0^t \eta_s \,\mathrm{d}s,$$

for $t \geq 0$.

14.2 Convergence of Traffic Process: Horizontal Aggregation

In this section we first focus on traffic generated by a single on–off source, by studying limit properties of $(J_{Tt})_t$, as $T \to \infty$ ('horizontal aggregation'). As it turns out, depending on the 'heaviness' of the tail distribution of the generic random variables T_{on} and T_{off}, one can distinguish two scenarios, leading to convergence either to Brownian motion or to α-stable Lévy motion (cf. the dichotomy presented in Chapter 5). We treat both scenarios separately. Later in this chapter we extend these results to multiple sources ('vertical aggregation').

Brownian approximation—Much of the (vast) literature on on–off models focuses on the case in which the successive on and off times are *light tailed.* Notice that the notion of 'light tailed' used here differs from the one used earlier in Chapter 8.

Definition 14.1 We say that the condition (LT) holds if

$$\mathbb{V}\text{ar}(T_{\text{on}}) = \sigma_{\text{on}}^2 < \infty \quad \text{and} \quad \mathbb{V}\text{ar}(T_{\text{off}}) = \sigma_{\text{off}}^2 < \infty.$$

Assume that (LT) holds and consider the sequence of scaled centered cumulative processes $(J_{Tt} - \mu Tt)_t$, in the regime that $T \to \infty$. It is convenient to decompose $J_{Tt} - \mu Tt$ in terms of the regeneration points, that is,

$$J_{Tt} - \mu Tt = 1_{\{T_0<Tt\}}\left(\int_0^{T_0} (\eta_s - \mu)\mathrm{d}s + \sum_{i=1}^{K_{Tt}} ((1-\mu)T_{\text{on},i} - \mu T_{\text{off},i}) + \int_{K_{Tt}}^{T} (\eta_s - \mu)\mathrm{d}s\right) + 1_{\{T_0\geq Tt\}}(J_{Tt} - \mu Tt).$$

Next, observe that after dividing the above by $\sqrt{T}$, some of the terms become negligible as T grows large. Indeed, it is a matter of straightforward checking that both

$$\frac{1}{\sqrt{T}}\int_0^{T_0}(\eta_s-\mu)\mathrm{d}s \quad \text{and} \quad \frac{1}{\sqrt{T}}\int_{K_{Tt}}^{t}(\eta_s-\mu)\mathrm{d}s$$

converge in probability to 0 as $T\to\infty$. Besides, for each $\varepsilon>0$, upon applying the Markov inequality, we have

$$\begin{aligned}\mathbb{P}\left(1_{\{T_0\geq Tt\}}(J_{Tt}-\mu Tt)/\sqrt{T}\geq\varepsilon\right) &\leq \mathbb{P}\left(1_{\{T_0\geq Tt\}}\geq\varepsilon/(t\sqrt{T})\right)\\ &\leq \frac{t\sqrt{T}\,\mathbb{P}(T_0\geq Tt)}{\varepsilon}\to 0,\end{aligned}$$

as $T\to\infty$; here (LT) is used. As a consequence, we have that

$$\begin{aligned}\lim_{T\to\infty}\frac{J_{Tt}-\mu Tt}{\sqrt{T}} &\stackrel{\mathrm{d}}{=} \lim_{T\to\infty}\frac{\sum_{i=1}^{K_{Tt}}((1-\mu)T_{\mathrm{on},i}-\mu T_{\mathrm{off},i})}{\sqrt{T}}\\ &\stackrel{\mathrm{d}}{=} \lim_{T\to\infty}\sqrt{\frac{K_{Tt}}{T}}\times\frac{\sum_{i=1}^{K_{Tt}}((1-\mu)T_{\mathrm{on},i}-\mu T_{\mathrm{off},i})}{\sqrt{K_{Tt}}}\\ &\stackrel{\mathrm{d}}{=} \lim_{T\to\infty}\sqrt{\frac{t}{\mu_{\mathrm{on}}+\mu_{\mathrm{off}}}}\times\frac{\sum_{i=1}^{[Tt]}((1-\mu)T_{\mathrm{on},i}-\mu T_{\mathrm{off},i})}{\sqrt{[Tt]}}\\ &\stackrel{\mathrm{d}}{=} \sqrt{\frac{t}{\mu_{\mathrm{on}}+\mu_{\mathrm{off}}}}\sqrt{\mathbb{V}\mathrm{ar}((1-\mu)T_{\mathrm{on},i}-\mu T_{\mathrm{off},i})}\,\mathscr{N}\stackrel{\mathrm{d}}{=}\breve{\sigma}\sqrt{t}\,\mathscr{N},\end{aligned}$$

where

$$\breve{\sigma}:=\sqrt{\frac{\mu_{\mathrm{on}}^2\sigma_{\mathrm{off}}^2+\mu_{\mathrm{off}}^2\sigma_{\mathrm{on}}^2}{(\mu_{\mathrm{on}}+\mu_{\mathrm{off}})^3}}$$

and $\mathscr{N}$ represents a standard normal random variable. We have just proved that single-dimensional distributions of $((J_{Tt}-\mu Tt)/\sqrt{T})_t$ converge to the single-dimensional distributions of $\mathbb{B}\mathrm{m}(0,\breve{\sigma}^2)$ as $T\to\infty$.

The convergence of the finite-dimensional distributions can be dealt with in a similar way. For example, to prove the convergence of two-dimensional distributions, it suffices to check that for each $b_1,b_2\in\mathbb{R}$ and $t_2>t_1\geq 0$, as $T\to\infty$,

$$b_1\frac{J_{Tt_1}-\mu Tt_1}{\sqrt{T}}+b_2\frac{J_{Tt_2}-J_{Tt_1}-\mu T(t_2-t_1)}{\sqrt{T}}\stackrel{\mathrm{d}}{\to}\left(\breve{\sigma}b_1\sqrt{t_1}\mathscr{N}_1+\breve{\sigma}b_2\sqrt{t_2-t_1}\mathscr{N}_2\right),$$

with mutually independent standard normal random variables $\mathscr{N}_1,\mathscr{N}_2$.

In order to justify the application of the limit model in communication networks it is crucial to establish convergence in a stronger sense, for instance the functional weak convergence in the space D equipped with the Skorokhod topology J_1—we refer to the book by Whitt [217] for a complete description of these notions and further background. The following result for vertical aggregation can be found in Taqqu et al. [210], but see also [217, Chapter 8].

Theorem 14.1 *Assume that (LT) holds. Then*

$$\left(\frac{J_{Tt} - \mu Tt}{\sqrt{T}}\right)_t \stackrel{\mathrm{d}}{\to} (X_t)_t,$$

as $T \to \infty$, where $X \in \mathbb{B}\mathrm{m}(0, \check{\sigma}^2)$. Here '$\stackrel{\mathrm{d}}{\to}$' denotes weak convergence in the $(D[0,\infty), J_1)$ Skorokhod space.

Example 14.1 A key model in traffic theory is the so-called *Anick–Mitra–Sondhi model* [10], that is, it is assumed that $T_{\mathrm{on}}, T_{\mathrm{off}}$ are exponentially distributed with means $\mu_{\mathrm{on}}, \mu_{\mathrm{off}} \in (0,\infty)$ respectively (i.e. exponentially distributed with hazard rates $\mu_{\mathrm{on}}^{-1}, \mu_{\mathrm{off}}^{-1}$ respectively). Then, with $(X_t)_t$ corresponding to $\mathbb{B}\mathrm{m}(0,1)$, we have that

$$\left(\frac{J_{Tt} - \mu Tt}{\sqrt{T}}\right)_t \stackrel{\mathrm{d}}{\to} \left(\sqrt{\frac{2\mu_{\mathrm{on}}^2 \mu_{\mathrm{off}}^2}{(\mu_{\mathrm{on}} + \mu_{\mathrm{off}})^3}} X_t\right)_t,$$

as $T \to \infty$, in the sense of weak convergence in $(D[0,\infty), J_1)$). ◇

Remark 14.1 Thm. 14.1 can immediately be extended to traffic processes consisting of superpositions of an arbitrary number (say, M) of i.i.d. on–off sources. Let $(J_t^{(1)})_t, (J_t^{(2)})_t, \ldots, (J_t^{(M)})_t$ be i.i.d. copies of $(J_t)_t$, and assume that (LT) holds. Then

$$\left(\frac{\sum_{i=1}^M (J_{Tt}^{(i)} - \mu Tt)}{\sqrt{MT}}\right)_t \stackrel{\mathrm{d}}{\to} \left(\sqrt{\frac{(\mathbb{E}T_{\mathrm{on}})^2 \sigma_{\mathrm{off}}^2 + (\mathbb{E}T_{\mathrm{off}})^2 \sigma_{\mathrm{on}}^2}{(\mathbb{E}T_{\mathrm{on}} + \mathbb{E}T_{\mathrm{off}})^3}} X_t\right)_t$$

as $T \to \infty$, in the sense of weak convergence in $(D[0,\infty), J_1)$. ◇

Stable Lévy approximation—Statistical measurements of data traffic in communication networks have shown that in specific situations assumption (LT) may be inadequate, in the sense that the tails of $T_{\mathrm{on}}, T_{\mathrm{off}}$ are potentially significantly heavier. The presence of such *heavy-tailed* phenomena strongly motivates the analysis of the on–off model under the following alternative assumption. Again, it is remarked that the notion of a random variable being 'heavy tailed' differs from the one used in Chapter 8. Let $\bar{F}_{\mathrm{on}}(\cdot)$ and $\bar{F}_{\mathrm{off}}(\cdot)$ be the complementary distribution functions of T_{on} and T_{off}, respectively.

Definition 14.2 We say that the condition (HT) holds if

$$\bar{F}_{\rm on}(x) = x^{-\alpha_{\rm on}} L_{\rm on}(x), \quad \bar{F}_{\rm off}(x) = x^{-\alpha_{\rm off}} L_{\rm off}(x) \ \text{ with } \ \alpha_{\rm on}, \alpha_{\rm off} \in (1,2),$$

where $L_{\rm on}(\cdot)$ and $L_{\rm off}(\cdot)$ are *slowly varying* at infinity.

To simplify notation we tacitly assume that $L_{\rm on}(\cdot) = L_{\rm off}(\cdot) = 1$ in the rest of this section. It can be verified that all results are valid for slowly varying $L_{\rm on}(\cdot)$ and $L_{\rm off}(\cdot)$. We note that condition (HT) ensures $\mathbb{E}T_{\rm on} = \mu_{\rm on} < \infty$ and $\mathbb{E}T_{\rm off} = \mu_{\rm off} < \infty$, while $\mathbb{V}{\rm ar}(T_{\rm on}) = \infty$ and $\mathbb{V}{\rm ar}(T_{\rm off}) = \infty$. In this sense, $T_{\rm on}, T_{\rm off}$ are heavy tailed.

Then, with $\alpha := \min\{\alpha_{\rm on}, \alpha_{\rm off}\}$, after switching the scaling from $\sqrt{T}$ to $T^{1/\alpha}$, one can repeat the argument given under the condition (LT). That is, we can restrict the analysis of the convergence of $(J_{Tt} - \mu Tt)/T^{1/\alpha}$, as $T \to \infty$, to the convergence of just

$$\frac{1}{T^{1/\alpha}} \sum_{i=1}^{K_{Tt}} ((1-\mu)T_{{\rm on},i} - \mu T_{{\rm off},i})$$

as $T \to \infty$ (cf. the proof of [210, Thm. 3]). The above expression, by virtue of (HT) being in place, tends to an α-stable law. The following result can be found in [210]. Recall $C_{\alpha,\sigma} := \sigma^\alpha(1-\alpha)/(\Gamma(2-\alpha)\cos(\pi\alpha/2))$, as in Prop. 2.1. Define

$$\sigma := \frac{(\mu_{\rm on} + \mu_{\rm off})^{1+1/\alpha}}{\mu_{\rm off}}, \quad \beta := \frac{\mu_{\rm off}^\alpha - \mu_{\rm on}^\alpha}{\mu_{\rm off}^\alpha + \mu_{\rm on}^\alpha}.$$

Theorem 14.2 *Assume that* (HT) *holds.*

(i) If $\alpha_{\rm on} < \alpha_{\rm off}$, *then, with* $X \in \mathbb{S}(\alpha, 1, 0)$, *as* $T \to \infty$,

$$\left(\frac{J_{Tt} - \mu Tt}{T^{1/\alpha}}\right)_t \xrightarrow{\rm d} \left((C_{\alpha,\sigma})^{-1/\alpha} X_t\right)_t.$$

(ii) If $\alpha_{\rm off} < \alpha_{\rm on}$, *then, with* $X \in \mathbb{S}(\alpha, -1, 0)$, *as* $T \to \infty$,

$$\left(\frac{J_{Tt} - \mu Tt}{T^{1/\alpha}}\right)_t \xrightarrow{\rm d} \left((C_{\alpha,\sigma})^{-1/\alpha} X_t\right)_t.$$

(iii) If $\alpha_{\rm on} = \alpha_{\rm off}$, *then, with* $X \in \mathbb{S}(\alpha, \beta, 0)$, *as* $T \to \infty$,

$$\left(\frac{J_{Tt} - \mu Tt}{T^{1/\alpha}}\right)_t \xrightarrow{\rm d} \left((C_{\alpha,\sigma})^{-1/\alpha} X_t\right)_t.$$

Here '$\xrightarrow{\rm d}$*' denotes weak convergence in the* $(D[0,\infty), M_1)$ *Skorokhod space.*

Observe that, although $(J_t)_t$ has continuous sample paths a.s., the obtained α-stable limit has non-continuous trajectories a.s. This contrasts with Thm. 14.1, where the limit process is continuous a.s. (as it is a Brownian motion). This observation is related to the fact that the obtained convergence in Thm. 14.2 is in the M_1 topology, and cannot be extended to J_1-convergence in the Skorokhod space $D[0,\infty)$. A comprehensive discussion of these issues can be found in [215, 216] and the book [217].

Remark 14.2 Thm. 14.2 can immediately be extended to traffic processes consisting of a superposition of i.i.d. on–off sources. Namely, for $(J_t^{(1)})_t, (J_t^{(2)})_t, \ldots, (J_t^{(M)})_t$ being i.i.d. copies of $(J_t)_t$, under (HT), one gets that

$$\left(\frac{\sum_{i=1}^{M}(J_{Tt}^{(i)} - \mu Tt)}{(MT)^{1/\alpha}}\right)_t$$

weakly converges in $(D[0,\infty), M_1)$, as $T \to \infty$, to α-stable Lévy motion, with the same regimes (in terms of α_{on} and α_{off}) as in Thm. 14.2. ◇

The infinite-source Poisson counterpart of the results given in this section can be found in [183].

14.3 Convergence of Traffic Process: Vertical Aggregation

In the previous section we observed that for a fixed number of on–off sources we obtain a limiting Lévy process (either Brownian motion or an α-stable Lévy motion), under a specific scaling, as $T \to \infty$. It is tempting to ask whether these limiting properties are preserved in the situation that the number of sources grows large, too. To this end, we study the model in a limiting regime in which one first lets the number of sources M go to ∞ and only then one sends T to ∞.

In order to answer this question we start from the analysis for light-tailed $T_{\text{on}}, T_{\text{off}}$, that is, it is assumed that (LT) holds. As previously, $(J_t^{(i)})_t$, for $i = 1, 2, \ldots$ are i.i.d. copies of $(J_t)_t$. First, by the central limit theorem, we observe that for given T and as $M \to \infty$, with

$$\check{J}_t^{(M)} := \frac{\sum_{i=1}^{M}(J_t^{(i)} - \mu t)}{\sqrt{M}},$$

$\check{J}_{Tt}^{(M)}$ converges in distribution to a centered normal random variable with variance

$$\mathbb{V}\text{ar}\left(\check{J}_{Tt_1}^{(M)}\right) = \mathbb{V}\text{ar}(J_{Tt}) = 2\int_0^{Tt}\int_0^{s} \mathbb{C}\text{ov}(\eta(v), \eta(0))\text{d}v\,\text{d}s.$$

Additionally, it is readily verified that for any given $t_1, t_2 > 0$ the scaling is such that the covariance

$$\mathbb{C}\text{ov}\left(\check{J}_{Tt_1}^{(M)}, \check{J}_{Tt_2}^{(M)}\right) = \mathbb{C}\text{ov}\left(J_{Tt_1}, J_{Tt_2}\right) = \frac{\mathbb{V}\text{ar}(J_{Tt_1}) + \mathbb{V}\text{ar}(J_{Tt_2}) - \mathbb{V}\text{ar}(J_{Tt_1} - J_{Tt_2})}{2}$$

is independent of the number of sources M. The above, given that we are able to prove tightness in $(D[0,\infty), J_1)$ of the considered sequence of processes (with respect to the parameter M), leads to the conclusion that $(\check{J}_{Tt}^{(M)})_t$ converges, as $M \to \infty$, in $(D[0,\infty), J_1)$ to a centered Gaussian process with the same covariance structure as $(J_{Tt})_t$.

In the second step we take the outer limit as $T \to \infty$. It is a matter of straightforward algebra to check that, by (LT),

$$\frac{\mathbb{V}\text{ar}(J_{Tt})}{T} \to \frac{2t}{\int_0^\infty \mathbb{C}\text{ov}(\eta(s), \eta(0))\text{d}s} = \frac{\mu_{\text{on}}^2 \sigma_{\text{off}}^2 + \mu_{\text{off}}^2 \sigma_{\text{on}}^2}{(\mu_{\text{on}} + \mu_{\text{off}})^3} t,$$

as $T \to \infty$. We thus arrive at the following result; cf. [210].

Theorem 14.3 *Assume that* (LT) *holds. Then*

$$\lim_{T\to\infty} \lim_{M\to\infty} \left(\frac{\sum_{i=1}^{M} (J_{Tt}^{(i)} - \mu T t)}{\sqrt{MT}} \right)_t \stackrel{\text{d}}{=} (X_t)_t,$$

where $X \in \mathbb{B}\text{m}(0, \check{\sigma}^2)$*; the weak convergence holds in the* $(D[0,\infty), J_1)$ *Skorokhod space.*

Interestingly, under the (HT) scenario, we still have convergence of $(\check{J}_{Tt}^{(M)})_t$, as $M \to \infty$, to a Gaussian process with the same covariance structure as the generic $(J_{Tt})_t$ process, as above. Then, after scaling by $T^{1/\alpha}$, the resulting Gaussian process converges, as $T \to \infty$, to a *fractional Brownian motion* (fBm) with Hurst parameter $H = 1/\alpha$, which is a non-Lévy self-similar Gaussian process with stationary increments and a long-range dependent structure of the increment process; see e.g. [210, 217].

An important contribution to the discussion on the validity of Lévy or fBm approximations (under (HT)) is given in Mikosch et al. [163], where dependence between the rates of growth of M and T is allowed. We briefly summarize these findings. Assume that $M = M_T$ is an integer-valued function, non-decreasing in T, such that $\lim_{T\to\infty} M_T = \infty$. For compactness we assume that $\alpha = \alpha_{\text{on}} < \alpha_{\text{off}}$; other cases can be dealt with in a similar way. Let $b(t) := (1/(1 - F_{\text{on}}))^{-1}(t)$; by (HT) the function $b(t)$ is regularly varying with index $1/\alpha$. Consider the process

$$\left(\frac{\sum_{i=1}^{M_T} (J_{Tt}^{(i)} - \mu T t)}{b(M_T T)} \right)_t$$

as $T \to \infty$. It turns out that, if $\lim_{t\to\infty} b(M_T T)/T = 0$, then the above converges in $(D[0,\infty), M_1)$, as $T \to \infty$, to an α-stable Lévy motion. If $\lim_{t\to\infty} b(M_T T)/T = \infty$, on the other hand, then the corresponding limit is a fractional Brownian motion with Hurst parameter $1/\alpha$. We refer to [163] or [164] for details. Infinite-source Poisson counterparts of the results given in this section can be found in [163].

14.4 Convergence of Workload Processes

In this section we validate the applicability of the limits obtained in previous sections in the context of communication networks. More specifically, we focus on showing that the convergence of the traffic processes to a given Lévy process (which we showed to apply under certain circumstances, as presented in the previous subsections) carries over to the workload process (in the sense that the workload process converges to the workload process corresponding to a queue fed by that specific limiting Lévy process). We primarily consider the analysis of the model with just a single source; the many-sources model can be dealt with in a similar way.

For given $T > 0$ consider a fluid queue, where the buffer is fed by an integrated on–off process $(J_{Tt})_t$. The buffer is emptied at rate $r_T > 0$ and it is assumed that $Q_0 = x_T$.

Then (for given T), the content of the fluid queue at time t equals, according to (2.4),

$$Q_t^{(T)} := X_{Tt} + \max\left\{x_T, -\inf_{s\in[0,Tt]} X_s\right\} = X_{Tt} + \max\left\{x_T, -\inf_{s\in[0,t]} X_{Ts}\right\},$$

with $X_{Tt} := J_{Tt} - r_T t$.

Reflected Brownian motion approximation—This case corresponds to the light-tailed scenario. Here it is assumed that condition (LT) is satisfied and

$$\lim_{T\to\infty} \frac{r_T - \mu T}{\sqrt{T}} = r, \quad \lim_{T\to\infty} \frac{x_T}{\sqrt{T}} = x.$$

Then, by Thm. 14.1,

$$\frac{X_{Tt}}{\sqrt{T}} = \frac{J_{Tt} - \mu T t}{\sqrt{T}} - \frac{r_T - \mu T}{\sqrt{T}} t$$

converges in $(D[0,\infty), J_1)$, as $T \to \infty$, to $\mathbb{B}\text{m}(-r, \check{\sigma}^2)$.

The reflection mapping, mapping a path $(y_v)_v$ for $v \in [0, s]$ onto $[0,\infty)$, given by

$$q[y](s) := y_s + \max\left\{x, -\inf_{v\in[0,s]} y_v\right\},$$

with $q[y](0) = x \geq 0$, is continuous in the space $D[0,t]$ equipped with the topology J_1. We refer for this result (as well as many related results) e.g. to Whitt [215, 217]. Hence we can apply the continuous mapping theorem, which straightforwardly implies that

$$\frac{Q_t^{(T)}}{\sqrt{T}} = \frac{1}{\sqrt{T}}\left(X_{Tt} + \max\left\{x_T, -\inf_{s\in[0,t]} X_{Ts}\right\}\right) \xrightarrow{\text{d}} Q_t,$$

as $T \to \infty$, where Q_t is a queue driven by $\mathbb{B}\text{m}(-r, \check{\sigma}^2)$, with $Q_0 = x$. The exact distribution of Q_t is given in (4.6).

Reflected α-stable approximation—This scenario corresponds to the heavy-tailed case. Assume that (HT) holds with

$$\lim_{T\to\infty} \frac{r_T - \mu T}{T^{1/\alpha}} = r, \quad \lim_{T\to\infty} \frac{x_T}{T^{1/\alpha}} = x.$$

The same reasoning as in the light-tailed case, combined with Thm. 14.2 and the fact that the reflection mapping is continuous in the space $(D[0,t], M_1)$, implies that

$$\frac{Q_t^{(T)}}{T^{1/\alpha}} \xrightarrow{\text{d}} Q_t,$$

as $T \to \infty$, where Q_t is a queue driven by α-stable Lévy motion (chosen according to Thm. 14.2) with drift $-rt$ and $Q_0 = x$. Distributional properties of Q_t were presented in e.g. Chapters 4 and 9.

Remark 14.3 Continuity of the reflection mapping both in the J_1 and the M_1 topology implies that the above findings extend to queues in which the input consists of a superposition of multiple i.i.d. on–off sources. ◇

Remark 14.4 Convergence of the transient workload process to an appropriately chosen reflected Lévy process does not imply directly the convergence of the corresponding stationary workload process, since the sup functional is *not* continuous in the $(D[0,\infty), J_1)$ space (or the $(D[0,\infty), M_1)$ space). Interesting results confirming such a convergence for the sequence of workload processes driven by Lévy inputs can be found in [139, 207]. ◇

Exercises

Exercise 14.1 Assume that $J_t = J_t^{(1)} + J_t^{(2)}$, where $(J_t^{(1)})_t, (J_t^{(2)})_t$ are independent integrated on–off processes.

(a) Find the counterpart of Thm. 14.1 under the assumption that both $(J_t^{(1)})_t, (J_t^{(2)})_t$ satisfy (LT).

(b) Find the counterpart of Thm. 14.2 under the assumption that both $(J_t^{(1)})_t$, $(J_t^{(2)})_t$ satisfy (HT).
(c) Assume that $(J_t^{(1)})_t$ satisfies (LT) and $(J_t^{(2)})_t$ satisfies (HT). What is the right scaling to get a non-trivial limit for $(J_t)_t$? Recognize the limit.

Exercise 14.2 Check that under (LT),

$$\int_0^\infty \mathbb{C}\mathrm{ov}(\eta(s), \eta(0))\mathrm{d}s = \frac{1}{2}\frac{(\mu_{\mathrm{on}} + \mu_{\mathrm{off}})^3}{\mu_{\mathrm{on}}^2\sigma_{\mathrm{off}}^2 + \mu_{\mathrm{off}}^2\sigma_{\mathrm{on}}^2}.$$

Hint: This follows by analyzing the transform of $\int_0^\infty \mathbb{C}\mathrm{ov}(\eta(s), \eta(0))\mathrm{d}s$.

Exercise 14.3 Check that the reflection mapping

$$q[y](s) := y_s + \max\left\{x, -\inf_{v\in[0,s]} y_v\right\}$$

is continuous in the space $D[0, t]$ with the topology J_1. Do the same with the topology M_1. For definitions see e.g. [217].

Exercise 14.4 Consider the space $D[0, \infty)$ with the topology J_1. For definitions see e.g. [217].

(a) Prove that the functional $\bar{g}(\cdot)$, defined by

$$\bar{g}(y) := \sup_{s\in[0,t]} y_s,$$

is continuous in the space $D[0, \infty)$.
(b) Show that the functional $\bar{h}(\cdot)$, defined by

$$\bar{h}(y) := \sup_{s\in[0,\infty)} y_s,$$

is *not* continuous in the space $D[0, \infty)$ (i.e. find a counterexample).
(c) Repeat tasks (a) and (b) with the topology M_1.

Chapter 15
Applications in Mathematical Finance

Lévy fluctuation theory is widely used in mathematical finance, primarily owing to its capability to model a wide range of path structures—as we saw earlier in the book, the class of Lévy models is rich, in that it includes processes with paths that are sometimes for a while seemingly continuous, then exhibit jumps, potentially both in the upward and the downward direction. The primary goal of this chapter is to illustrate how the theory of the previous chapters can be used in various subareas within the broader domain of mathematical finance.

In e.g. Cont and Tankov [63] it is argued that one could model the price evolution of various risky assets by the process $(S_t)_t$, where

$$S_t = S_0 e^{X_t},$$

with $(X_t)_t$ being a Lévy process. The payoff structures of popular options are typically expressed in terms of the value S_T at the maturity time T, or possibly the associated running maximum $\bar{S}_T$ (or running minimum). This explains why Lévy fluctuation theory is a useful tool when pricing such options, and why we choose *option pricing* as one of the leading examples in the chapter.

The second leading example that we include in this chapter is *non-life insurance*. Modeling the cumulative claim process as a Lévy process, it turns out that a significant subset of the results presented in this book can be used when quantifying the insurer's ruin probability.

For an extensive treatment of applications in finance, and a general account of the use of Lévy modeling in this context, we refer e.g. to [63]; see e.g. [194] for an extensive account of applications in *credit risk*.

In this chapter, we first consider a number of specific Lévy processes that are frequently used in the financial literature. Then we give a brief account of methods that have been developed to estimate the parameters of the Lévy process from time series data. We continue by providing a number of examples in which we indicate how (exotic) options can be priced relying on the theory presented in this book. The

K. Dębicki, M. Mandjes, *Queues and Lévy Fluctuation Theory*, Universitext,
DOI 10.1007/978-3-319-20693-6_15

chapter is concluded by presenting a number of applications of Lévy fluctuation theory in non-life insurance and a short account of other applications in finance.

15.1 Specific Lévy Processes in Finance

In Chapter 2, as well as Sections 3.4 and 3.5, we introduced a series of standard Lévy processes. Several other Lévy processes have been developed specifically for financial applications; we review them in this section.

Jump diffusion processes—In the 'classical' literature, the models proposed are typically Brownian motion with compound Poisson jumps, referred to as *jump diffusion processes*. In the classical paper by Merton [160] these jumps are assumed to have a normal distribution, whereas in the paper by Kou [141] the jumps have a Lévy measure of the form (with $\lambda_-, \lambda_+ > 0$ and $p \in (0,1)$)

$$\Pi(\mathrm{d}x) = \left(p\lambda_- e^{\lambda_- x} 1_{\{x<0\}} + (1-p)\lambda_+ e^{-\lambda_+ x} 1_{\{x>0\}}\right) \mathrm{d}x,$$

meaning that with probability p there is an exponentially distributed downward jump with mean $1/\lambda_-$, and with probability $1-p$ there is an exponentially distributed upward jump with mean $1/\lambda_+$.

Normal inverse Gaussian process—Among several other references, [34, 191] advocate the use of normal inverse Gaussian processes in financial modeling. These processes are constructed as follows.

With X being $\mathbb{B}\mathrm{m}(d,\sigma^2)$ and Y being an increasing Lévy process, it can be shown that $(Z_t)_t := (X_{Y_t})_t$ is again a Lévy process. Indeed, an elementary computation yields that

$$\log \mathbb{E} e^{\mathrm{i}sZ_t} = t \log \mathbb{E} e^{(-(\sigma^2 s^2/2)+d\,\mathrm{i}s)Y_1}.$$

Now take for $(Y_t)_t$ an inverse Gaussian process, as introduced in Chapter 2, with parameters $-\bar{d}$ and $\bar{\sigma}^2$; more specifically, Y_t is the first time that a Brownian motion with parameters $-\bar{d}$ (with $\bar{d} > 0$) and $\bar{\sigma}^2$ reaches the level $-t$, for $t \geq 0$. Recall that, due to its very definition, $(Y_t)_t$ is increasing. After some calculus, it follows that

$$\log \mathbb{E} e^{\mathrm{i}sZ_1} = \frac{1}{\kappa} - \frac{1}{\kappa}\sqrt{1 + \frac{\kappa\sigma^2 s^2}{\bar{d}} - 2\frac{\kappa d\,\mathrm{i}s}{\bar{d}}},$$

with $\kappa := \bar{\sigma}^2/\bar{d}$. In the sequel we refer to Z as a *normal inverse Gaussian* process, denoted by $\mathbb{N}\mathrm{IG}(d,\bar{d},\sigma^2,\kappa)$; cf. [63, Table 4.5].

Variance gamma process—We have observed that with X being $\mathbb{B}\mathrm{m}(d,\sigma^2)$ and Y being an increasing Lévy process, $(Z_t)_t := (X_{Y_t})_t$ is again a Lévy process. Now taking $Y \in \mathbb{G}(\gamma, \beta)$ (being an increasing process!), we obtain

$$\log \mathbb{E}e^{isZ_1} = -\beta \log \left(1 + \frac{(\frac{1}{2}\sigma^2 s^2 - d\, is)}{\gamma}\right).$$

Picking $\gamma = \beta = 1/\kappa$, we call the resulting Lévy process $(Z_t)_t$ a *variance gamma process* [59, 154, 155, 196] with parameters d, σ^2, and κ; cf. expression [63, Eqn. (4.23)]; we use the notation $\mathbb{VG}(d, \sigma^2, \kappa)$.

There are at least two alternative ways to construct the variance gamma process. The first construction is the following. With a substantial amount of calculus, we can rewrite the Lévy exponent $\log \mathbb{E}e^{isZ_1}$ as

$$-\frac{1}{\kappa}\log\left(1+\frac{1}{2}\sigma^2\kappa s^2 - \kappa d\, is\right) = -\frac{1}{\kappa}\log\left(1 - \frac{i\cdot(-s)}{A_-}\right)$$
$$-\frac{1}{\kappa}\log\left(1-\frac{is}{A_+}\right), \tag{15.1}$$

with

$$A_- := \frac{W - d\kappa}{\sigma^2\kappa}, \quad A_+ := \frac{W + d\kappa}{\sigma^2\kappa}, \quad W := \sqrt{d^2\kappa^2 + 2\sigma^2\kappa}.$$

Conclude that $\mathbb{VG}(d, \sigma^2, \kappa)$ can be written as the difference between two gamma processes; with a bit of abuse of notation,

$$\mathbb{VG}(d, \sigma^2, \kappa) \stackrel{d}{=} \mathbb{G}(A_+, 1/\kappa) - \mathbb{G}(A_-, 1/\kappa). \tag{15.2}$$

An alternative way to represent the variance gamma process is as follows. Using the above representation in conjunction with the expression for the Lévy measure of the gamma process, we can characterize $\mathbb{VG}(d, \sigma^2, \kappa)$ through

$$\Pi(\mathrm{d}x) = \left(-\frac{1}{\kappa x}e^{A_- x}1_{\{x<0\}} + \frac{1}{\kappa x}e^{-A_+ x}1_{\{x>0\}}\right)\mathrm{d}x;$$

recall that A_+ and A_- are positive. It is concluded that the variance gamma process inherits from the gamma process the property of the 'small jumps': there are infinitely many jumps in a finite amount of time; as opposed to the gamma process, the variance gamma process has both positive and negative jumps. Observe that the Lévy measure behaves as $1/|x|$ for small x, and as a result the condition for the Brownian approximation (3.12) does not hold. In addition, it is verified that variance gamma processes are light tailed (due to the fact that the upper tail of the Lévy density is essentially exponential).

Generalized tempered stable processes—As we saw in Chapter 2, the class of α-stable Lévy motions has 'small jumps' (i.e. infinitely many jumps in a finite time interval), but also regularly varying tails. If we want to keep the former effect, but mitigate the latter, one could choose *tempered stable processes*. They are characterized by the following Lévy measure, for positive A_+, A_-, C_+, and C_-, and $\alpha_+, \alpha_- < 2$:

$$\Pi(\mathrm{d}x) = \left(\frac{C_-}{(-x)^{1+\alpha_-}} e^{A_- x} 1_{\{x<0\}} + \frac{C_+}{x^{1+\alpha_+}} e^{-A_+ x} 1_{\{x>0\}} \right) \mathrm{d}x; \tag{15.3}$$

the resulting process is light tailed. In the literature, the process obtained when choosing $\alpha := \alpha_+ = \alpha_-$ and $C := C_+ = C_-$ is usually referred to as the *CGMY process* [58], denoted by $\mathbb{C}\mathrm{GMY}(\alpha, C, A_+, A_-)$; an early reference for the case $\alpha_- = \alpha_+$ is [138]. It is noted that we have 'small jumps' when choosing α positive, while we have a compound Poisson process for negative α. It is seen that for $\alpha > 0$ the condition for the Brownian approximation (3.12) is met. The evaluation of the Lévy exponent is a routine calculation (recognize the gamma function!); it is noted that the cases $\alpha = 0$ and $\alpha = 1$ should be treated separately.

Processes in the β-class—The following model was introduced in Kuznetsov [143]; it has 'small jumps' (for certain parameter values) and light tails as well. As it belongs to the class $\mathscr{M}$ of meromorphic Lévy processes (see Section 3.5), it allows relatively easy numerical evaluation, relying on the techniques presented in Section 3.5. The β-class is characterized by as many as 10 parameters: obviously the deterministic drift $d \in \mathbb{R}$ and the variance σ^2 corresponding to the Brownian component, but also $\alpha_+ > 0$, $\beta_+ > 0$, $c_+ > 0$, and $\lambda_+ \in (0,3) \setminus \{1,2\}$ corresponding to the positive jumps, and $\alpha_- > 0$, $\beta_- > 0$, $c_- > 0$, and $\lambda_- \in (0,3) \setminus \{1,2\}$ corresponding to the negative jumps. More concretely, the Lévy measure is given by

$$\Pi(\mathrm{d}x) = \left(\frac{c_-}{(1 - e^{\beta_- x})^{\lambda_-}} e^{\alpha_- \beta_- x} 1_{\{x<0\}} + \frac{c_+}{(1 - e^{-\beta_+ x})^{\lambda_+}} e^{-\alpha_+ \beta_+ x} 1_{\{x>0\}} \right) \mathrm{d}x.$$

The corresponding Lévy exponent can be found by performing a standard computation; it involves the beta function. This family of processes has a relatively large number of parameters, and therefore offers a large amount of flexibility. For instance, one obtains 'small jumps' (infinite activity) by picking λ_- or λ_+ in the interval $(1,3)$.

Various other models have been proposed; see e.g. [144] for several other examples in the class of meromorphic Lévy processes $\mathscr{M}$.

15.2 Estimation

We now give a brief account of methods to estimate the parameters of the Lévy process based on N periodic observations $r_k := X_{k\Delta} - X_{(k-1)\Delta}$, for $k = 1, \ldots, N$. For a more complete treatment in the specific context of financial modeling we refer e.g. to Cont and Tankov [63, Chapter VII] or Sueishi and Nishiyama [204]. A review from a more statistical perspective can be found in Gugushvili [105].

The first method that is often used is *maximum likelihood estimation.* Then a class of Lévy processes should be picked, leaving us with estimation of the parameters corresponding to that class, for instance the vector $\boldsymbol{\theta} = (\alpha, C, A_+, A_-)$ in the case of the CGMY process. Suppose the density of X_Δ is known explicitly in terms of $\boldsymbol{\theta}$; call it $f(\cdot \mid \boldsymbol{\theta})$. Then the idea is to maximize the likelihood of the observations, that is,

$$\max_{\boldsymbol{\theta} \in \Theta} \prod_{k=1}^{N} f(r_k \mid \boldsymbol{\theta}),$$

or, equivalently, maximize the logarithm of the likelihood function $\ell(\boldsymbol{\theta}) := \sum_{k=1}^{N} \log f(r_k \mid \boldsymbol{\theta})$. This method has a number of obvious disadvantages.

- First, a parametric class of Lévy processes should be picked. If there is little prior knowledge of which model would fit well, it may sound reasonable to take a class with many parameters, for instance a generalized tempered stable process (6 parameters), or a process in the β-class proposed in Kuznetsov [143] (10 parameters). To allow a fair comparison between various models, however, one needs to include a penalty for the number of parameters, perhaps in the spirit of Akaike [4].
- For several classes of Lévy processes the density $f(\cdot \mid \boldsymbol{\theta})$ of X_Δ is not available in closed form, or it is available only in terms of special functions (whose evaluation may be costly).
- Third, there are numerical issues. For instance, it is not guaranteed that $\ell(\boldsymbol{\theta})$ is concave in $\boldsymbol{\theta}$. As a result, standard optimization procedures may end up in a local maximum.

An alternative to maximum likelihood is the *generalized method of moments* [106]. In this method, we need to have a vector-valued function $\boldsymbol{g}(\boldsymbol{\theta})$ such that

$$\boldsymbol{m}(\boldsymbol{\theta}^\star) := \mathbb{E}\, \boldsymbol{g}(X_\Delta \mid \boldsymbol{\theta}^\star) = 0,$$

where $\boldsymbol{\theta}^\star$ is the true value of the parameter (these are often referred to as the *moment conditions*). Now define the following estimator of $\boldsymbol{m}(\boldsymbol{\theta})$:

$$\hat{\boldsymbol{m}}(\boldsymbol{\theta}) := \frac{1}{N} \sum_{k=1}^{N} \boldsymbol{g}(r_k \mid \boldsymbol{\theta}).$$

The idea is now to estimate the parameters by performing the following minimization, for some positive-definite weight matrix W:

$$\min_{\boldsymbol{\theta} \in \Theta} [\boldsymbol{m}(\boldsymbol{\theta})]' \, W \boldsymbol{m}(\boldsymbol{\theta}). \tag{15.4}$$

Conditions have been established under which the resulting estimator (i.e. the minimizer of the optimization program (15.4)) has specific nice properties, e.g. consistency. Evidently, this procedure has drawbacks as well. In addition to some of the aspects mentioned above (with respect to the maximum likelihood estimator), it is mentioned that the choice of the function $\boldsymbol{g}(\boldsymbol{\theta})$ in the moment conditions has potentially a substantial impact on the efficiency of the estimator.

In specific cases alternative algorithms have been developed. For instance, in the case that X is a compound Poisson process (and hence the r_k correspond to a Poisson number of i.i.d. terms), interesting techniques have been developed; see e.g. [55, 56, 107, 212] and references therein.

15.3 Distribution of Running Maximum

As we will see later in this section, in financial applications knowledge of the distribution of the running maximum of a Lévy process X, that is,

$$\bar{X}_t := \sup_{0 \le s \le t} X_s,$$

in some cases jointly with the value of X_t, is of utmost importance. For general Lévy processes, this knowledge is immediately available from the Wiener–Hopf results. Apart from reviewing these results, we also present approximations that can be used when X is a subordinated Brownian motion.

We first consider the joint distribution of $\bar{X}_t$ and X_t. To this end, we trivially write

$$\alpha \bar{X}_t + \beta X_t = (\alpha + \beta)\bar{X}_t + \beta(X_t - \bar{X}_t).$$

The key step is that, with T being exponentially distributed with mean ϑ^{-1}, according to Wiener–Hopf theory, the random variables $\bar{X}_T$ and $X_T - \bar{X}_T$ are independent [43, 146]; see also Section 3.3. It is also seen that, with as usual $X'_t := -X_t$,

$$X_T - \bar{X}_T \stackrel{\mathrm{d}}{=} -\bar{X}'_T.$$

Focusing for the moment on $X \in \mathscr{S}_+$, the above findings give after elementary calculations (using the explicit expressions for $X \in \mathscr{S}_+$ as given in Section 3.3) that, for $\alpha, \beta \ge 0$,

$$\mathbb{E}e^{-\alpha \bar{X}_T - \beta X_T} = \mathbb{E}e^{-(\alpha+\beta)\bar{X}_T}\,\mathbb{E}e^{\beta \bar{X}'_T} = \frac{\vartheta}{\vartheta - \varphi(\alpha+\beta)}\,\frac{\psi(\vartheta) - (\alpha+\beta)}{\psi(\vartheta) - \beta}, \tag{15.5}$$

where we use that $X' \in \mathscr{S}_-$. Likewise, for $X \in \mathscr{S}_-$, for $\alpha, \beta \ge 0$, with T being exponentially distributed with mean q^{-1}, again relying on the findings of

Section 3.3,

$$\mathbb{E}e^{\alpha\bar{X}_T+\beta X_T} = \frac{\Psi(q)-\beta}{\Psi(q)-(\alpha+\beta)}\frac{q}{q-\Phi(\beta)}, \tag{15.6}$$

using that $X' \in \mathscr{S}_+$.

In the case that X is not spectrally one sided, we have results in terms of just Wiener–Hopf factors. As before, we first use the independence of $\bar{X}_T$ and $X_T - \bar{X}_T$, to obtain

$$\mathbb{E}e^{\mathrm{i}\alpha\bar{X}_T+\mathrm{i}\beta X_T} = \mathbb{E}e^{\mathrm{i}(\alpha+\beta)\bar{X}_T}\mathbb{E}e^{\mathrm{i}\beta(X_T-\bar{X}_T)}.$$

From Thm. 3.4 we know how to compute these quantities.

As is the case for the stationary and transient workloads, this result does not lend itself for further evaluation in terms of the Lévy exponent of X in the spectrally two-sided case, except when either the upward jumps or the downward jumps are of phase type. This observation suggests that evaluation of the distribution of $\bar{X}_t$ is hard for the spectrally two-sided processes introduced in Section 15.1 (normal inverse Gaussian, variance gamma, CGMY), but there turns out to be an interesting approximation.

Subordinated Brownian motion—Let X be $\mathbb{B}\mathrm{m}(d,\sigma^2)$ and Y be an increasing Lévy process (or *subordinator*), and define, as before, $(Z_t)_t := (X_{Y_t})_t$; we call the process Z a *subordinated Brownian motion*. Then, due to the fact that Y is increasing, we obviously have

$$X_{Y_t} = Z_t \le \bar{Z}_t = \sup_{s\in[0,t]} Z_s = \sup_{s\in[0,t]} X_{Y_s} \le \sup_{s\in[0,Y_t]} X_s = \bar{X}_{Y_t},$$

and hence

$$\mathbb{P}(X_{Y_t} \ge x) \le \mathbb{P}(\bar{Z}_t \ge x) \le \int_0^\infty \mathbb{P}(\bar{X}_y \ge x)\mathbb{P}(Y_t \in \mathrm{d}y).$$

Relying on the explicit result for the transient workload of a Brownian-motion-driven queue (4.6), we thus have

$$\mathbb{P}(\bar{Z}_t \ge x) \le \int_0^\infty \left(\Phi_{\mathrm{N}}\left(\frac{-x+dy}{\sigma\sqrt{y}}\right) + e^{2dx/\sigma^2}\Phi_{\mathrm{N}}\left(\frac{-x-dy}{\sigma\sqrt{y}}\right)\right)\mathbb{P}(Y_t \in \mathrm{d}y)$$

and

$$\mathbb{P}(\bar{Z}_t \ge x) \ge \int_0^\infty \Phi_{\mathrm{N}}\left(\frac{-x+dy}{\sigma\sqrt{y}}\right)\mathbb{P}(Y_t \in \mathrm{d}y).$$

We conclude that if we can write Z as a subordinated Brownian motion, this elementary procedure can be applied to find bounds on the distribution of $\bar{Z}_t$. Clearly, the upper bound will be tight if the jumps of the subordinator are typically small.

In the case that the drift d is negative, we can find logarithmic asymptotics of $\mathbb{P}(\bar{Z}_t \geq x)$ for a large class of subordinating processes Y. Following Dębicki et al. [73], we sketch the idea behind the derivation of these logarithmic asymptotics for the case that Y_t has a density function of the form $L(x)/x^{\lambda+1}$, where $L(\cdot)$ is slowly varying at infinity and $\lambda > 0$. For the upper bound we use that

$$\mathbb{P}(\bar{Z}_t \geq x) \leq \mathbb{P}\left(\sup_{s\in[0,\infty)} X_s \geq x\right) = \exp\left(\frac{2dx}{\sigma^2}\right).$$

To get the lower bound, we first observe that

$$\begin{aligned} \mathbb{P}(\bar{Z}_t \geq x) &\geq \mathbb{P}(X_{Y_t} \geq x) \\ &\geq \min_{s\in[-x/d-\sqrt{x},-x/d+\sqrt{x}]} \mathbb{P}(X_s \geq x)\mathbb{P}\left(Y_s \in \left[-\frac{x}{d} - \sqrt{x}, -\frac{x}{d} + \sqrt{x}\right]\right). \end{aligned}$$

Using that $\mathbb{P}(Y_s \in [-x/d - \sqrt{x}, -x/d + \sqrt{x}])$ is regularly varying at infinity and

$$\lim_{x\to\infty} \frac{\log\left(\min_{s\in[-x/d-\sqrt{x},-x/d+\sqrt{x}]} \mathbb{P}(X_s \geq x)\right)}{x} = \frac{2d}{\sigma^2},$$

we obtain the following result [73].

Theorem 15.1 *Let X be $\mathbb{B}\mathrm{m}(d, \sigma^2)$, with $d < 0$ and Y be an increasing Lévy process. Assume that Y_t has an absolutely continuous distribution with a probability density function of the form $L(x)/x^{\lambda+1}$, where $L(\cdot)$ is slowly varying at infinity and $\lambda > 0$. Then, with $(Z_t)_t = (X_{Y_t})_t$,*

$$\lim_{x\to\infty} \frac{\log \mathbb{P}(\bar{Z}_t \geq x)}{x} = \frac{2d}{\sigma^2}.$$

The case that the subordinating process Y is light tailed can be addressed as well; it needs a more subtle approach, however, leading to several scenarios. We refer to [13, 73] for more results on asymptotics of a subordinated Brownian motion and extensions to more general subordinated Gaussian processes.

Likewise, we can also find an upper bound on $\mathbb{P}(\bar{Z}_t \geq x, Z_t \in \mathrm{d}z)$. Relying on explicit results for the Brownian bridge, we can find the joint distribution of $\bar{X}_t$ and X_t:

$$\mathbb{P}(\bar{X}_t \geq x, X_t \in \mathrm{d}y) = \frac{1}{\sqrt{2\pi t}\sigma} \exp\left(-\frac{(y-dt)^2}{2\sigma^2 t}\right) \mathbb{P}(\bar{X}_t \geq x \,|\, X_t = y)\,\mathrm{d}y,$$

where

$$\mathbb{P}(\bar{X}_t \geq x \,|\, X_t \in \mathrm{d}y) = \exp\left(-\frac{2x}{\sigma^2 t}(x-y)\right)$$

for $x \geq y$ and 1 otherwise. We thus find

$$\mathbb{P}(\bar{Z}_t \geq x, Z_t \in \mathrm{d}z) \leq \int_{y=0}^{\infty} \mathbb{P}(\bar{X}_y \geq x, X_y \in \mathrm{d}z)\mathbb{P}(Y_t \in \mathrm{d}y).$$

This leaves us with the question of how we can verify whether a Lévy process corresponds to a subordinated Brownian motion. In this respect, the following equivalence is of crucial importance [63, Thm. 4.3]. Without loss of generality we take the variance σ^2 of the Brownian motion equal to 1.

Theorem 15.2 *A Lévy process Z is a subordinated Brownian motion (i.e.* $(Z_t)_t := (X_{Y_t})_t$ *with X being* $\mathbb{B}\mathrm{m}(d,1)$ *and Y being an increasing Lévy process) if and only if the following three conditions hold:*

- *the Lévy measure* Π *of Z is absolutely continuous, with associated density* $\pi(x)$*;*
- $\pi(x)e^{-dx} = \pi(-x)e^{dx}$ *for all x;*
- $\pi(\sqrt{x})e^{-d\sqrt{x}}$ *is completely monotone on* $(0,\infty)$*.*

This result can be used to show that normal inverse Gaussian, variance gamma, and CGMY are all subordinated Brownian motions. For normal inverse Gaussian and variance gamma we already know that property from their very construction, but it can now easily be re-proved by the above result. CGMY can also be shown to be a subordinated Brownian motion [176, 221], with $\Pi(x)$ containing a parabolic cylinder function [218, p. 347]; for the broader class of tempered stable processes this is not necessarily true (in other words, one should choose $\alpha_- = \alpha_+$ and $C_- = C_+$ in (15.3)).

15.4 Option Pricing: Payoff Structures

In this and the next section we demonstrate how the results on the running maximum facilitate option pricing; for more background we refer e.g. to [63, 170, 171]. We start with simpler options, and gradually look at increasingly complicated variants. As indicated earlier, we have that $S_t = S_0 e^{X_t}$ for a Lévy process $(X_t)_t$ and a known value S_0.

Vanilla options—The simplest of all options is the so-called *vanilla call option*. The call variant is defined through its *payoff function*

$$P^{(c)}_{\mathrm{van}}(T,K) := \max\{S_T - K, 0\} = (S_T - K)^+;$$

informally, this means that the holder has the right to buy the asset for K at the maturity date T. Similarly, $P_{\mathrm{van}}^{(\mathrm{p})}(T,K) := (K - S_T)^+$ is the price of the right to sell, referred to as the vanilla *put* option.

Lookback options—Vanilla options have a payoff structure that depends on the price evolution of the underlying asset only through the price at expiration. There is an abundance of exotic options that are traded nowadays, however, with payoff structures that are substantially more involved. Lookback options are examples of derivatives for which the payoff depends on the maximum (or minimum) price over the life of the option, and possibly the price of the underlying asset at maturity as well. They come in two flavors: lookback options with a *fixed* strike, and those with a *floating* strike.

With the stochastic process S_t representing the evolution of the stock prices and $\bar{S}_T := \sup_{0\leq t\leq T} S_t$ the associated *running maximum process*, the payoff of the fixed-strike call option is

$$P_{\mathrm{fix}}^{(\mathrm{c})}(T,K) := \max\{\bar{S}_T - K, 0\} = (\bar{S}_T - K)^+,$$

with strike price K and maturity time T; analogously, the payoff of the put counterpart is given by $P_{\mathrm{fix}}^{(\mathrm{p})}(T,K) := (K - \underline{S}_T)^+$, with $(\underline{S}_t)_t$ the running minimum process. As indicated by these payoffs, this type of option has a fixed, a priori known strike price, but as opposed to the 'traditional' European option, the underlying trigger is not the price at maturity but rather the maximum (or minimum) of the underlying asset price over the life of the option.

The payoff of the floating-strike call option is

$$P_{\mathrm{fl}}^{(\mathrm{c})}(T,L) := \max\{S_T - L\underline{S}_T, 0\} = (S_T - L\underline{S}_T)^+;$$

in the case $L \leq 1$ the payoff is always non-negative, and reduces to $S_T - L\underline{S}_T$. This means that the strike price is fixed at the asset's minimal price during the option's life, multiplied by a specified constant L. The payoff of the put counterpart is defined by $P_{\mathrm{fl}}^{(\mathrm{p})}(T,L) := \max\{L\bar{S}_T - S_T, 0\}$, which reduces to $L\bar{S}_T - S_T$ if $L \geq 1$.

Importantly, unlike vanilla options, the lookback options discussed above, as well as other exotic options, have a *path-dependent* payoff. This means that their payoff does not depend on S_T alone, but also involves a certain functional of the process S_t, for $0 \leq t \leq T$ (i.e. the maximum or minimum value attained). As a consequence it is highly non-trivial to price such options, or to numerically assess the sensitivities of the price with respect to the various model parameters such as the maturity and the initial price of the underlying asset (the 'Greeks').

Barrier options—Another type of option involves both the running maximum (or minimum), and the process' value at the maturity time. Consider for instance the so-called *Up-and-In barrier* call option, where it is remarked that other flavors

(Up-and-Out, Down-and-In, Down-and-Out, and the put variants) can be defined analogously [171]. The Up-and-In barrier call option has payoff

$$P^{(c)}_{uib}(T,K,H) := (S_T - K)^+ 1_{\{\bar{S}_T \geq H\}};$$

we are interested in the more challenging case that $\max\{S_0, K\} < H$ (noting that if this condition is not fulfilled the payoff is non-negative with certainty).

Conclude from the above that all payoff functions depend on S_t, $\bar{S}_t$, $\underline{S}_t$, or a subset of these. In Section 15.3 we found expressions for the joint distribution of $\bar{X}_t$ and X_t (which were quite implicit, i.e. in terms of transforms); expressions for the joint distribution of $\underline{X}_t$ and X_t can be found analogously. This means that, in principle, we can evaluate the prices of the options. The next section further elaborates on this.

15.5 Option Pricing: Transforms of Prices

In this section we present results on the transforms of option prices, and also pay some attention to corresponding sensitivities ('Greeks'). We do so for the vanilla option, lookback option, and barrier option, as introduced in the previous section. We consider the usual setup, as introduced in more detail e.g. in Nguyen-Ngoc [170] and Asghari and Mandjes [17]: a market with two basic assets, that is, the usual bank account with an interest rate $r > 0$, and the option associated with an underlying asset whose evolution in time is represented by the stochastic process S_t.

Locally, we use short notation for the Laplace transforms of $\bar{X}_{T(q)}$ and $\underline{X}_{T(q)}$, with $T(q)$ being an exponentially distributed random variable with mean q^{-1}. We introduce

$$\kappa^+(\alpha,q) := \mathbb{E}\, e^{-\alpha \bar{X}_{T(q)}} = \frac{k(q,\alpha)}{k(q,0)}, \quad \kappa^-(\alpha,q) := \mathbb{E}\, e^{-\alpha \underline{X}_{T(q)}} = \frac{\bar{k}(q,-\alpha)}{\bar{k}(q,0)}.$$

In addition,

$$\begin{aligned}
\kappa^+(\alpha,q)\kappa^-(\alpha,q) &= \mathbb{E}e^{-\alpha \bar{X}_{T(q)}}\, \mathbb{E}e^{-\alpha \underline{X}_{T(q)}} = \mathbb{E}e^{-\alpha \bar{X}_{T(q)}}\, \mathbb{E}e^{-\alpha X_{T(q)} + \alpha \bar{X}_{T(q)}} \\
&= \mathbb{E}e^{-\alpha X_{T(q)}} = \int_0^\infty q e^{-qt} \mathbb{E}e^{-\alpha X_t} \mathrm{d}t \\
&= \int_0^\infty q \left(\exp(-q + \log \mathbb{E}e^{-\alpha X_1})\right)^t \mathrm{d}t \\
&= \frac{q}{q - \log \mathbb{E}e^{-\alpha X_1}} =: \mathscr{K}(\alpha,q). \qquad (15.7)
\end{aligned}$$

Vanilla options—Our goal is to compute the price of the vanilla option, that is, for given numbers K and T,

$$V_{\rm van}^{(c)}(T,K) := \mathbb{E}\left[e^{-rT}P_{\rm van}^{(c)}(T,K)\right];$$

the analysis of the put counterpart works similarly. It requires some elementary algebra to verify that, with $k := \log(K/S_0)$,

$$V_{\rm van}^{(c)}(T,K) = S_0\, e^{-rT}\int_k^{\infty}(e^x - e^k)\mathbb{P}(X_T \in {\rm d}x).$$

Let $\hat{V}_{\rm van}^{(c)}(T,\alpha)$ be the Fourier transform with respect to k:

$$\hat{V}_{\rm van}^{(c)}(T,\alpha) := S_0 e^{-rT}\int_{-\infty}^{\infty} e^{{\rm i}\alpha k}e^{\eta k}\int_k^{\infty}\left(e^x - e^k\right)\mathbb{P}(X_T \in {\rm d}x){\rm d}k,$$

where $\eta > 0$ is a damping factor. By changing the integration order it is readily found that

$$\begin{aligned}\hat{V}_{\rm van}^{(c)}(T,\alpha) &= \frac{S_0 e^{-rT}}{({\rm i}\alpha+\eta)({\rm i}\alpha+\eta+1)}\mathbb{E}e^{({\rm i}\alpha+\eta+1)X_T}\\ &= \frac{S_0}{({\rm i}\alpha+\eta)({\rm i}\alpha+\eta+1)}\left(e^{-r}\,\mathbb{E}e^{({\rm i}\alpha+\eta+1)X_1}\right)^T, \end{aligned} \tag{15.8}$$

where the last step is due to the Lévy nature of X_t. We have expressed the transform $\hat{V}_{\rm van}^{(c)}(T,\alpha)$ in terms of the Lévy exponent corresponding to X_t and the maturity T.

We now determine the transforms of a set of Greeks, that is, sensitivities. We focus on the sensitivities with respect to the initial price of the underlying asset S_0 and the maturity T; in the sequel we refer to these Greeks as Δ and Θ. Regarding the former, it is elementary to verify that

$$\Delta_{\rm van}^{(c)}(T,K) := \frac{\partial V_{\rm van}^{(c)}(T,K)}{\partial S_0} = e^{-rT}\int_{\log(K/S_0)}^{\infty} e^x\mathbb{P}(X_T \in {\rm d}x).$$

Writing $k := \log(K/S_0)$ and transforming to k in the same way as above, we obtain the transform

$$\begin{aligned}\hat{\Delta}_{\rm van}^{(c)}(T,\alpha) &:= e^{-rT}\int_{-\infty}^{\infty} e^{{\rm i}\alpha k}e^{\eta k}\int_k^{\infty} e^x\,\mathbb{P}(X_T \in {\rm d}x){\rm d}k\\ &= \frac{1}{{\rm i}\alpha+\eta}\left(e^{-r}\,\mathbb{E}e^{({\rm i}\alpha+\eta+1)X_1}\right)^T. \end{aligned}$$

Realize that the expression in the right-hand side implicitly depends on S_0, as $k = \log(K/S_0)$.

We now concentrate on the Greek with respect to the maturity time. With

$$\Theta^{(c)}_{\rm van}(T,K) := \frac{\partial V^{(c)}_{\rm van}(T,K)}{\partial T},$$

we have that the corresponding transform equals

$$\hat{\Theta}^{(c)}_{\rm van}(T,\alpha) := \frac{S_0}{(\mathrm{i}\alpha+\eta)(\mathrm{i}\alpha+\eta+1)}\left(e^{-r}\,\mathbb{E}e^{(\mathrm{i}\alpha+\eta+1)X_1}\right)^T\left(\log\mathbb{E}e^{(\mathrm{i}\alpha+\eta+1)X_1}-r\right).$$

It is noted that transforms of second-order Greeks can be determined similarly.

The vanilla options are *path independent*, in the sense that their prices depend on the asset price process only through the asset price at maturity time T, and are independent of the specific shape of the path during the time interval $(0,T)$. The lookback options and barrier options, which we are going to study now, *are* path dependent.

Fixed-strike lookback options—We now focus on pricing fixed-strike lookback options; again we present our analysis for the call option, but the put variant is dealt with analogously. In our derivations, we follow the same line of reasoning as in Nguyen-Ngoc [170]. Our goal is to evaluate, in terms of transforms,

$$V^{(c)}_{\rm fix}(T,K) := \mathbb{E}\left[e^{-rT}P^{(c)}_{\rm fix}(T,K)\right],$$

as well as its Greeks with respect to S_0 and T; recall the definition of the payoff $P^{(c)}_{\rm fix}(T,K)$ from Section 15.4. If $K\leq S_0$, it automatically follows that $P^{(c)}_{\rm fix}(T,K)=\bar{S}_T-K$. Realize that this case corresponds to a 'riskless' option, for which it is guaranteed that the payoff is non-negative. Let us therefore turn to the more realistic setting in which $K>S_0$.

We again parameterize $k=\log(K/S_0)$, which is now necessarily positive. Let $\hat{V}^{(c)}_{\rm fix}(\vartheta,\alpha)$ be the transform with respect to k and T:

$$\hat{V}^{(c)}_{\rm fix}(\vartheta,\alpha) := \int_0^\infty \vartheta e^{-\vartheta T}\int_0^\infty e^{-\alpha k}\,V^{(c)}_{\rm fix}(T,K)\,\mathrm{d}k\,\mathrm{d}T.$$

The idea of including the maturity T as an exponential random variable was first proposed in Geman and Yor [98] for barrier options, but just for the Black–Scholes model. This expression can be rewritten as the threefold integral

$$S_0\int_0^\infty \vartheta e^{-(r+\vartheta)T}\int_0^\infty e^{-\alpha k}\int_k^\infty (e^x-e^k)\,\mathbb{P}(\bar{X}_T\in\mathrm{d}x)\,\mathrm{d}k\,\mathrm{d}T,$$

which we in the sequel assume to converge. Now change the order of integration: first integrate over $k \in [0, x]$, so as to obtain

$$S_0 \frac{\vartheta}{r+\vartheta} \int_0^\infty (r+\vartheta) e^{-(r+\vartheta)T}$$
$$\int_0^\infty \left(\frac{1}{\alpha} \left(e^x - e^{(1-\alpha)x} \right) - \frac{1}{\alpha - 1} \left(1 - e^{(1-\alpha)x} \right) \right) \mathbb{P}(\bar{X}_T \in \mathrm{d}x) \, \mathrm{d}T.$$

This quantity can be expressed in term of transforms related to the running maximum after an exponentially distributed time with mean $(r+\vartheta)^{-1}$:

$$S_0 \frac{\vartheta}{r+\vartheta} \left(\frac{1}{\alpha} \left(\mathbb{E} e^{\bar{X}_{T(r+\vartheta)}} - \mathbb{E} e^{(1-\alpha)\bar{X}_{T(r+\vartheta)}} \right) - \frac{1}{\alpha - 1} \left(1 - \mathbb{E} e^{(1-\alpha)\bar{X}_{T(r+\vartheta)}} \right) \right).$$

This expression can be written in terms of the transform $\kappa^+(\alpha, q)$ introduced earlier:

$$S_0 \frac{\vartheta}{r+\vartheta} \left(\frac{\kappa^+(-1, r+\vartheta) - \kappa^+(\alpha - 1, r+\vartheta)}{\alpha} - \frac{1 - \kappa^+(\alpha - 1, r+\vartheta)}{\alpha - 1} \right).$$

The Greek related to the initial asset price S_0 is

$$\Delta_{\text{fix}}^{(c)}(T, K) := \frac{\partial V_{\text{fix}}^{(c)}(T, K)}{\partial S_0} = e^{-rT} \int_{\log(K/S_0)}^\infty e^x \mathbb{P}(\bar{X}_T \in \mathrm{d}x).$$

With the usual transformation $k = \log(K/S_0)$, we find

$$\hat{\Delta}_{\text{fix}}^{(c)}(\vartheta, \alpha) := \int_0^\infty \vartheta e^{-(r+\vartheta)T} \int_{-\infty}^\infty e^{-\alpha k} \int_k^\infty e^x \, \mathbb{P}(\bar{X}_T \in \mathrm{d}x) \mathrm{d}k \, \mathrm{d}T$$
$$= \frac{\vartheta}{r+\vartheta} \frac{\kappa^+(-1, r+\vartheta) - \kappa^+(\alpha - 1, r+\vartheta)}{\alpha}.$$

Now consider the Greek with respect to the maturity T. Interchanging the order of the integrals and integration by parts yields

$$\hat{\Theta}_{\text{fix}}^{(c)}(\vartheta, \alpha) := \int_0^\infty \vartheta e^{-\vartheta T} \int_0^\infty e^{-\alpha k} \frac{\partial V_{\text{fix}}^{(c)}(T, K)}{\partial T} \, \mathrm{d}k \, \mathrm{d}T = \vartheta \hat{V}_{\text{fix}}^{(c)}(\vartheta, \alpha).$$

Floating-strike lookback options—In this subsection the focus lies on floating-strike lookback options, presenting, as usual, the results for the call variant. We characterize, in terms of transforms,

$$V_{\text{fl}}^{(c)}(T, L) := \mathbb{E}\left[e^{-rT} P_{\text{fl}}^{(c)}(T, L) \right],$$

as well as its Greeks with respect to S_0 and T; the payoff function $P_{\rm fl}^{(c)}(T,L)$ is defined in the previous section. If $L \leq 1$, this payoff equals $S_T - L\underline{S}_T$, being non-negative, and allowing for relatively easy evaluation. We therefore focus on the more realistic (and challenging) setting in which $L > 1$.

We parameterize $\ell := \log L$ (which is positive), and define

$$\hat{V}_{\rm fl}^{(c)}(\vartheta,\alpha) = \int_0^\infty \vartheta e^{-\vartheta T} \int_0^\infty e^{-\alpha\ell} V_{\rm fl}^{(c)}(T,e^\ell)\,{\rm d}\ell\,{\rm d}T.$$

After some algebra, it is seen that this expression equals

$$S_0 \int_0^\infty \vartheta e^{-(r+\vartheta)T} \int_0^\infty e^{-\alpha\ell} \int_{y=-\infty}^0 \int_{x=\ell+y}^\infty (e^x - e^{\ell+y})\mathbb{P}(X_T \in {\rm d}x, \underline{X}_T \in {\rm d}y)\,{\rm d}\ell\,{\rm d}T.$$

Interchange the order of the integrals, such that first the integral over $\ell \in [0, x-y]$ is evaluated. With the inner integral corresponding to the variable y and the 'middle' integral to x, this reduces to

$$S_0 \int_0^\infty \vartheta e^{-(r+\vartheta)T} \int_{-\infty}^\infty \int_{-\infty}^0$$
$$\left(e^x \frac{(1-e^{-\alpha(x-y)})}{\alpha} - e^y \frac{(1-e^{-(\alpha-1)(x-y)})}{\alpha-1}\right)\mathbb{P}(X_T \in {\rm d}x, \underline{X}_T \in {\rm d}y){\rm d}T.$$

We thus find

$$\hat{V}_{\rm fl}^{(c)}(\vartheta,\alpha) = S_0 \frac{\vartheta}{r+\vartheta}\left(\frac{1}{\alpha(\alpha-1)}\mathbb{E}\left[e^{-(\alpha-1)X_{T(r+\vartheta)}+\alpha\underline{X}_{T(r+\vartheta)}}\right] + \frac{1}{\alpha}\mathbb{E}e^{X_{T(r+\vartheta)}} - \frac{1}{\alpha-1}\mathbb{E}e^{\underline{X}_{T(r+\vartheta)}}\right).$$

Consider the first term between the brackets in the previous display. By virtue of (i) the trivial identity $-(\alpha-1)x + \alpha\underline{x} = (\alpha-1)(\underline{x}-x) + \underline{x}$, (ii) the fact that (due to Wiener–Hopf theory) $\underline{X}_{T(r+\vartheta)}$ and $X_{T(r+\vartheta)} - \underline{X}_{T(r+\vartheta)}$ are independent, and (iii) the fact that $\underline{X}_{T(r+\vartheta)} - X_{T(r+\vartheta)}$ is distributed as $-\bar{X}_{T(r+\vartheta)}$, we have that

$$\mathbb{E}\left[e^{-(\alpha-1)X_{T(r+\vartheta)}+\alpha\underline{X}_{T(r+\vartheta)}}\right] = \mathbb{E}e^{-(\alpha-1)\bar{X}_{T(r+\vartheta)}}\,\mathbb{E}e^{\underline{X}_{T(r+\vartheta)}}.$$

Using the notation introduced in Eqn. (15.7), these considerations eventually lead to the identity

$$\hat{V}_{\text{fl}}^{(\text{c})}(\vartheta,\alpha) = S_0 \frac{\vartheta}{r+\vartheta}\left(\frac{\kappa^+(\alpha-1,r+\vartheta)\kappa^-(-1,r+\vartheta)}{\alpha(\alpha-1)} + \frac{\mathscr{K}(-1,r+\vartheta)}{\alpha} - \frac{\kappa^-(-1,r+\vartheta)}{\alpha-1}\right).$$

We now turn to the Greeks. The quantity $\Delta_{\text{fl}}^{(\text{c})}(T,L)$, defined in the obvious way, is simply $\hat{V}_{\text{fl}}^{(\text{c})}(\vartheta,\alpha)/S_0$, which is independent of S_0; it is evident from the definition of the payoff that $\hat{V}_{\text{fl}}^{(\text{c})}(\vartheta,\alpha)$ is linear in S_0. Regarding, in self-evident notation, $\Theta_{\text{fl}}^{(\text{c})}(T,L)$, it is seen that, with the same line of reasoning as used for the fixed-strike lookback option, $\hat{\Theta}_{\text{fl}}^{(\text{c})}(\vartheta,\alpha) = \vartheta\hat{V}_{\text{fl}}^{(\text{c})}(\vartheta,\alpha)$.

Barrier options—In this situation, we are interested in the more challenging case that $\max\{S_0,K\} < H$ (noting that if this condition is not fulfilled the payoff is non-negative with certainty). Putting $k := \log(K/S_0)$ and $h := \log(H/S_0)$, we wish to evaluate

$$V_{\text{uib}}^{(\text{c})}(T,K,H) := \mathbb{E}\left[e^{-rT}P_{\text{uib}}^{(\text{c})}(T,K,H)\right].$$

Now let $\hat{V}_{\text{uib}}^{(\text{c})}(\vartheta,\alpha,\beta)$ be the transform with respect to k, h, and T:

$$S_0\int_0^\infty \vartheta e^{-(r+\vartheta)T}\int_{-\infty}^\infty\int_0^\infty e^{\alpha i k}e^{-\beta h}\int_{y=k}^\infty\int_{x=h}^\infty (e^y-e^k)\mathbb{P}(\bar{X}_T\in \mathrm{d}x, X_T\in\mathrm{d}y)\mathrm{d}h\,\mathrm{d}k\,\mathrm{d}T.$$

This expression reduces, when interchanging the order of integration, to

$$\frac{S_0}{\mathrm{i}\alpha(\mathrm{i}\alpha+1)\beta}\int_0^\infty \vartheta e^{-(r+\vartheta)T}\int_{y=-\infty}^\infty\int_{x=0}^\infty e^{(\mathrm{i}\alpha+1)y}(1-e^{-\beta x})\mathbb{P}(\bar{X}_T\in\mathrm{d}x, X_T\in\mathrm{d}y)\mathrm{d}T$$
$$= \frac{S_0}{\mathrm{i}\alpha(\mathrm{i}\alpha+1)\beta}\frac{\vartheta}{r+\vartheta}\mathbb{E}\left(e^{(\mathrm{i}\alpha+1)X_{T(r+\vartheta)}}\left(1-e^{-\beta\bar{X}_{T(r+\vartheta)}}\right)\right),$$

which can be expressed in terms of the functions $\kappa^+(\alpha,q)$, $\kappa^-(\alpha,q)$, and $\mathscr{K}(\alpha,q)$ (as we did for the floating-strike lookback option):

$$S_0\frac{\vartheta}{r+\vartheta}\left(\frac{\mathscr{K}(-\mathrm{i}\alpha-1,r+\vartheta)-\kappa^+(-\mathrm{i}\alpha-1+\beta,r+\vartheta)\kappa^-(-\mathrm{i}\alpha-1,r+\vartheta)}{\mathrm{i}\alpha(\mathrm{i}\alpha+1)\beta}\right).$$

The Greeks can be characterized (in terms of transforms) as before.

We conclude this section with a few remarks. Above we demonstrated how to compute the first moment of the option prices (where a discount factor r is imposed). In making portfolio decisions, it may be that one needs to have information about higher moments as well, for instance the second moment (to be able to compute the variance). Transforms of higher moments can be derived analogously.

Besides the options mentioned above, there is a plethora of alternatives. One variant is the so-called American option, in which the holder decides the moment of executing the option. This means that, in order to compute the price of such an option (for instance in terms of a first moment), an optimal stopping problem needs to be solved; see e.g. Mordecki [167].

15.6 Applications in Non-life Insurance

In this section we treat a selection of problems arising in non-life insurance that can be analyzed using Lévy fluctuation theory. An important role is played by a duality relation between ruin models and specific associated queueing models.

Lévy risk processes—In collective risk theory the risk reserve process $(U_t)_t$, describing the dynamics of the insurer's capital balance in time, is modeled as

$$U_t = u + ct - S_t, \tag{15.9}$$

where u is the initial capital (at time 0) of the insurance company, $c > 0$ is the (constant) premium rate per unit of time, and $(S_t)_t$ is a stochastic process that models the cumulative claims.

The celebrated Cramér–Lundberg model, to be found in e.g. [21, 162], assumes that successive claims $Y_1, Y_2, \ldots$ arrive according to a Poisson process $(N_t)_t$ with constant intensity, say $\lambda > 0$. The claims form a sequence of i.i.d. non-negative random variables with finite mean $\mu > 0$, which is also assumed to be independent of the arrival process. As such, the cumulative claim process $(S_t)_t$ is a compound Poisson process, and can be represented as

$$S_t = \sum_{k=1}^{N_t} Y_k.$$

Although the Cramér–Lundberg model has a natural and elegant interpretation, in many cases the framework appears to be too restrictive. This explains why recent studies focus on more flexible and richer classes of processes to model the cumulative claim process. In particular, there is a substantial body of work that considers the idea of representing $(S_t)_t$ by a Lévy process with finite mean.

The theoretical justification behind approximating the risk reserve process by a Lévy motion goes back to work of Iglehart [113]. He found, under the assumption that $\mathbb{E}Y_1^2 < \infty$, that an appropriately scaled sequence of Cramér–Lundberg risk reserve processes may be approximated by Brownian motion with linear drift. In the case that the condition $\mathbb{E}Y_1^2 < \infty$ does not hold, an α-stable approximation was established by Furrer et al. [97]. The idea that underlies these approximations is that time is sped up, but simultaneously the claim sizes are made smaller; this is done in such a way that the corresponding risk processes converge to a non-degenerate limit.

More specifically, the scaling considered is the following. Let $S_1^{(\alpha)} \stackrel{d}{=} S_\alpha(1, \beta, 0)$ for $\alpha < 2$, and let $S_1^{(2)}$ be distributed as a standard normal random variable. We suppose that the partial sums of the claims $Y_1, Y_2, \ldots$ obey the following limiting law as $n \to \infty$:

$$\frac{1}{d(n)} \sum_{k=1}^{n} (Y_k - \mu) \xrightarrow{d} S_1^{(\alpha)},$$

where $d(n) := An^{1/\alpha}$ with $\alpha \in (1, 2]$, and A is some positive constant.

Then we consider the following sequence of Cramér–Lundberg risk processes:

$$U_t^{(n)} := u + c_n t - \frac{1}{d(n)} \sum_{k=1}^{N_{nt}} Y_k, \tag{15.10}$$

for $n = 1, 2, \ldots$. Here the premium rates are equal to $c_n := c + \lambda\mu A^{-1} n^{1-1/\alpha}$, and $(N_t)_t$ denotes a Poisson process with intensity $\lambda > 0$. It is readily verified that

$$U_t^{(n)} = u + ct - \frac{N_{nt} - \lambda nt}{d(n)} \mu - \frac{1}{d(n)} \sum_{k=1}^{N_{nt}} (Y_k - \mu).$$

From this we draw the following two conclusions. First, it follows that for $\alpha \in (1, 2)$, as $n \to \infty$,

$$\frac{N_{nt} - \lambda nt}{d(n)} \to 0$$

in probability in the Skorokhod topology. Second, if $\alpha = 2$, then

$$\left(\frac{N_{nt} - \lambda nt}{d(n)} \right)_t \xrightarrow{d} (S_t)_t$$

in the sense of weak convergence in $(D[0, \infty), J_1)$, where $S \in \mathbb{B}\mathrm{m}\left(0, \lambda/A^2\right)$. These observations straightforwardly lead to the following result; see Iglehart [113] and Furrer et al. [97] for detailed proofs.

Theorem 15.3 *Let* $(U_t^{(n)})_t$, $n = 1, 2, \ldots$ *be a sequence of Cramér–Lundberg risk processes as defined through* (15.10).

(i) If $\alpha \in (1, 2)$, *then with* $S^{(\alpha)} \in \mathbb{S}(\alpha, \beta, 0)$, *as* $n \to \infty$,

$$(U_t^{(n)})_t \xrightarrow{d} (u + ct - \lambda^{1/\alpha} S_t^{(\alpha)})_t.$$

(ii) If $\alpha = 2$, then with $S \in \mathbb{B}\mathrm{m}\left(0, \lambda(\mu^2 + A^2)/A^2\right)$, as $n \to \infty$,

$$(U_t^{(n)})_t \stackrel{\mathrm{d}}{\to} (u + ct - S_t)_t.$$

Here '$\stackrel{\mathrm{d}}{\to}$' denotes weak convergence in the $(D[0,\infty), J_1)$ Skorokhod space.

Dualities between ruin and overflow probabilities—One of the central problems in collective risk theory concerns the evaluation of the ruin probability. For a risk process $U_t = u + ct - S_t$, where $(S_t)_t$ is a Lévy process with $\mathbb{E}S_1 < c$, we define the *ruin time* $\tau(u)$ as the first time the risk process is negative, that is,

$$\tau(u) := \inf\{t > 0 : U_t < 0\}.$$

Since $\tau(u) = \inf\{t > 0 : ct - S_t < -u\}$, the ruin time coincides with the passage time of the process $(ct - S_t)_t$ below level $-u$, as introduced in Chapter 6. As an immediate consequence, the probability that the risk process $(U_t)_t$ ever becomes negative, sometimes referred to as the *infinite-time* ruin probability, equals

$$\mathbb{P}(\tau(u) < \infty).$$

Now it follows directly that

$$\mathbb{P}(\tau(u) < \infty) = \mathbb{P}\left(\inf_{t\geq 0} U_t < 0\right) = \mathbb{P}\left(\sup_{t\geq 0}(S_t - ct) > u\right).$$

Thus, by Reich's identity, as given by Eqn. (2.5), the following theorem holds. It establishes a useful duality property between ruin and queueing models.

Theorem 15.4 *For $u > 0$,*

$$\mathbb{P}(\tau(u) < \infty) = \mathbb{P}(Q > u),$$

where Q is the stationary workload of a queue with net input process $(S_t - ct)_t$.

As a straightforward application of Thm. 15.4, we may translate all the results derived for the stationary workload in terms of infinite-time ruin probabilities. In particular, upon combining Thm. 3.1 with Remark 3.1, we obtain an exact formula for the distribution of the infinite-time ruin probability in the Cramér–Lundberg model.

In an analogous way we can find the counterpart of Thm. 15.4 for the *finite-time* ruin probability

$$\mathbb{P}(\tau(u) \leq t),$$

which represents the probability that the risk process $(U_t)_t$ becomes negative before time t. Using that

$$\mathbb{P}(\tau(u) \leq t) = \mathbb{P}\left(\inf_{s\in[0,t]} U_s < 0\right) = \mathbb{P}\left(\sup_{s\in[0,t]} (S_s - cs) > u\right),$$

in combination with the representation that we derived for the transient workload $(Q_t \mid Q_0 = 0)$ in Chapter 4, we obtain the following relation between the finite-time ruin probability and the exceedance probability in the corresponding transient queue.

Theorem 15.5 *For $u > 0$,*

$$\mathbb{P}(\tau(u) \leq t) = \mathbb{P}(Q_t > u \mid Q_0 = 0),$$

where $(Q_t)_t$ is the workload process of a queue with net input process $(S_t - ct)_t$.

It is clear that, by applying Thm. 15.5 to the results derived in Chapters 4 and 9, we can obtain both exact and asymptotic results for such finite-time ruin probabilities.

Lévy insurance risk process with tax—Consider the Lévy risk model under the additional assumption that tax payments are deducted from the premium income. Following Albrecher and Hipp [5], and also [6], we suppose that taxes are paid at a constant proportion $\gamma \in [0, 1)$, whenever the risk process is at its running maximum. This leads to the *modified* risk process

$$\tilde{U}_t := u + ct - S_t - \gamma \sup_{s\in[0,t]} (cs - S_s),$$

where the ruin time is given by

$$\tau_\gamma(u) := \inf\{t \geq 0 : \tilde{U}_t < 0\}.$$

Thus, the infinite-time ruin probability takes the following form:

$$\mathbb{P}\left(\tau_\gamma(u) < \infty\right) = \mathbb{P}\left(\sup_{t\geq 0}\left(S_t - ct - \gamma \inf_{s\in[0,t]} (S_s - cs)\right) > u\right).$$

Interestingly, there is an intimate relation between the ruin probability under tax constraints and the ruin probability in the classical tax-free model (corresponding to $\gamma = 0$). Albrecher et al. [6] established the following remarkably simple identity, which allows us to reduce the analysis of the ruin probabilities with tax to those related to the classical 'non-tax model'.

Theorem 15.6 *Suppose that $S \in \mathscr{S}_-$, $\gamma \in [0, 1)$, and $\mathbb{E}S_1 < c$. Then, for $u > 0$,*

$$\mathbb{P}\left(\tau_\gamma(u) < \infty\right) = 1 - \left(1 - \mathbb{P}\left(\tau_0(u) < \infty\right)\right)^{1/(1-\gamma)}.$$

Now the duality presented in Thm. 15.4, combined with the distributional properties of the stationary workload that were derived in Chapters 3 and 8, provides a tool that facilitates the analysis of ruin probabilities associated with the Lévy insurance risk process with tax.

As an implication of the *tax identity* formula given in Thm. 15.6, we have, as $u \to \infty$,

$$\mathbb{P}\left(\tau_\gamma(u) < \infty\right) \sim \frac{1}{1-\gamma}\mathbb{P}\left(\tau_0(u) < \infty\right).$$

Reinsurance models—We now consider the extension of the standard risk model (15.9) to the situation in which several insurance companies share the same risk portfolio. The model aims to compactly represent a system with an insurer and reinsurer. We refer to Avram et al. [32, 33] and references therein for more background on this model.

In order to make the notation transparent, we focus on a two-dimensional risk model with a so-called proportional reinsurance scheme. Suppose that two companies, to be interpreted as the insurance company and the reinsurance company, share the payout of each claim in proportions $p_1, p_2 > 0$, where $p_1 + p_2 = 1$, and receive premiums at rates $c_1, c_2 > 0$, respectively. Then the risk process for the ith company is given by

$$U_t^{(i)} = u_i + c_i t - p_i S_t,$$

where $u_i > 0$ is the initial capital (reserve) of ith company. As previously, the claim process $(S_t)_t$ is a Lévy process, where it is throughout assumed that

$$\mathbb{E}S_1 < \min\left\{\frac{c_1}{p_1}, \frac{c_2}{p_2}\right\}.$$

In what follows we assume that the second company, that is, the reinsurer, receives less premium per amount paid out, that is, it holds that

$$\frac{c_1}{p_1} > \frac{c_2}{p_2}.$$

Consider the infinite-time ruin probability that at least one insurance company will go bankrupt, that is,

$$\mathbb{P}\left(\tau_{\text{or}}(u_1, u_2) < \infty\right) \tag{15.11}$$

with

$$\tau_{\text{or}}(u_1, u_2) := \inf\left\{t \ge 0 : U_t^{(1)} < 0 \quad \text{or} \quad U_t^{(2)} < 0\right\}.$$

It takes a bit of elementary algebra to verify that

$$\begin{aligned}\mathbb{P}\left(\tau_{\text{or}}(u_1, u_2) < \infty\right) &= \mathbb{P}\left(\exists t \ge 0 : U_t^{(1)} < 0 \text{ or } U_t^{(2)} < 0\right) \\ &= \mathbb{P}\left(\exists t \ge 0 : S_t - \frac{c_1}{p_1}t > \frac{u_1}{p_1} \text{ or } S_t - \frac{c_2}{p_2}t > \frac{u_2}{p_2}\right).\end{aligned}$$

Now relate the model that we currently consider with the Lévy-driven tandem queueing model, as introduced in Chapter 12. It is then a straightforward exercise to check that

$$\mathbb{P}\left(\tau_{\text{or}}(u_1, u_2) < \infty\right) = \mathbb{P}\left(\left\{Q^{(1)} > \frac{u_1}{p_1}\right\} \cup \left\{Q^{(1)} + Q^{(2)} > \frac{u_2}{p_2}\right\}\right), \quad (15.12)$$

where, using the notation introduced in Chapter 12, $(Q^{(1)}, Q^{(2)})$ is the joint stationary workload in the tandem system with

$$X_t^{(1)} := S_t - \frac{c_1}{p_1}t, \quad X_t^{(2)} := S_t - \frac{c_2}{p_2}t.$$

Recall that $(X_t^{(2)})_t$ is the net input process of the total queue $Q^{(1)} + Q^{(2)}$.

Relation (15.12), combined with results that were derived in Chapter 8 as well as the straightforward observation that

$$\begin{aligned}&\max\left\{\mathbb{P}\left(Q^{(1)} > \frac{u_1}{p_1}\right), \mathbb{P}\left(Q^{(1)} + Q^{(2)} > \frac{u_2}{p_2}\right)\right\} \\ &\quad \le \mathbb{P}\left(\left\{Q^{(1)} > \frac{u_1}{p_1}\right\} \cup \left\{Q^{(1)} + Q^{(2)} > \frac{u_2}{p_2}\right\}\right) \\ &\quad \le \mathbb{P}\left(Q^{(1)} > \frac{u_1}{p_1}\right) + \mathbb{P}\left(Q^{(1)} + Q^{(2)} > \frac{u_2}{p_2}\right), \quad (15.13)\end{aligned}$$

in many cases allows us to identify the asymptotic behavior of the infinite-time ruin probability (15.11).

Example 15.1 Consider the case of $S \in \mathbb{B}\text{m}(0, 1)$. By virtue of (15.13) and Example 3.1, we distinguish two scenarios. If $c_1/p_1^2 < c_2/p_2^2$, then, as $u \to \infty$,

$$\mathbb{P}\left(\tau_{\text{or}}(u, u) < \infty\right) \sim \mathbb{P}\left(Q^{(1)} > \frac{u}{p_1}\right) = \exp\left(-\frac{2c_1}{p_1^2}u\right).$$

If $c_1/p_1^2 > c_2/p_2^2$, then, as $u \to \infty$,

$$\mathbb{P}\left(\tau_{\text{or}}(u,u) < \infty\right) \sim \mathbb{P}\left(Q^{(1)} + Q^{(2)} > \frac{u}{p_2}\right) = \exp\left(-\frac{2c_2}{p_2^2}u\right).$$

The case $c_1/p_1^2 = c_2/p_2^2$ requires a more refined analysis. ◇

So far we have concentrated on the probability of bankruptcy of *at least one* of the insurance companies. Relying on the same arguments, however, we can analyze the infinite-time ruin probability of *both* insurance companies. To this end, we define

$$\tau_{\text{and}}(u_1,u_2) := \min\left\{\tau^{(1)}(u_1), \tau^{(2)}(u_2)\right\},$$

where $\tau^{(i)}(u_i)$ is the ruin time for the ith company, where $i = 1, 2$. Then,

$$\mathbb{P}\left(\tau_{\text{and}}(u_1,u_2) < \infty\right) = \mathbb{P}\left(\left\{Q^{(1)} > \frac{u_1}{p_1}\right\} \cap \left\{Q^{(1)} + Q^{(2)} > \frac{u_2}{p_2}\right\}\right). \quad (15.14)$$

Despite the relation

$$\mathbb{P}\left(\tau_{\text{and}}(u_1,u_2) < \infty\right) = \mathbb{P}(\tau^{(1)}(u_1) < \infty) + \mathbb{P}(\tau^{(2)}(u_2) < \infty) - \mathbb{P}\left(\tau_{\text{or}}(u_1,u_2) < \infty\right)$$

which holds for all $u_1, u_2 > 0$, the asymptotic analysis of (15.14) is typically more involved than the one corresponding to the 'or case', and needs to deal with several special cases; see e.g. [151]. Complications arise when the probability $\mathbb{P}\left(\tau_{\text{or}}(u_1,u_2) < \infty\right)$ is of the same order as the largest of the $\mathbb{P}(\tau^{(i)}(u_i) < \infty)$, with $i = 1, 2$; due to the minus sign in the above relation, a refined analysis is required in order to identify the limiting behavior of $\mathbb{P}\left(\tau_{\text{and}}(u_1,u_2) < \infty\right)$.

For the special case of $(S_t)_t$ being a Brownian motion, however, a formula is derived in [151] for the exact distribution of (15.14), which translated to the reinsurance model gives the following result. As usual $\Psi_{\text{N}}(\cdot)$ denotes the tail distribution function of a standard normal random variable. We also use the following notation:

$$\chi := \frac{p_1u_2 - p_2u_1}{p_2c_1 - p_1c_2}, \quad k(x,y) := \frac{x + y\chi}{\sqrt{\chi}}.$$

Theorem 15.7 *Suppose that $S \in \mathbb{B}\text{m}(0,1)$ and $u_1, u_2 > 0$.*

(i) If $u_1/p_1 \geq u_2/p_2$, then

$$\mathbb{P}\left(\tau_{\text{and}}(u_1,u_2) < \infty\right) = \exp\left(-\frac{2c_1}{p_1^2}u_1\right).$$

(ii) If $u_1/p_1 < u_2/p_2$, *then*

$$\begin{aligned}\mathbb{P}\left(\tau_{\text{and}}(u_1,u_2)<\infty\right) &= \Psi_{\mathrm{N}}\left(k\left(-\frac{u_1}{p_1},\frac{c_1}{p_1}\right)\right)\exp\left(-\frac{2c_1}{p_1^2}u_1\right)\\ &+\Psi_{\mathrm{N}}\left(k\left(\frac{u_1}{p_1},\frac{c_1}{p_1}-\frac{2c_2}{p_2}\right)\right)\exp\left(-\frac{2c_2}{p_2^2}u_2\right)\\ &+\left(1-\Psi_{\mathrm{N}}\left(k\left(-\frac{u_1}{p_1},\frac{c_1}{p_1}-\frac{2c_2}{p_2}\right)\right)\right)\exp\left(-2\left(\frac{u_1}{p_1}\left(\frac{c_1}{p_1}-\frac{2c_2}{p_2}\right)+\frac{c_2}{p_2^2}u_2\right)\right)\\ &-\Psi_{\mathrm{N}}\left(k\left(\frac{u_1}{p_1},\frac{c_1}{p_1}\right)\right).\end{aligned}$$

Related asymptotic results for light-tailed claim processes $(S_t)_t$ can be found in Avram et al. [32, 33].

For more results on insurance models and their relations with queueing theory we refer to textbooks by Asmussen and Albrecher [21], Asmussen [19], Mikosch [162], and Rolski et al. [189].

15.7 Other Applications in Finance

Above we pointed out how Lévy-based fluctuation theory can be used when pricing options, but the material presented in this book is actually applicable in many other contexts within mathematical finance. In this monograph we primarily considered single-dimensional Lévy processes, but in many of these applications the *joint* evolution of asset prices plays a role. The question of how the correlation between the asset prices should be incorporated, however, is still largely unsolved.

One of the proposed techniques relies on the concept of *copulas*, as advocated in Jaworski et al. [115] or Nelsen [168]; then the joint distribution can be written in terms of the marginal distributions and a copula. These copulas have been predominantly applied in risk/portfolio management and the pricing of derivatives (for instance *collateralized debt obligations*, CDOs). In the latter domain the use of copulas has become controversial since the financial crisis of 2008–2009. An alternative Lévy-based model was proposed in den Iseger et al. [80]: there, the *distances to default* of the individual obligors are modeled as the sum of a common component (through which the obligors become correlated) and an obligor-specific component, where both the common and the obligor-specific components follow variance gamma processes. The resulting model has attractive computational properties and allows for easy calibration. We also mention that the textbook of Schoutens and Cariboni [194] addresses the use of Lévy processes in credit risk.

Exercises

Exercise 15.1 Verify Eqns. (15.1) and (15.2).

Exercise 15.2 Evaluate the Lévy exponent for $X \in \mathbb{C}\text{GMY}(\alpha, C, A_+, A_-)$.

Exercise 15.3 Verify Eqns. (15.5) and (15.6).

Exercise 15.4 Find the exact asymptotics of $\mathbb{P}(\bar{X}_t > u)$ as $t \to \infty$, where X is standard Brownian motion, that is, $X \in \mathbb{B}\text{m}(0, 1)$.

Exercise 15.5 Consider $\bar{X}_T$, where X is a standard Brownian motion, and T is exponentially distributed with mean $1/\vartheta$ (independently of X).

(a) Check that $\bar{X}_T \stackrel{\text{d}}{=} T^{1/2}\bar{X}_1$.
(b) Derive the distribution of $\bar{X}_T$.

Exercise 15.6 Suppose that $X_t = X_t^{(1)} + X_t^{(2)}$ where the process $X^{(1)}$ corresponds to $\mathbb{B}\text{m}(d, \sigma^2)$ and the process $X^{(2)} \in \mathbb{S}(\alpha, 1, -m)$, with $\alpha \in (1, 2)$ and $m > 0$. Find the exact asymptotics of $\mathbb{P}(\bar{X}_t > u)$ as $u \to \infty$.

Exercise 15.7 Determine the transforms of the *second-order Greeks* of the vanilla option

$$\frac{\partial^2 V_{\text{van}}^{(\text{c})}(T, K)}{\partial S_0^2}, \quad \frac{\partial^2 V_{\text{van}}^{(\text{c})}(T, K)}{\partial T^2};$$

do the same for the lookback and barrier options.

Exercise 15.8 Consider the situation of a claim process $S \in \mathbb{S}(\alpha, \beta, 0)$ with $\alpha \in (1, 2)$, $\beta \in (-1, 1]$, and let the premium rate be $c > 0$. Find the asymptotics of $\mathbb{P}(\tau(u) < \infty)$ and $\mathbb{P}(\tau(u) < t)$, as $u \to \infty$.

Exercise 15.9 Consider the proportional reinsurance problem with $U_t^{(i)} = u_i + c_i t - p_i S_t$, for $i = 1, 2$. It is assumed that $S \in \mathbb{S}(\alpha, \beta, 0)$ with $\alpha \in (1, 2)$, $\beta \in (-1, 1]$; as usual p_1, p_2 are positive and add up to 1. In addition, we assume that $c_1/p_1 > c_2/p_2$.

Develop estimates for the probabilities $\mathbb{P}(\tau_{\text{or}}(u, u) < \infty)$ and $\mathbb{P}(\tau_{\text{and}}(u, u) < \infty)$, as $u \to \infty$.

Chapter 16
Computational Aspects: Inversion Techniques

In this chapter we focus on the issue of numerically computing

$$\mathbb{P}\left(\bar{X}_t > x\right) = \mathbb{P}\left(\sup_{0\le s\le t} X_s > x\right),$$

for given $x, t \geq 0$, and for arbitrary Lévy processes $(X_t)_t$, or the more ambitious goal of accurately evaluating the corresponding density $f_{\bar{X}_t}(x) := \mathbb{P}(\bar{X}_t \in \mathrm{d}x)$. We focus on two techniques, with their own specific pros and cons.

Before providing brief sketches of each of these techniques in Sections 16.1 and 16.2, we first recall that for any Lévy process in principle the full distribution of $\bar{X}_t$ is described by Thm. 3.4: with $k(\cdot\,,\cdot)$ as defined in (3.10), for $\alpha \geq 0$, and with T exponentially distributed with mean ϑ^{-1},

$$\begin{aligned}\mathbb{E}e^{-\alpha\bar{X}_T} &= \frac{k(\vartheta,\alpha)}{k(\vartheta,0)}\\ &= \exp\left(-\int_0^\infty\int_0^\infty \frac{1}{t}\left(e^{-\vartheta t} - e^{-\vartheta t}e^{-\alpha x}\right)\mathbb{P}(X_t\in \mathrm{d}x)\mathrm{d}t\right).\end{aligned}$$

A first, admittedly naïve, idea is to evaluate $f_{\bar{X}_t}(x)$ by first numerically computing $\mathbb{E}e^{-\alpha\bar{X}_T}$, and then performing inversion with respect to α and ϑ. While fast and accurate techniques exist for such a double inversion [79], a major complication lies in the fact that in many situations we know the Lévy exponent corresponding to $(X_t)_t$, but unfortunately hardly any explicit expressions for the density $\mathbb{P}(X_t \in \mathrm{d}x)$ are available (think of variance gamma, CGMY, normal inverse Gaussian, etc.).

The first ('approximation and inversion') of the two approaches that we present in this chapter, based on Asghari et al. [16], follows the line of reasoning pointed out in Section 3.3: approximate the Lévy process by a Lévy process for which the function $k(\vartheta, \alpha)$ *can* be evaluated. This means that the small jumps are approximated

K. Dębicki, M. Mandjes, *Queues and Lévy Fluctuation Theory*, Universitext,
DOI 10.1007/978-3-319-20693-6_16

by a Brownian motion, whereas the distribution of the jumps in one direction is approximated by that of a phase-type random variable.

The second approach ('repeated inversion'), based on Gruntjes et al. [104], (i) first uses numerical inversion to find $\mathbb{P}(X_t \in dx)$ from the Lévy exponent, (ii) then numerically evaluates $k(\vartheta, \alpha)$, and (iii) finally performs a second inversion to obtain $f_{\bar{X}_t}(x) = \mathbb{P}(\bar{X}_t \in dx)$. We describe both methods in greater detail now.

16.1 Approach 1: Approximation and Inversion

As mentioned above, in this approach we replace the jumps in one direction (say the upward direction) by phase-type jumps, and the 'small jumps' by a Brownian motion. The Lévy process we thus obtain allows (semi-)explicit evaluation of $\mathbb{E}e^{-\alpha \bar{X}_T}$; see e.g. Lewis and Mordecki [150]. We now provide more detail regarding both approximation ideas.

Replace upward jumps by a phase-type counterpart—There is a relatively large set of papers dealing with approximating a distribution on $(0, \infty)$ by a phase-type distribution; see e.g. [91, 112]. Here we rely on the approach developed in Asmussen et al. [28], which is based on the *EM algorithm*, and Thümmler et al. [211], which proposes a comparable approach that focuses primarily on mixtures of Erlangs.

For a precise definition of phase-type distributions, see e.g. Asmussen [19, Chapter III]; they can be thought of as distributions of absorption times in finite-state continuous-time Markov chains, as we described in Section 3.3 of this textbook. More precisely, with $n + 1$ denoting the dimension of the state space, where n states are transient and the remaining state is absorbing, a phase-type distribution corresponds to the entrance time of the absorbing state. This class covers mixtures and sums of exponential distributions (and hence also the Erlang distribution, being distributed as the sum of independent exponential random variables with the same mean). As we discussed in Section 3.3, the class of phase-type distributions is dense, in that any distribution on $(0, \infty)$ can, in principle, be approximated arbitrarily well; the price to be paid, though, is that the dimension n of the associated Markov chain may become prohibitively large.

The performance of the EM-based algorithm proposed is assessed in detail in [28]—it was shown that quite a large class of distributions can be accurately approximated by phase-type distributions of relatively low dimension d. From this it is, however, not a priori clear what the impact is of replacing the upward jumps by an appropriate phase-type random variable when evaluating $\mathbb{P}(\bar{X}_t \leq x)$ in the way described above—there are no explicit bounds available on the error introduced by replacing the jump distribution by its phase-type counterpart.

Replace small jumps by Brownian motion—As pointed out in Section 3.3, in the case of small jumps (i.e. $\int_{-\infty}^{\infty} \Pi(dx) = \infty$), the Lévy process under study can be accurately approximated by the sum of an appropriately chosen compound Poisson process and Brownian motion. We first write the jump part of the Lévy exponent in the form

$$\int_{-\infty}^{\infty}(e^{isx}-1-isx1_{\{|x|<\varepsilon\}})\Pi(dx)=\int_{-\varepsilon}^{\varepsilon}(e^{isx}-1-isx)\Pi(dx)$$
$$+\int_{\mathbb{R}\setminus[-\varepsilon,\varepsilon]}(e^{isx}-1)\Pi(dx);$$

let the first term correspond to a Lévy process, say, $X_t^{(1,\varepsilon)}$, and the second term (which is a compound Poisson process) correspond to, say, $X_t^{(2,\varepsilon)}$. Then the 'small jump component' $X_t^{(1,\varepsilon)}$ can be approximated by (for some small value of ε)

$$\mu_\varepsilon t+\sigma_\varepsilon B_t, \tag{16.1}$$

where B_t is a standard Brownian motion, and

$$\mu_\varepsilon:=\int_{-\varepsilon}^{\varepsilon}x\Pi(dx),\ \ \sigma_\varepsilon^2:=\int_{-\varepsilon}^{\varepsilon}x^2\Pi(dx).$$

This approximation is motivated by the limit result (3.12). As pointed out in Section 3.3, a sufficient condition for (3.12) to hold is that, with $L(\cdot)$ a slowly varying function at 0, $\Pi(\cdot)$ has a density of the form $L(x)/|x|^{\alpha+1}$ for $x\downarrow 0$, with $\alpha\in(0,2)$. It is noted that this condition applies for e.g. stable Lévy processes and CGMY processes, but not for e.g. variance gamma processes (as these correspond to $\alpha=0$). We also mention that the use of (16.1) is advocated for variance gamma in Fu [95]—see in particular the third algorithm [95, p. 25].

Numerical inversion—Now that we can approximate the (double) Laplace transform $k(\vartheta,\alpha)$, the next step is to use a numerical technique to perform Laplace inversion, so as to obtain an approximation for $f_{\bar{X}_t}(x)=\mathbb{P}(\bar{X}_t\in dx)$. We advocate the use of the method developed by den Iseger [79]; see also [81]. It is in the spirit of approaches developed earlier [2, 86], in the sense that it relies on the Poisson summation formula. This Poisson summation formula relates an infinite sum of Laplace transform values to the z-transform of the function values $f(k\Delta)$, with $k=0,\dots,M-1$, that we wish to evaluate, from which the $f(k\Delta)$ can be computed relying on the well-known *fast Fourier transform*; see e.g. [64].

A first complication is that the above-mentioned infinite sum tends to converge slowly. Abate and Whitt [2] remedy this using a so-called Euler summation, but in general the convergence remains prohibitively slow unless knowledge of the location of singularities is available. One of den Iseger's contributions [79] is to approximate the infinite sum by a finite sum by using Gaussian quadrature. The resulting algorithm is a substantial improvement over earlier algorithms in the sense that (i) it can handle a larger class of Laplace transforms (e.g. no knowledge of the location of discontinuities or singularities is needed), (ii) the algorithm needs only numerical values of the Laplace transform, is fast (i.e. the function values $f(k\Delta)$, with $k = 0, \ldots, M-1$, are computed at once, in order $M \log M$ time), and is of nearly machine precision, (iii) can be extended to multiple dimensions. It is stressed that this last feature is of crucial importance to us, as in our setting we are often dealing with two-dimensional transforms.

Numerical results for this approach, as well as implementation details, are provided in [16]. There the algorithm is tested for a broad variety of driving Lévy processes and parameters. Unless the scenarios considered are 'extreme', the performance of the algorithm is excellent. The major drawback of the approach lies in the fact that, in the case that the upward jumps are not of phase type, we have to identify a well-fitting phase-type distribution, which is not trivial to automate. In addition, the techniques for evaluating $k(\vartheta, \alpha)$ for Lévy processes with phase-type upward jumps require the computation of a set of roots, which can be time consuming. The same approach has been used in the context of option pricing in [17].

16.2 Approach 2: Repeated Inversion

We now propose an alternative algorithm, based on Gruntjes et al. [104], which settles the drawbacks of the approach described above. We have observed that in the analysis a key role is played by the double transform $k(\vartheta, \alpha)/k(\vartheta, 0)$. In this approach, we first write $K(\vartheta, \alpha) := \log(k(\vartheta, \alpha)/k(\vartheta, 0))$ in a more convenient form (in terms of ϑ, α, and the Lévy exponent of the underlying Lévy process), and then we point out how this new form can be used to develop an algorithm for fast and accurate evaluation of the density $f_{\bar{X}_t}(x)$.

In this section we write $\mathbb{E}e^{-sX_t} = \exp(-t\zeta(s))$. The main idea is that we want an algorithm that does not require us to approximate the driving Lévy process $(X_t)_t$ (as *was* the case for the algorithm of Section 16.1); instead, we aim to devise a procedure that has ϑ, α, and $\zeta(\cdot)$ as inputs, and provides us with an accurate approximation of $K(\vartheta, \alpha)$, which can then be numerically inverted.

Writing the Wiener–Hopf factors in a convenient form—Our first objective is to rewrite the function $K(\vartheta, \alpha)$. To this end, observe that

$$K(\vartheta, \alpha) = \int_0^\infty \int_{(0,\infty)} \frac{1}{t} (1 - e^{-\alpha x}) \, e^{-\vartheta t} f_{X_t}(x) \mathrm{d}x \, \mathrm{d}t,$$

where $f_{X_t}(\cdot)$ is the density of X_t (assumed to exist). We recall that we do not have an explicit expression for $f_{X_t}(\cdot)$ at our disposal (instead, the probabilistic properties of $(X_t)_t$ are captured by $\zeta(\cdot)$).

Now denote by $\mathscr{F}$ the Fourier transform, and by $\mathscr{F}^{-1}$ the inverse Fourier transform; in the sequel, $\mathscr{F}[f](s) \equiv \mathscr{F}[f(\cdot)](s)$ is the Fourier transform of f, evaluated in s. We then have the following obvious relations:

$$\mathscr{F}[f_{X_t}(\cdot)](s) = e^{-t\zeta(s)}, \quad f_{X_t}(x) = \mathscr{F}^{-1}[e^{-t\zeta(\cdot)}](x),$$

and

$$\mathscr{F}[g_t(\cdot)](s) = t\zeta'(s)e^{-t\zeta(s)}, \quad \text{with } g_t(x) := xf_{X_t}(x).$$

It follows that

$$\mathscr{F}^{-1}[t\zeta'(\cdot)e^{-t\zeta(\cdot)}](x) = x\mathscr{F}^{-1}[e^{-t\zeta(\cdot)}](x).$$

From the above, it is concluded that

$$K(\vartheta,\alpha) = \int_0^\infty \int_{(0,\infty)} \frac{1}{xt}(1-e^{-\alpha x})\, e^{-\vartheta t}\, \mathscr{F}^{-1}[t\zeta'(\cdot)e^{-t\zeta(\cdot)}](x)\mathrm{d}x\,\mathrm{d}t.$$

Using Fubini's theorem, in conjunction with the fact that $\mathscr{F}^{-1}$ is a linear operator, this in turn equals

$$K(\vartheta,\alpha) = \int_{(0,\infty)} \left(\frac{1}{x}(1-e^{-\alpha x})\, \mathscr{F}^{-1}\left[\int_0^\infty \zeta'(\cdot)e^{-t(\zeta(\cdot)+\vartheta)}\,\mathrm{d}t\right](x)\right)\mathrm{d}x.$$

Now define

$$F_\vartheta(x) := \mathscr{F}^{-1}\left[\int_0^\infty \zeta'(\cdot)e^{-t(\zeta(\cdot)+\vartheta)}\,\mathrm{d}t\right](x) = \frac{1}{x}\mathscr{F}^{-1}\left[\frac{\zeta'(\cdot)}{\zeta(\cdot)+\vartheta}\right](x).$$

By $F_\vartheta^+(x)$ we denote $F_\vartheta(x)$ if $x \geq 0$ and 0 otherwise. With $T_\vartheta(\cdot)$ the Fourier transform of $F_\vartheta^+(\cdot)$, it then follows immediately that

$$K(\vartheta,\alpha) = T_\vartheta(0) - T_\vartheta(\alpha).$$

In this way we have found a compact expression purely in terms of ϑ, α, and $\zeta(\cdot)$: we can plug in ϑ, α, and $\zeta(\cdot)$, and obtain $K(\vartheta,\alpha)$.

It is clear, however, that there may be numerical issues if

$$\lim_{x\downarrow 0} \mathscr{F}^{-1}\left[\frac{\zeta'(\cdot)}{\zeta(\cdot)+\vartheta}\right](x) \neq 0.$$

To remedy this issue, let $\bar{F}_\vartheta^+(x)$ equal $F_\vartheta(x) - e^{-x}F_\vartheta(0)$ if $x \geq 0$ and 0 otherwise. Then obviously, with $\bar{T}_\vartheta(\cdot)$ the Fourier transform of $\bar{F}_\vartheta^+(\cdot)$,

$$\begin{aligned} K(\vartheta,\alpha) &= \bar{T}_\vartheta(0) - \bar{T}_\vartheta(\alpha) + F(0)\int_0^\infty \frac{1}{x}(1-e^{-\alpha x})e^{-x}\mathrm{d}x \\ &= \bar{T}_\vartheta(0) - \bar{T}_\vartheta(\alpha) + F(0)\log(1+\alpha), \end{aligned}$$

where in the last step the Frullani integral equality has been used. We arrive at the following pseudocode to determine $K(\vartheta,\alpha) = \log \mathbb{E}e^{-\alpha\bar{X}_T}$, with T, as usual, an exponentially distributed random variable with mean $1/\vartheta$. It requires a routine to perform Fourier inversion to evaluate $F_\vartheta(\cdot)$, and a routine to perform the Fourier transform to evaluate $\bar{T}_\vartheta(\cdot)$, so as to compute $K(\vartheta,\alpha)$ from ϑ, α, and $\zeta(\cdot)$.

Pseudocode 16.1 Input: ϑ, α, and $\zeta(\cdot)$. Output: $K(\vartheta,\alpha) = \log \mathbb{E}e^{-\alpha X_T}$.

1. Compute the function $F_\vartheta(\cdot)$.
2. Compute the function $\bar{T}_\vartheta(\cdot)$.
3. Set $K(\vartheta,\alpha) = \log \mathbb{E}e^{-\alpha X_T} := \bar{T}_\vartheta(0) - \bar{T}_\vartheta(\alpha) + F_\vartheta(0)\log(1+\alpha)$.

Implementation issues—Above we mentioned that the input of the procedure consists of ϑ, α, and $\zeta(\cdot)$, but, as we have seen, in principle also $\zeta'(\cdot)$ is needed. One could either evaluate $\zeta'(\cdot)$ numerically, or use an explicit expression for $\zeta'(\cdot)$. Evidently, from a numerical standpoint the latter option is preferred.

Once we have $K(\vartheta,\alpha)$, in order to find the density $f_{\bar{X}_t}(\cdot)$, we have to perform a double Laplace inversion (with respect to α and ϑ) to

$$\int_0^\infty \int_{(0,\infty)} e^{-\vartheta t} e^{-\alpha x} f_{\bar{X}_t}(x)\mathrm{d}x\,\mathrm{d}t = \frac{1}{\vartheta}\cdot \mathbb{E}e^{-\alpha\bar{X}_T} = \frac{e^{K(\vartheta,\alpha)}}{\vartheta}.$$

We refer for more detailed implementation issues to [104]; see also Section 16.1 for more background on the Laplace inversion.

The experiments reported in [104] show that, for a broad set of scenarios, the performance of the algorithm is excellent. In contrast with the first approach, as described in Section 16.1, however, this second approach does not work well when the driving Lévy process has small jumps.

16.3 Other Applications

Above we showed how to write the transform of the random quantity $\bar{X}_T$ in a form that facilitates numerical evaluation; we presented two approaches. In this section, we consider various other random quantities that can be assessed in a similar way.

Overshoots—By virtue of Lemma 6.4, with $\sigma(x)$ defined as the first passage time over level x, that is, $\inf\{t \geq 0 : X_t \geq x\}$,

$$\int_0^\infty e^{-\beta x}\mathbb{E}\left(e^{-\vartheta\sigma(x)-\bar{\vartheta}(X_{\sigma(x)}-x)}1_{\{\sigma(x)<\infty\}}\right)\mathrm{d}x = \frac{1}{\beta-\bar{\vartheta}}\left(1-\frac{k(\vartheta,\beta)}{k(\vartheta,\bar{\vartheta})}\right). \quad (16.2)$$

Now both approaches, as presented in the previous sections, can be followed.

(i) We could follow the approach of Section 16.1: approximate X by a process with phase-type upward jumps and the small jumps part replaced by Brownian motion; evaluate (16.2) for this process; and perform the numerical inversion to obtain the joint density of the overshoot and the first passage time.

(ii) Alternatively, we could follow the second approach, as presented in Section 16.2. To this end, realize that expression (16.2) equals

$$\frac{1}{\beta-\bar{\vartheta}}\left(1-\frac{e^{K(\vartheta,\beta)}}{e^{K(\vartheta,\bar{\vartheta})}}\right).$$

It is easily checked that, inserting $\vartheta = 0$, this formula gives $1/(\Psi(0)+\beta)$ for $X \in \mathscr{S}_-$, as desired (why?). The above expressions effectively show that, using Pseudocode 16.1, we can also evaluate the triple transform (16.2). Then numerical inversion has to be performed to obtain the joint density.

Joint distribution of running maximum and position—We found earlier that

$$\mathbb{E}e^{\mathrm{i}\alpha_1\bar{X}_T+\mathrm{i}\alpha_2 X_T} = \mathbb{E}e^{\mathrm{i}(\alpha_1+\alpha_2)\bar{X}_T}\mathbb{E}e^{\mathrm{i}\alpha_2(X_T-\bar{X}_T)} = \frac{k(\vartheta,-\mathrm{i}(\alpha_1+\alpha_2))}{k(\vartheta,0)}\frac{\bar{k}(\vartheta,\mathrm{i}\alpha_2)}{\bar{k}(\vartheta,0)}.$$

Again both approaches can be followed. Regarding the second approach, note that it is elementary to express this quantity in terms of $K(\cdot\,,\cdot)$ and $\bar{K}(\cdot\,,\cdot)$ (the latter function defined in the obvious way), and as we are able to evaluate these, we can also evaluate the joint transform under consideration.

Joint distribution of running maximum and corresponding epoch—In Thm. 3.4, we found the joint transform of the running maximum and the epoch at which the maximum was attained:

$$\mathbb{E}e^{-\beta G_T-\alpha\bar{X}_T} = \frac{k(\vartheta+\beta,\alpha)}{k(\vartheta,0)}. \tag{16.3}$$

It is clear how to evaluate the joint distribution using the first approach.

Regarding the second approach, it is noted that unfortunately this transform cannot be expressed in terms of $K(\cdot\,,\cdot)$. We now point out how to evaluate $L(\vartheta) := \log k(\vartheta,0)$; it is easily seen that if one can compute $K(\cdot\,,\cdot)$ and $L(\cdot)$, then one can evaluate (16.3) as well. Using 'Fubini' and 'Frullani',

$$\begin{aligned}
k(\vartheta,0) &= -\int_0^\infty\int_{(0,\infty)}\frac{1}{t}\left(e^{-t}-e^{-\vartheta t}\right)f_{X_t}(x)\mathrm{d}x\,\mathrm{d}t\\
&= -\int_0^\infty\int_{(0,\infty)}\frac{1}{t}\left(e^{-t}-e^{-\vartheta t}\right)\mathscr{F}^{-1}[e^{-\zeta(\cdot)t}](x)\mathrm{d}x\,\mathrm{d}t\\
&= -\int_{(0,\infty)}\mathscr{F}^{-1}\left[\int_0^\infty\frac{1}{t}\left(e^{-t}-e^{-\vartheta t}\right)e^{-\zeta(\cdot)t}\mathrm{d}t\right](x)\,\mathrm{d}x\\
&= -\int_{(0,\infty)}\mathscr{F}^{-1}\left[\log\left(\frac{\zeta(\cdot)+\vartheta}{\zeta(\cdot)+1}\right)\right](x)\,\mathrm{d}x.
\end{aligned}$$

Least concave majorant—The methodology presented in this section clearly facilitates numerical computations regarding quantities related to the transient workload distribution of a Lévy-driven queue (as in Thm. 4.3); in addition, it can be used when pricing options (as in Section 15.4). Now we show how our tools can be used to numerically evaluate the *least concave majorant* $\hat{X}_t$ of a Lévy process X_t. For the Brownian case, for instance Carolan and Dykstra [57] and Groeneboom [103] performed explicit calculations; the results below address the general Lévy case. The following lemma applies to any stochastic process X_t.

Lemma 16.1 *Let $(\hat{X}_t)_t$ be the concave majorant of $(X_t)_t$ over the interval $[0,T]$, with T possibly equal to ∞. With $t \in [0,T]$, the event $\{\hat{X}_t \le x\}$ is equivalent to*

$$\left\{ \inf_{0\le s\le t} \frac{x - X_s}{t-s} \ge \sup_{t\le r\le T} \frac{X_r - x}{r-t} \right\}.$$

Proof Realize that $(x - X_s)/(t-s)$ is the slope of the line through (t,x) and (s, X_s), so that

$$B_- := \inf_{0\le s\le t} \frac{x - X_s}{t-s}$$

is the slope of the 'steepest' line through (t,x) that majorizes X_s for any $s \in [0,t]$. Likewise,

$$B_+ := \sup_{t\le r\le T} \frac{X_r - x}{r-t}$$

is the slope of the 'flattest' line through (t,x) that majorizes X_s for any $s \in [t,T]$. Then the stated result follows immediately. □

From the above lemma it is evident that

$$\mathbb{P}(\hat{X}_t \le x) = \int_{b=x/t}^{\infty} \int_{y=-\infty}^{x} \mathbb{P}(B_- \ge b, X_t \in \mathrm{d}y)\mathbb{P}(B_+ \in \mathrm{d}b \,|\, X_t = y)\mathrm{d}y.$$

We now point out how the probabilities and densities in the integrand can be determined. It is straightforward to verify that

$$\mathbb{P}(B_- \ge b, X_t \in \mathrm{d}y) = \mathbb{P}\left(\sup_{0\le s\le t} X_s - bs \le x - bt, X_t \in \mathrm{d}y \right). \tag{16.4}$$

It is evident that we can evaluate (16.4) with the techniques described above; rephrase it in terms of the joint distribution of $\bar{Y}_t$ and Y_t, with $Y_t := X_t - bt$. Furthermore, observe that

$$\mathbb{P}(B_+ \le b \,|\, X_t = y) = \mathbb{P}(\forall r \in [t,T] : X_r \le x - bt + br \,|\, X_t = y),$$

which due to the Markov property equals $\mathbb{P}(\bar{Y}_{T-t} \leq x - y)$. Also this probability can be evaluated relying on the approach proposed earlier in this section.

Exercises

Exercise 16.1 Use the EM technique advocated in [28] to find a phase-type approximation of a standard normal random variable (by splitting this into a positive and negative part). Repeat this for the dimension d of the associated continuous-time Markov chain taking the values 3, 4, and 5, respectively.

Exercise 16.2 Let $X \in \mathbb{C}\mathrm{GMY}(\alpha, C, A_+, A_-)$. Suppose we wish to replace the small jumps by Brownian motion. Evaluate μ_ε and σ_ϵ^2.

Chapter 17
Concluding Remarks

In this textbook we have highlighted a set of important results on queues with Lévy input, and explicitly drawn the connection with fluctuation theory. An obvious disclaimer is in place here: with this field being large, some relevant contributions may have been overlooked. Also, given the connection between Lévy-driven queues and risk theory in a Lévy environment, compactly reflected by Eqn. (2.5), perhaps not all relations with the vast finance and insurance literature have been fully exploited.

Despite the fact that the field develops rapidly, there are still many open problems; we mention here just a few challenging directions.

(i) In the first place, quite a number of results presented in this book are restricted to spectrally one-sided cases, whereas in practical situations the underlying Lévy process often has *two-sided jumps*; see however [22, 149, 150].
(ii) Another domain in which still only partial results are known is that of Lévy-driven *networks*: hardly any results are available when the underlying network does not satisfy conditions (T_1)–(T_5); see however the novel contribution [166].
(iii) Also, in the area of *numerical evaluation* (by either simulation or numerical inversion) there is still substantial scope for improvement.
(iv) Finally, there are still many open problems related to various *functionals* of the workload process: for instance, one would like to uniquely characterize the full distribution of $V(t, u)$, as defined in Section 4.3, and only partial results are available for the area under the workload graph [14, 48].

The variety of open questions, which emerge from analyzing Lévy-driven queueing systems and Lévy fluctuation theory, stimulates the current research to lie at the interface of such areas as extreme value theory, stochastic geometry, large deviations, stochastic simulation theory, etc. This fuels the expectation that Lévy-driven queueing theory and fluctuation theory will increasingly become a key subdiscipline of applied and theoretical probability.

K. Dębicki, M. Mandjes, *Queues and Lévy Fluctuation Theory*, Universitext,
DOI 10.1007/978-3-319-20693-6_17

Chapter 17
Concluding Remarks

[illegible]

References

1. Abate, J., Whitt, W.: The correlation functions of RBM and M/M/1. Stoch. Mod. **4**, 315–359 (1988)
2. Abate, J., Whitt, W.: Numerical inversion of Laplace transforms of probability distributions. ORSA J. Comput. **7**, 36–43 (1995)
3. Abate, J., Whitt, W.: Asymptotics for M/G/1 low-priority waiting-time tail probabilities. Queueing Syst. **25**, 173–223 (1997)
4. Akaike, H.: A new look at the statistical model identification. IEEE Trans. Autom. Contr. **19**, 716–723 (1974)
5. Albrecher, H., Hipp, C.: Lundberg's risk process with tax. Blätter DGVFM **28**, 13–28 (2007)
6. Albrecher, H., Renaud, J., Zhou, X.: A Lévy insurance risk process with tax. J. Appl. Probab. **45**, 363–375 (2008)
7. Alili, L., Kyprianou, A.: Some remarks on first passage of Lévy processes, the American put and smooth pasting. Ann. Appl. Probab. **15**, 2062–2080 (2004)
8. Andersen, L.N., Asmussen, S., Glynn, P., Pihlsgård, M.: Lévy processes with two-sided reflection. Lévy Matters (accepted)
9. Andersen, L.N., Mandjes, M.: Structural properties of reflected Lévy processes. Queueing Syst. **63**, 301–322 (2009)
10. Anick, D., Mitra, D., Sondhi, M.: Stochastic theory of a data-handling system with multiple sources. Bell Syst. Tech. J. **61**, 1871–1894 (1982)
11. Applebaum, D.: Lévy Processes and Stochastic Calculus. Cambridge University Press, Cambridge (2004)
12. Applebaum, D.: Lévy processes—from probability to finance and quantum groups. Not. Am. Math. Soc. **51**, 1336–1347 (2004)
13. Arendarczyk, M., Dębicki, K.: Asymptotics of supremum distribution of a Gaussian process over a Weibullian time. Bernoulli **17**, 194–210 (2011)
14. Arendarczyk, M., Dębicki, K., Mandjes, M.: On the tail asymptotics of the area swept under the Brownian storage graph. Bernoulli **20**, 395–415 (2014)
15. Asghari, N., Dębicki, K., Mandjes, M.: Exact tail asymptotics of the supremum attained by a Lévy process. Stat. Probab. Lett. **96**, 180–184 (2015)
16. Asghari, N., den Iseger, P., Mandjes, M.: Numerical techniques in Lévy fluctuation theory. Methodol. Comput. Appl. Probab. **16**, 31–52 (2014)
17. Asghari, N., Mandjes, M.: Transform-based evaluation of prices and Greeks of lookback options driven by Lévy processes. J. Comput. Financ. (accepted)
18. Asmussen, S.: Subexponential asymptotics for stochastic processes: extremal behavior, stationary distributions and first passage probabilities. Ann. Appl. Probab. **8**, 354–374 (1998)

K. Dębicki, M. Mandjes, *Queues and Lévy Fluctuation Theory*, Universitext,
DOI 10.1007/978-3-319-20693-6

19. Asmussen, S.: Applied Probability and Queues, 2nd edn. Springer, New York (2003)
20. Asmussen, S.: Lévy processes, phase-type distributions and martingales. Stoch. Mod. **30**, 443–468 (2014)
21. Asmussen, S., Albrecher, H.: Ruin Probabilities, 2nd edn. World Scientific, Singapore (2010)
22. Asmussen, S., Avram, F., Pistorius, M.: Russian and American put options under exponential phase-type Lévy models. Stoch. Proc. Appl. **109**, 79–111 (2004)
23. Asmussen, S., Binswanger, K.: Simulation of ruin probabilities for subexponential claims. ASTIN Bull. **27**, 297–318 (1997)
24. Asmussen, S., Glynn, P.: Stochastic Simulation: Algorithms and Analysis. Springer, New York (2007)
25. Asmussen, S., Kella, O.: A multi-dimensional martingale for Markov additive processes and its applications. Adv. Appl. Probab. **32**, 376–393 (2000)
26. Asmussen, S., Klüppelberg, C., Sigman, K.: Sampling at subexponential times, with queueing applications. Stoch. Proc. Appl. **79**, 265–286 (1999)
27. Asmussen, S., Kroese, D.: Improved algorithms for rare event simulation with heavy tails. Adv. Appl. Probab. **38**, 545–558 (2006)
28. Asmussen, S., Nerman, O., Olsson, M.: Fitting phase-type distributions via the EM algorithm. Scand. J. Stat. **23**, 419–441 (1996)
29. Asmussen, S., Pihlsgård, M.: Loss rates for Lévy processes with two reflecting barriers. Math. Oper. Res. **32**, 308–321 (2007)
30. Asmussen, S., Rosiński, J.: Approximations of small jumps of Lévy processes with a view towards simulation. J. Appl. Probab. **38**, 482–493 (2001)
31. Avi-Itzhak, B.: A sequence of service stations with arbitrary input and regular service times. Man. Sci. **11**, 565–571 (1965)
32. Avram, F., Palmowski, Z., Pistorius, M.: Exit problem of a two-dimensional risk process from the quadrant: exact and asymptotic results. Ann. Appl. Probab. **18**, 2124–2149 (2008)
33. Avram, F., Palmowski, Z., Pistorius, M.: A two-dimensional ruin problem on the positive quadrant. Insur. Math. Econ. **42**, 227–234 (2008)
34. Barndorff-Nielsen, O.: Processes of normal inverse Gaussian type. Financ. Stoch. **2**, 41–68 (1998)
35. Baxter, G., Donsker, M.: On the distribution of the supremum functional for the processes with stationary independent increments. Trans. Am. Math. Soc. **85**, 73–87 (1957)
36. Bekker, R., Borst, S., Boxma, O., Kella, O.: Queues with workload-dependent arrival and service rates. Queueing Syst. **46**, 537–556 (2004)
37. Bekker, R., Boxma, O., Kella, O.: Queues with delays in two-state strategies and Lévy input. J. Appl. Probab. **45**, 314–332 (2008)
38. Bekker, R., Boxma, O., Resing, J.: Lévy processes with adaptable exponent. Adv. Appl. Probab. **41**, 177–205 (2009)
39. Beneš, V.: On queues with Poisson arrivals. Ann. Math. Stat. **3**, 670–677 (1957)
40. Berman, A., Plemmons, R.: Nonnegative Matrices in the Mathematical Sciences. Academic, New York (1979)
41. Bernstein, S.: Sur les fonctions absolument monotones. Acta Math. **52**, 1–66 (1929)
42. Bertoin, J.: Regularity of the half-line for Lévy processes. Bull. Sci. Math. **121**, 345–354 (1997)
43. Bertoin, J.: Lévy Processes. Cambridge University Press, Cambridge (1998)
44. Bertoin, J., Doney, R.: Cramér's estimate for Lévy processes. Stat. Probab. Lett. **21**, 363–365 (1994)
45. Billingsley, P.: Convergence of Probability Measures. Wiley, New York (1999)
46. Bingham, N., Doney, R.: Asymptotic properties of subcritical branching processes I: the Galton–Watson process. Adv. Appl. Probab. **6**, 711–731 (1974)
47. Bingham, N., Goldie, C., Teugels, J.: Regular Variation. Cambridge University Press, Cambridge (1987)
48. Blanchet, J., Mandjes, M.: Asymptotics of the area under the Lévy-driven storage graph. Oper. Res. Lett. **41**, 730–736 (2013)

49. Borovkov, A.: Stochastic Processes in Queueing Theory. Springer, New York (1976)
50. Boxma, O., Cohen, J.: Heavy-traffic analysis for the GI/G/1 queue with heavy-tailed distributions. Queueing Syst. **33**, 177–204 (1999)
51. Boxma, O., Ivanovs, J., Kosiński, K., Mandjes, M.: Lévy-driven polling systems and continuous-state branching processes. Stoch. Syst. **1**, 411–436 (2011)
52. Boxma, O., Mandjes, M., Kella, O.: On a queueing model with service interruptions. Probab. Eng. Inf. Sci. **22**, 537–555 (2008)
53. Breuer, L.: A quintuple law for Markov additive processes with phase-type jumps. J. Appl. Probab. **47**, 441–458 (2010)
54. Brockwell, P., Resnick, S., Tweedie, R.: Storage processes with general release rule and additive inputs. Adv. Appl. Probab. **14**, 392–433 (1982)
55. Buchmann, B., Grübel, R.: Decompounding: an estimation problem for Poisson random sums. Ann. Stat. **31**, 1054–1074 (2003)
56. Buchmann, B., Grübel, R.: Decompounding Poisson random sums: recursively truncated estimates in the discrete case. Ann. Inst. Stat. Math. **56**, 743–756 (2004)
57. Carolan, C., Dykstra, R.: Characterization of the least concave majorant of Brownian motion, conditional on a vertex point, with application to construction. Ann. Inst. Stat. Math. **55**, 487–497 (2003)
58. Carr, P., Geman, H., Madan, D., Yor, M.: The fine structure of asset returns: an empirical investigation. J. Bus. **75**, 305–332 (2002)
59. Carr, P., Madan, D.: Option valuation using the fast Fourier transform. J. Comput. Financ. **2**, 61–73 (1999)
60. Chaumont, L.: On the law of the supremum of Lévy processes. Ann. Probab. **41**, 1191–1217 (2013)
61. Çinlar, E.: Markov additive processes, II. Probab. Theory. Relat. Flds. **24**, 95–121 (1972)
62. Cohen, J.: Some results on regular variation for distributions in queueing and fluctuation theory. J. Appl. Probab. **10**, 343–353 (1973)
63. Cont, R., Tankov, P.: Financial Modelling with Jump Processes. Chapman & Hall/CRC, Boca Raton (2003)
64. Cooley, J., Tukey, J.: An algorithm for the machine calculation of complex Fourier series. Math. Comput. **19**, 297–301 (1965)
65. Cox, D., Smith, W.: Queues. Methuen, London (1961)
66. Darling, D., Liggett, T., Taylor, H.: Optimal stopping for partial sums. Ann. Math. Stat. **43**, 1363–1368 (1972)
67. D'Auria, B., Ivanovs, J., Kella, O., Mandjes, M.: First passage process of a Markov additive process, with applications to reflection problems. J. Appl. Probab. **47**, 1048–1057 (2010)
68. D'Auria, B., Ivanovs, J., Kella, O., Mandjes, M.: Two-sided reflection of Markov-modulated Brownian motion. Stoch. Mod. **28**, 316–332 (2012)
69. De Acosta, A.: Large deviations for vector-valued Lévy processes. Stoch. Proc. Appl. **51**, 75–115 (1994)
70. De Finetti, B.: Sulle funzioni ad incremento aleatorio. Rend. Acc. Naz. Lincei. **10**, 163–168 (1929)
71. Dębicki, K., Dieker, T., Rolski, T.: Quasi-product forms for Lévy-driven fluid networks. Math. Oper. Res. **32**, 629–647 (2007)
72. Dębicki, K., Es-Saghouani, A., Mandjes, M.: Transient asymptotics of Lévy-driven queues. J. Appl. Probab. **47**, 109–129 (2010)
73. Dębicki, K., Hashorva, E., Ji, L.: Tail asymptotics of supremum of certain Gaussian processes over threshold dependent random intervals. Extremes **17**, 411–429 (2014)
74. Dębicki, K., Kosiński, K., Mandjes, M.: On the infimum attained by a reflected Lévy process. Queueing Syst. **70**, 23–35 (2012)
75. Dębicki, K., Mandjes, M., Sierpińska-Tułacz, I.: Transient analysis of Lévy-driven tandem queues. Stat. Probab. Lett. **83**, 1776–1781 (2013)
76. Dębicki, K., Mandjes, M., van Uitert, M.: A tandem queue with Lévy input: a new representation of the downstream queue length. Probab. Eng. Inf. Sci. **21**, 83–107 (2007)

77. Dembo, A., Zeitouni, O.: Large Deviations Techniques and Applications, 2nd edn. Springer, New York (1998)
78. de Meyer, A., Teugels, J.: On the asymptotic behaviour of the distributions of the busy period and service time in M/G/1. J. Appl. Probab. **17**, 802–813 (1980)
79. den Iseger, P.: Numerical transform inversion using Gaussian quadrature. Probab. Eng. Inf. Sci. **20**, 1–44 (2006)
80. den Iseger, P., Gruntjes, P., Mandjes, M.: Modeling dynamic default correlation in a Lévy world, with applications to CDO pricing (2011, submitted)
81. den Iseger, P., Oldenkamp, E.: Pricing guaranteed return rate products and discretely sampled Asian options. J. Comput. Financ. **9**, 1–39 (2006)
82. Dieker, T.: Applications of factorization embeddings for Lévy processes. Adv. Appl. Probab. **38**, 768–791 (2006)
83. Dieker, T., Mandjes, M.: Extremes of Markov-additive processes with one-sided jumps, with queueing applications. Methodol. Comput. Appl. Probab. **13**, 221–267 (2009)
84. Doney, R.: Some excursion calculations for spectrally one-sided Lévy processes. Séminaire de Probabilités **XXXVIII**, 5–15 (2005)
85. Dube, P., Guillemin, F., Mazumdar, R.: Scale functions of Lévy processes and busy periods of finite-capacity M/GI/1 queues. J. Appl. Probab. **41**, 1145–1156 (2004)
86. Dubner, H., Abate, J.: Numerical inversion of Laplace transforms by relating them to the finite Fourier cosine transform. J. ACM **15**, 115–123 (1968)
87. Elwalid, A., Mitra, D.: Effective bandwidth of general Markovian traffic sources and admission control of high speed networks. IEEE/ACM Trans. Netw. **1**, 329–343 (1993)
88. Elwalid, A., Mitra, D.: Analysis, approximations and admission control of a multi-service multiplexing system with priorities. Proceedings of IEEE INFOCOM, Boston, pp. 463–472 (1995)
89. Embrechts, P., Klüppelberg, C., Mikosch, T.: Modelling Extremal Events. Springer, New York (1997)
90. Es-Saghouani, A., Mandjes, M.: On the correlation structure of a Lévy-driven queue. J. Appl. Probab. **45**, 940–952 (2008)
91. Feldmann, A., Whitt, W.: Fitting mixtures of exponentials to long-tail distributions to analyze network performance models. Perform. Eval. **31**, 245–279 (1998)
92. Feller, W.: An Introduction to Probability Theory and its Applications, 2nd edn. Wiley, New York (1971)
93. Foss, S., Korshunov, D., Zachary, S.: An Introduction to Heavy-Tailed and Subexponential Distributions. Springer, New York (2011)
94. Friedman, H.: Reduction methods for tandem queueing systems. Oper. Res. **13**, 121–131 (1965)
95. Fu, M.: Variance-Gamma and Monte Carlo. In: Fu, M., Jarrow, R., Yen, J., Elliott, R. (eds.) Advances in Mathematical Finance, pp. 21–35. Birkhäuser, Boston (2007)
96. Furrer, H.: Risk Theory and Heavy-Tailed Lévy Processes. Ph.D. thesis, Eidgenössische Technische Hochschule, Zürich (1997). http://e-collection.ethbib.ethz.ch/eserv/eth:22556/eth-22556-02.pdf.
97. Furrer, H., Michna, Z., Weron, A.: Stable Lévy motion approximation in collective risk theory. Insur. Math. Econ. **20**, 97–114 (1997)
98. Geman, H., Yor, M.: Pricing and hedging double barrier options: a probabilistic approach. Math. Financ. **6**, 365–378 (1996)
99. Gibbens, R., Hunt, P.: Effective bandwidths for the multi-type UAS channel. Queueing Syst. **9**, 17–28 (1991)
100. Glynn, P.: Diffusion approximations. In: Heyman, D., Sobel, M. (eds.) Handbooks on Operations Research & Management Science, vol. 2, pp. 145–198. Elsevier, Amsterdam (1990)
101. Glynn, P., Mandjes, M.: Simulation-based computation of the workload correlation function in a Lévy-driven queue. J. Appl. Probab. **48**, 114–130 (2011)

102. Greenwood, P., Pitman, J.: Fluctuation identities for Lévy processes and splitting at the maximum. Adv. Appl. Probab. **12**, 839–902 (1979)
103. Groeneboom, P.: The concave majorant of Brownian motion. Ann. Probab. **11**, 1016–1027 (1983)
104. Gruntjes, P., den Iseger, P., Mandjes, M.: A Wiener–Hopf based approach to numerical computations in fluctuation theory for Lévy processes. Math. Methods Oper. Res. **78**, 101–118 (2013)
105. Gugushvili, S.: Nonparametric Inference for Partially Observed Lévy Processes. Ph.D. thesis, Universiteit van Amsterdam. http://dare.uva.nl/document/95633 (2008)
106. Hansen, L.: Large sample properties of generalized method of moments estimators. Econometrica **50**, 1029–1054 (1982)
107. Hansen, M., Pitts, S.: Decompounding random sums: a nonparametric approach. Ann. Inst. Stat. Math. **62**, 855–872 (2010)
108. Harrison, J.: Brownian Motion and Stochastic Flow Systems. Wiley, New York (1985)
109. Harrison, J., Reiman, M.: Reflected Brownian motion on an orthant. Ann. Probab. **9**, 302–308 (1981)
110. Harrison, J., Resnick, S.: The recurrence classification of risk and storage processes. Math. Oper. Res. **3**, 57–66 (1978)
111. Heath, D., Resnick, S., Samorodnitsky, G.: Heavy tails and long range dependence in on/off processes and associated fluid models. Math. Oper. Res. **23**, 145–165 (1998)
112. Horváth, A., Telek, M.: Approximating heavy tailed behavior with phase type distributions. In: Proceedings of 3rd International Conference on Matrix-Analytic Methods in Stochastic Models, Leuven (2000)
113. Iglehart, D.: Diffusion approximations in collective risk theory. J. Appl. Probab. **6**, 285–292 (1969)
114. Ivanovs, J., Boxma, O., Mandjes, M.: Singularities of the generator of a Markov additive process with one-sided jumps. Stoch. Proc. Appl. **120**, 1776–1794 (2010)
115. Jaworski, P., Durante, F., Härdle, W.-K., Rychlik, T.: Copula Theory and its Applications. Lecture Notes in Statistics. Springer, New York (2010)
116. Jeannin, M., Pistorius, M.: A transform approach to compute prices and Greeks of barrier options driven by a class of Lévy processes. Quant. Financ. **10**, 629–644 (2010)
117. Karatzas, I., Shreve, S.: Brownian Motion and Stochastic Calculus, 2nd edn. Springer, New York (1991)
118. Kaspi, H.: On the symmetric Wiener–Hopf factorization for Markov-additive processes. Probab. Theory Relat. Flds. **59**, 179–196 (1982)
119. Kella, O.: Concavity and reflected Lévy processes. J. Appl. Probab. **29**, 209–215 (1992)
120. Kella, O.: Parallel and tandem fluid networks with dependent Lévy inputs. Ann. Appl. Probab. **3**, 682–695 (1993)
121. Kella, O.: Stability and nonproduct form of stochastic fluid networks with Lévy inputs. Ann. Appl. Probab. **6**, 186–199 (1996)
122. Kella, O.: Stochastic storage networks: stationarity and the feedforward case. J. Appl. Probab. **34**, 498–507 (1997)
123. Kella, O.: Non-product form of two-dimensional fluid networks with dependent Lévy inputs. J. Appl. Probab. **37**, 1117–1122 (2000)
124. Kella, O.: Reflecting thoughts. Stat. Probab. Lett. **76**, 1808–1811 (2006)
125. Kella, O., Boxma, O., Mandjes, M.: A Lévy process reflected at a Poisson age process. J. Appl. Probab. **43**, 221–230 (2006)
126. Kella, O., Mandjes, M.: Transient analysis of reflected Lévy processes. Stat. Probab Lett. **83**, 2308–2315 (2013)
127. Kella, O., Sverchkov, M.: On concavity of the mean function and stochastic ordering for reflected processes with stationary increments. J. Appl. Probab. **31**, 1140–1142 (1994)
128. Kella, O., Whitt, W.: Queues with server vacations and Lévy processes with secondary jump input. Ann. Appl. Probab. **1**, 104–117 (1991)

129. Kella, O., Whitt, W.: A tandem fluid network with Lévy input. In: Bhat, U., Basawa, I., (eds.) Queueing and Related Models, pp. 112–128. Oxford University Press, Oxford (1992)
130. Kella, O., Whitt, W.: Useful martingales for stochastic storage processes with Lévy input. J. Appl. Probab. **33**, 1169–1180 (1992)
131. Kelly, F.: (1996) Notes on effective bandwidths. In: Kelly, F., Zachary, S., Ziedins, I. (eds.) Stochastic Networks: Theory and Applications, pp. 141–168. Oxford University Press, Oxford
132. Kesidis, G., Walrand, J., Chang, C.: Effective bandwidths for multiclass Markov fluids and other ATM sources. IEEE/ACM Trans. Netw. **1**, 424–428 (1993)
133. Khintchine, A.: Matematicheskaya teoriya statsionarnoi ocheredi. Mat. Sb. **30**, 73–84 (1932)
134. Kingman, J.: The single server queue in heavy traffic. Proc. Camb. Philos. Soc. **57**, 902–904 (1961)
135. Kingman, J.: On queues in heavy traffic. J. R. Stat. Soc. B **24**, 383–392 (1962)
136. Kingman, J.: The heavy traffic approximation in the theory of queues. In: Smith, W.L., Wilkinson, W.E. (eds.) Proceedings of Symposium on Congestion Theory, Chapel Hill, pp. 137–159. University of North Carolina Press, Chapel Hill (1965)
137. Klüppelberg, C., Kyprianou, A., Maller, R.: Ruin probabilities and overshoots for general Lévy insurance risk processes. Ann. Appl. Probab. **14**, 1766–1801 (2004)
138. Koponen, I.: Analytic approach to the problem of convergence of truncated Lévy flights towards the Gaussian stochastic process. Phys. Rev. E **52**, 1197–1199 (1995)
139. Kosiński, K., Boxma, O., Zwart, B.: Convergence of the all-time supremum of a Lévy process in the heavy-traffic regime. Queueing Syst. **67**, 295–304 (2011)
140. Kosten, L.: Stochastic theory of data-handling systems with groups of multiple sources. In: Rudin, H., Bux, W. (eds.) Proceedings of Performance of Computer-Communication Systems, Zurich, pp. 321–331. Elsevier, Amsterdam (1984)
141. Kou, S.: A jump-diffusion model for option pricing. Man. Sci. **48**, 1086–1101 (2002)
142. Kruk, L., Lehoczky, J., Ramanan, K., Shreve, S.: An explicit formula for the Skorokhod map on $[0, a]$. Ann. Probab. **35**, 1740–1768 (2007)
143. Kuznetsov, A.: Wiener–Hopf factorization and distribution of extrema for a family of Lévy processes. Ann. Appl. Probab. **2**, 1801–1830 (2010)
144. Kuznetsov, A.: Wiener–Hopf factorization for a family of Lévy processes related to theta functions. J. Appl. Probab. **47**, 1023–1033 (2010)
145. Kuznetsov, A., Kyprianou, A., Pardo, J.: Meromorphic Lévy processes and their fluctuation identities. Ann. Appl. Probab. **22**, 1101–1135 (2012)
146. Kyprianou, A.: Introductory Lectures on Fluctuations of Lévy Processes with Applications. Springer, Berlin (2006)
147. Kyprianou, A.: The Wiener–Hopf decomposition. In: Cont, R. (ed.) Encyclopedia of Quantitative Finance. Wiley, Chichester (2010)
148. Kyprianou, A., Palmowski, Z.: Fluctuations of spectrally negative Markov additive processes. Séminaire de Probabilité XLI. Lect. Notes Math. **1934**, 121–135 (2008)
149. Lewis, A., Mordecki, E.: Wiener–Hopf factorization for Lévy processes having negative jumps with rational transforms. J. Appl. Probab. **45**, 118–134 (2005)
150. Lewis, A., Mordecki, E.: Wiener–Hopf factorization for Lévy processes having positive jumps with rational transforms. J. Appl. Probab. **45**, 118–134 (2008)
151. Lieshout, P., Mandjes, M.: Tandem Brownian queues. Math. Methods Oper. Res. **66**, 275–298 (2007)
152. Lieshout, P., Mandjes, M.: Asymptotic analysis of Lévy-driven tandem queues. Queueing Syst. **60**, 203–226 (2008)
153. Lindley, D.: Discussion of a paper of C.B. Winsten. R. J. Stat. Soc. B **21**, 22–23 (1959)
154. Madan, D., Carr, P., Chang, E.: The variance gamma process and option pricing. Eur. Financ. Rev. **2**, 79–105 (1998)
155. Madan, D., Seneta, E.: The variance gamma model for share market returns. J. Bus. **63**, 511–524 (1990)

156. Mandjes, M.: Packet models revisited: tandem and priority systems. Queueing Syst. **47**, 363–377 (2004)
157. Mandjes, M.: Large Deviations of Gaussian Queues. Wiley, Chichester (2007)
158. Mandjes, M., Palmowski, Z., Rolski, T.: Quasi-stationary workload of a Lévy-driven queue. Stoch. Mod. **28**, 413–432 (2012)
159. Mandjes, M., Ridder, A.: Finding the conjugate of Markov fluid processes. Probab. Eng. Inf. Sci. **9**, 297–315(1995)
160. Merton, R.: Option pricing when underlying stock returns are discontinuous. J. Financ. Econ. **3**, 125–144 (1976)
161. Michna, Z.: Formula for the supremum distribution of a spectrally positive Lévy process. ArXiv:1104.1976 [math.PR] (2012, submitted)
162. Mikosch, T.: Non-Life Insurance Mathematics. Springer, New York (2004)
163. Mikosch, T., Resnick, S., Rootzé, H., Stegeman, A.: Is network traffic approximated by stable Lévy motion or fractional Brownian motion? Ann. Appl. Probab. **12**, 23–68 (2002)
164. Mikosch, T., Samorodnitsky, G.: Scaling limits for cumulative input processes. Math. Oper. Res. **32**, 890–919 (2007)
165. Mitra, D.: Stochastic theory of a fluid model of producers and consumers coupled by a buffer. Adv. Appl. Probab. **20**, 646–676 (1988)
166. Miyazawa, M., Rolski, T.: Tail asymptotics for a Lévy-driven tandem queue with an intermediate input. Queueing Syst. **63**, 323–353 (2009)
167. Mordecki, E.: Optimal stopping and perpetual options for Lévy processes. Finac. Stoch. **6**, 473–493 (2002)
168. Nelsen, R.: An Introduction to Copulas. Springer, New York (1999)
169. Neveu, J.: Une généralisation des processus à accroissements positifs indépendents. Abh. Math. Sem. Univ. Hamburg **25**, 36–61 (1961)
170. Nguyen-Ngoc, L.: Exotic options in general exponential Lévy models. Prépublication du Laboratoire de Probabilités et Modèles Aléatoires (2003)
171. Nguyen-Ngoc, L., Yor, M.: Lookback and barrier options under general Lévy processes. In: Aït-Sahalia, Y., Hansen, L. (eds.) Handbook of Financial Econometrics. North-Holland, Amsterdam (2007)
172. Ott, T.: The covariance function of the virtual waiting-time process in an M/G/1 queue. Adv. Appl. Probab. **9**, 158–168 (1977)
173. Palmowski, Z., Rolski, T.: On the exact asymptotics of the busy period in GI/G/1 queues. Adv. Appl. Probab. **38**, 792–803 (2006)
174. Pecherskii, E., Rogozin, B.: On the joint distribution of random variables associated with fluctuations of a process with independent increments. Theory Probab. Appl. **14**, 410–423 (1969)
175. Pistorius, M.: On exit and ergodicity of the spectrally one-sided Lévy process reflected at its infimum. J. Theor. Probab. **17**, 183–220 (2004)
176. Poirot, J., Tankov, P.: Monte Carlo option pricing for tempered stable (CGMY) processes. Asia Pac. Financ. Markets **13**, 327–344 (2006)
177. Pollaczek, F.: Über eine Aufgabe der Wahrscheinlichkeitsrechnung. Math. Zeit. **32**, 64–100 and 729–850 (1930)
178. Port, S.: Stable processes with drift on the line. Trans. Am. Math. Soc. **313**, 805–841 (1989)
179. Prabhu, N.: Stochastic Storage Processes, 2nd edn. Springer, New York (1998)
180. Prohorov, V.: Transition phenomena in queueing processes, I. Litovsk. Mat. Sb. **3**, 199–205 (1963)
181. Pruitt, E.W. The growth of random walks and Lévy processes. Ann. Probab. **9**, 948–956 (1981)
182. Reich, E.: On the integrodifferential equation of Takács I. Ann. Math. Stat. **29**, 563–570 (1958)
183. Resnick, S., van den Berg, E.: Weak convergence of high-speed network traffic models. J. Appl. Probab. **37**, 575–597 (2000)

184. Robbins, H., Siegmund, D.: Boundary crossing probabilities for the Wiener process and sample sums. Ann. Math. Stat. **41**, 1410–1429 (1970)
185. Robert, P.: Stochastic Networks and Queues. Springer, Berlin (2003)
186. Rogers, L.: Wiener–Hopf factorization of diffusions and Lévy processes. Proc. Lond. Math. Soc. **47**, 177–191 (1983)
187. Rogers, L.: Fluid models in queueing theory and Wiener–Hopf factorisation of Markov chains. Ann. Appl. Probab. **4**, 390–413 (1994)
188. Rogers, L.: Evaluating first-passage probabilities for spectrally one-sided Lévy processes. J. Appl. Probab. **37**, 1173–1180 (2000)
189. Rolski, T., Schmidli, H., Schmidt, V., Teugels, J.: Stochastic Processes for Insurance and Finance. Wiley, New York (2009)
190. Rubin, I.: Path delays in communication networks. Appl. Math. Opt. **1**, 193–221 (1974)
191. Rydberg, T.: The normal inverse Gaussian Lévy process: simulation and approximation. Stoch. Mod. **13**, 887–910 (1997)
192. Samorodnitsky, G., Taqqu, M.: Stable Non-Gaussian Random Processes: Stochastic Models with Inifinite Variance. Chapman and Hall, New York (1994)
193. Sato, K.: Lévy Processes and Infinitely Divisible Distributions. Cambridge University Press, Cambridge (1999)
194. Schoutens, W., Cariboni, J.: Lévy Processes in Credit Risk. Wiley, New York (2009)
195. Seneta, E.: Nonnegative Matrices. Wiley, New York (1973)
196. Seneta, E.: The early years of the variance gamma process. In: Fu, M., Jarrow, R., Yen, J., Elliott, R. (eds.) Advances in Mathematical Finance, pp. 3–19, Birkhäuser, Boston (2000)
197. Shneer, S., Wachtel, V.: Heavy-traffic analysis of the maximum of an asymptotically stable random walk. Teor. Verojatn. i Primenen. **55**, 335–344 (2010). (English translation in Theor. Probab. Appl. **55**, 332–341, 2011)
198. Siegmund, D.: Importance sampling in the Monte Carlo study of sequential tests. Ann. Stat. **4**, 673–684 (1976)
199. Siegmund, D.: The equivalence of absorbing and reflecting barrier problems for stochastically monotone Markov processes. Ann. Probab. **4**, 914–924 (1976)
200. Siegmund, D.: Sequential Analysis. Springer, New York (1985)
201. Skorokhod, A.: Stochastic equations for diffusion processes in a bounded region, part I. Teor. Verojatn. i Primenen. **6**, 264–274 (1961)
202. Skorokhod, A.: Stochastic equations for diffusion processes in a bounded region, part II. Teor. Verojatn. i Primenen. **7**, 3–23 (1962)
203. Sonneveld, P.: Some properties of the generalized eigenvalue problem $Mx = A(\Gamma - cI)x$, where M is the infinitesimal generator of a Markov process and Γ is a real diagonal matrix. Report 04-02, Delft University of Technolog (2004)
204. Sueishi, N., Nishiyama, Y.: Estimation of Lévy processes in mathematical finance: a comparative study. In: Proceedings of International Congress on Modelling and Simulation, Melbourne, pp. 953–959 (2005)
205. Suprun, V.: Ruin problem and the resolvent of a terminating process with independent increments. Ukr. Mat. Zh. **28**, 53–61 (1976)
206. Surya, B.: Evaluating scale functions of spectrally negative Lévy processes. J. Appl. Probab. **45**, 135–149 (2008)
207. Szczotka, W., Woyczyński, W.: Distributions of suprema of Lévy processes via the heavy-traffic invariance principle. Probab. Math. Stat. **23**, 251–272 (2003)
208. Takács, L.: Investigations of waiting time problems by reduction to Markov processes. Acta Math. Acad. Sci. Hung. **6**, 101–129 (1955)
209. Takács, L.: On the distribution of the supremum of stochastic processes with exchangeable increments. Trans. Am. Math. Soc. **119**, 367–379 (1965)
210. Taqqu, M., Willinge, W., Sherman, R.: Proof of a fundamental result in self-similar traffic modeling. Comput. Commun. Rev. **27**, 5–23 (1997)
211. Thümmler, A., Buchholz, P., Telek, M.: A novel approach for phase-type fitting with the EM Algorithm. IEEE Trans. Dependable Sec. Comput. **3**, 245–258 (2006)

212. van Es, B., Gugushvili, S., Spreij, P.: A kernel type nonparametric density estimator for decompounding. Bernoulli **13**, 672–694 (2007)
213. Weron, R.: Lévy-stable distributions revisited: tail index > 2 does not exclude the Lévy-stable regime. Int. J. Mod. Phys. C **12**, 209–223 (2001)
214. Whitt, W.: Tail probabilities with statistical multiplexing and effective bandwidths in multi-class queues. Telecomm. Syst. **2**, 71–107 (1993)
215. Whitt, W.: The reflection map is Lipschitz with appropriate Skorokhod M-metrics. AT&T Labs (1999, preprint)
216. Whitt, W.: Limits for cumulative input processes to queues. Probab. Eng. Inf. Sci. **14**, 123–150 (2000)
217. Whitt, W.: Stochastic-Process Limits. Springer, New York (2002)
218. Whittaker, E., Watson, G.: A Course in Modern Analysis, 4th edn. Cambridge University Press, Cambridge (1990)
219. Willekens, E.: On the supremum of an infinitely divisible process. Stoch. Proc. Appl. **26**, 173–175 (1987)
220. Williams, D.: Probability with Martingales. Cambridge University Press, Cambridge (1991)
221. Yor, M., Madan, D.: CGMY and Meixner subordinators are absolutely continuous with respect to one-sided stable subordinators. Prépublication du Laboratoire de Probabilités et Modèles Aléatoires (2005)
222. Zolotarev, V.: The first passage time of a level and the behaviour at infinity for a class of processes with independent increments. Theor. Probab. Appl. **9**, 653–661 (1964)
223. Zwart, B.: Queueing systems with Heavy Tails. Ph.D. thesis, Eindhoven University of Technology. http://alexandria.tue.nl/extra2/200112999.pdf. (2001)
224. Zwart, B., Borst, S., Dębicki, K.: Subexponential asymptotics of hybrid fluid and ruin models. Ann. Appl. Probab. **15**, 500–517 (2005)

Printed in the United States
By Bookmasters